Meeting Health Information Needs Outside of Healthcare

CHANDOS

INFORMATION PROFESSIONAL SERIES

Series Editor: Ruth Rikowski
(email: Rikowskigr@aol.com)

Chandos' new series of books is aimed at the busy information professional. They have been specially commissioned to provide the reader with an authoritative view of current thinking. They are designed to provide easy-to-read and (most importantly) practical coverage of topics that are of interest to librarians and other information professionals. If you would like a full listing of current and forthcoming titles, please visit www.chandospublishing.com.

New authors: we are always pleased to receive ideas for new titles; if you would like to write a book for Chandos, please contact Dr Glyn Jones on g.jones.2@elsevier.com or telephone +44 (0) 1865 843000.

Meeting Health Information Needs Outside of Healthcare

Opportunities and Challenges

Edited by

Catherine Arnott Smith
and
Alla Keselman

AMSTERDAM • BOSTON • CAMBRIDGE • HEIDELBERG
LONDON • NEW YORK • OXFORD • PARIS • SAN DIEGO
SAN FRANCISCO • SINGAPORE • SYDNEY • TOKYO
Chandos Publishing is an imprint of Elsevier

Chandos Publishing is an imprint of Elsevier
225 Wyman Street, Waltham, MA 02451, USA
Langford Lane, Kidlington, OX5 1GB, UK

Copyright © C. Arnott Smith, A. Keselman, 2015. Published by Elsevier Ltd. All rights reserved.

No part of this publication may be reproduced or transmitted in any form or by any means, electronic or mechanical, including photocopying, recording, or any information storage and retrieval system, without permission in writing from the publisher. Details on how to seek permission, further information about the Publisher's permissions policies and our arrangements with organizations such as the Copyright Clearance Center and the Copyright Licensing Agency, can be found at our website: www.elsevier.com/permissions.

This book and the individual contributions contained in it are protected under copyright by the Publisher (other than as may be noted herein).

Notices
Knowledge and best practice in this field are constantly changing. As new research and experience broaden our understanding, changes in research methods, professional practices, or medical treatment may become necessary.

Practitioners and researchers must always rely on their own experience and knowledge in evaluating and using any information, methods, compounds, or experiments described herein. In using such information or methods they should be mindful of their own safety and the safety of others, including parties for whom they have a professional responsibility.

To the fullest extent of the law, neither the Publisher nor the authors, contributors, or editors, assume any liability for any injury and/or damage to persons or property as a matter of products liability, negligence or otherwise, or from any use or operation of any methods, products, instructions, or ideas contained in the material herein.

ISBN: 978-0-08-100248-3 (print)
ISBN: 978-0-08-100259-9 (online)

British Library Cataloguing-in-Publication Data
A catalogue record for this book is available from the British Library

Library of Congress Control Number: 2015939550

For information on all Chandos Publishing publications
visit our website at http://store.elsevier.com/

Contents

About the authors	ix
Editors' foreword	xvii
Acknowledgments	xix

Overview 1

1 Designing health information programs to promote the health and well-being of vulnerable populations: the benefits of evidence-based strategic health communication   3
Gary L. Kreps and Linda Neuhauser
- 1.1 Introduction   3
- 1.2 Barriers   4
- 1.3 Lessons learned: improving health communication for vulnerable populations   7
- 1.4 Strategies to develop strategic communication   8
- 1.5 Evaluating health communication   12
- 1.6 Practice implications   13
- References   14

2 Health literacy research's growth, challenges, and frontiers   19
Robert A. Logan
- 2.1 Introduction   19
- 2.2 Four milestones in health literacy research   20
- 2.3 Health literacy's evolving definition and conceptual underpinnings   26
- 2.4 The range and vitality of health literacy research   29
- 2.5 Health literacy research's current needs and frontiers   30
- 2.6 Conclusions   33
- References   34

Libraries 39

3 Medical information for the consumer before the World Wide Web   41
Catherine Arnott Smith
- 3.1 Introduction: "Closed to the Public"   41
- 3.2 Background: the beginnings of consumer health information   44

	3.3	Libraries	52
	3.4	Librarian	58
	3.5	The patron	63
	3.6	Content	67
	3.7	Conclusions	72
		References	73

4 Ethical health information: Do it well! Do it right! Do no harm! — 77
Michelynn McKnight

4.1	Introduction	77
4.2	Responsibility for the best possible information service	78
4.3	The right to privacy and responsibility for confidentiality	83
4.4	Providing fair and equitable access	85
4.5	Intellectual property rights and access to information	87
4.6	Advocacy for information access	88
4.7	Providing information versus giving advice	88
4.8	Conflicting values, dilemmas, and tough decisions	90
4.9	Keep learning	91
	References	92

5 Health information resource provision in the public library setting — 97
Mary Grace Flaherty

5.1	Background	97
5.2	Challenges	100
5.3	Case study: embedded consumer health librarians in Delaware	110
5.4	Conclusions	112
	References	113

6 Who needs a health librarian? Ethical reference transactions in the consumer health library — 117
Nancy C. Seeger

6.1	Introduction	117
6.2	The reference transaction: asking the right questions, avoiding the wrong answers	120
6.3	Looking for the answers: symptom-checkers and self-diagnosing	126
6.4	What did the doctor say? Health literacy and deciphering a whole new language	133
6.5	When the answers have questions: experimental treatments and integrative medicine	136
6.6	Conclusions	139
	References	140

7 Consumer health information: the community college conundrum — 143
Anne Chernaik

7.1	The community college setting	144
7.2	Health information needs at the community college	147

7.3	Issues in health information provision	149
7.4	Health literacy in the community college setting	156
7.5	The future for community colleges and health information	157
7.6	Conclusions	163
	Appendix A: Community and Junior College Libraries Section (CJCLS) of the Association of College and Research Libraries	163
	References	163

Contexts   167

8   Health information delivery outside the clinic in a developing nation: The Qatar Cancer Society in the State of Qatar   169
Ellen N. Sayed and Alan S. Weber

8.1	Introduction and background	169
8.2	Qatar	170
8.3	Methods	172
8.4	Sources of consumer health information in the GCC	173
8.5	Barriers to health care in Qatar	175
8.6	The Qatar Cancer Society	177
8.7	Cancer information delivery outside the clinical setting	183
8.8	Conclusions	186
	Appendix 1: Questionnaire	187
	References	187

9   Health information and older adults   191
Kay Hogan Smith

9.1	Introduction	191
9.2	Background	191
9.3	Settings: where do older adults go for information?	193
9.4	Health information format considerations	194
9.5	Format summary	197
9.6	Health information comprehension among older adults: barriers and solutions	197
9.7	Comprehension summary	203
	References	204

10 Re-envisioning the health information-seeking conversation: insights from a community center   209
Prudence W. Dalrymple and Lisl Zach

10.1	Introduction	209
10.2	Understanding information behaviors	210
10.3	Health information seeking in a local context	218
10.4	Conclusions	228
	References	231

11 For the mutual benefit: health information provision in the science classroom 235
Albert Zeyer, Daniel M. Levin and Alla Keselman
11.1 Background 235
11.2 The science classroom as a setting for health literacy 239
11.3 Challenges and opportunities for bringing health education into the science classroom 248
11.4 Conclusions and implications 256
References 257

12 "You will be glad you hung onto this quit": sharing information and giving support when stopping smoking online 263
Marie-Thérèse Rudolf von Rohr
12.1 Introduction 263
12.2 Interpersonal aspects of advice-giving and showing support online 265
12.3 Methodology 270
12.4 Results and discussion 275
12.5 Conclusions 287
References 289

13 Health information in bits and bytes: considerations and challenges of digital health communication 291
Clare Tobin Lence and Korey Capozza
13.1 Introduction 291
13.2 The health programs 291
13.3 The digital divide 294
13.4 Don't make me think 297
13.5 Humanizing technology 306
13.6 Know your audience 309
13.7 Data dilemmas 311
13.8 Conclusions 314
References 315

14 Does specialization matter? How journalistic expertise explains differences in health-care coverage 321
Michael W. Wagner
14.1 Introduction 321
14.2 Why specialization should matter 323
14.3 Methodology 324
14.4 Results 330
14.5 Discussion 336
14.6 Conclusions 338
References 338

Afterword 341
Index 347

About the authors

Korey Capozza

Korey Capozza is Director of Consumer and Community Engagement at HealthInsight, a nonprofit organization based in Salt Lake City that works to improve the quality of health care in Utah. She has 15 years of experience in health program development and policy analysis. Her research focus is on testing and improving health-care innovations for patients in the community setting. Ms Capozza was principal investigator for the Utah Diabetes Mobile Health Pilot, a study to test the clinical and quality-of-life impact of a two-way text-messaging program for diabetes self-management (Care4Life) and leads the Crowdsource project, an effort to aggregate, analyze, and understand patient comments from peer-to-peer online communities. She is an appointed member of the federal Health Information Technology, Consumer Empowerment Workgroup and a standing reviewer for the Patient Centered Outcomes Research Institute (PCORI). Previously, she served as a consumer advocate in the Utah legislature and was appointed by the Governor of Utah to the Utah Health Exchange advisory board in 2010. A former Pew Scholar and Knight Fellow, Ms Capozza holds an undergraduate degree from the University of Pennsylvania and a master's in health policy and management from University of California, Berkeley.

Anne Chernaik

Anne Chernaik has a master's of science in library and information science from Syracuse University and is an associate professor and reference librarian at the College of Lake County (CLC) in Grayslake, IL, USA, a community college. Since 2006 she has been the health and life sciences librarian and works closely with faculty and students to expand the library's role in nursing and allied health programs across all three campuses. She is also department chair for the Library Technical Assistant (LTA) program, a course of study offering academic certificate and Associates of Arts Degree options for library support staff. Over the past few years she also completed the master of online teaching professional certificate available through the Illinois Online Network and used that knowledge in instructional design and online education to launch all but one of the core LTA courses in the online environment.

Prudence W. Dalrymple

Prudence Dalrymple is a research and teaching professor at Drexel University's College of Computing & Informatics where she directs its Institute for Health Informatics, an interdisciplinary initiative preparing professionals to meet the public's information and data-related needs. In addition to her early work in cognitive models of information retrieval, she has more than 25 years' experience in the field of health information and communications. She uses mixed methods to examine how both consumers and health professionals seek and use information to make decisions. She holds a master's degree in library and information science from Simmons College and a PhD from the University of Wisconsin-Madison. While a National Library of Medicine (NLM) Fellow, she received a master's degree in health informatics from the Johns Hopkins University School of Medicine and a certificate in health communication from its Bloomberg School of Public health. She is a Fellow of the Medical Library Association.

Mary Grace Flaherty

Mary Grace Flaherty is currently an assistant professor at the School of Information & Library Science at the University of North Carolina—Chapel Hill. She received her PhD from Syracuse University's School of Information Studies where she was an IMLS fellow. She received her MLS from the University of Maryland, and her MS in applied behavioral science from Johns Hopkins University and is a current member of the Academy of Health Information Professionals. Dr Flaherty has over 20 years of experience working in a variety of library settings, including academic, medical research, special, and public libraries. Dr Flaherty's research interests include health information, health literacy, health promotion, and public libraries.

Alla Keselman

Alla Keselman holds a doctorate in human cognition and learning and a master's in biomedical informatics from Columbia University. She is a senior social science analyst in the Division of Specialized Information Services at the US NLM, National Institutes of Health. Dr Keselman conducts research into lay conceptual understanding of health, as well as the relationship between formal science education and health reasoning in everyday contexts. Her publications have appeared in many science education and informatics journals. She has contributed to several foundational Science|Environment|Health publications, including *Science|Environment|Health: Towards a Renewed Pedagogy for Science Education* and Science|Environment|Health special issue of the *International Journal of Science Education*. At NLM, Alla Keselman leads a team that develops science education Web sites, lesson plans, games, and activities. Together with coauthor Albert Zeyer, she is a co-coordinator of the Special

Interest Group Science|Environment|Health at ESERA (European Science Education Research Association).

Gary L. Kreps

Gary L. Kreps is a University Distinguished Professor of Communication and Director of the Center for Health and Risk Communication at George Mason University in Fairfax, Virginia, USA. His research examines health communication, health promotion, health informatics, multicultural relations, social organization, and applied research methods, with a major focus on reducing health disparities. He publishes widely (more than 400 articles, books, and monographs) concerning the applications of communication knowledge to address important health issues. Before joining the faculty at George Mason University in 2004, he served as the founding Chief of the Health Communication and Informatics Research Branch at the National Cancer Institute, NIH.

Clare Tobin Lence

Clare Tobin Lence is a project coordinator at HealthInsight Utah. Her projects primarily leverage technology-based tools to help health-care consumers make informed choices. Ms Lence has served as the strategic lead for the public reporting Web site UtahHealthScape.org since 2012, which has received attention from multiple national agencies, including the Institute of Medicine, for its consumer-friendly design; she also manages HealthInsight's current work to add health-care pricing information to UtahHealthScape, based on data from Utah's All Payer Claims Database. Her expertise in consumer engagement lies in understanding of health literacy issues and how to communicate complex information, and particularly data, to nonexpert audiences. She has further utilized this expertise in developing another consumer-oriented Web site, Leaving-Well.org, that supports end-of-life decision making. Ms Lence holds an undergraduate degree in human biology from Stanford University and master of public health and master of public policy degrees from the University of Utah. She is the 2014 winner of Policy Solutions Challenge USA, a national public policy analysis competition.

Daniel M. Levin

Daniel M. Levin is clinical assistant professor in the Department of Teaching and Learning, Policy and Leadership at the University of Maryland, College Park, where he teaches science pedagogy, teacher inquiry, and biology education courses, supervises student teaching interns, and coordinates middle school math and science teacher education programs. His research focuses on responsive science teaching, students' participation in scientific practices, and teaching and learning of socioscientific issues. He is the author or coauthor of peer-reviewed articles in *Journal of Research in Science Teaching, Science Education, Science Educator, Journal of Teacher Education, International Journal of Science Education, American Biology Teacher*, and *The Science Teacher*. He has also coauthored a book, *Becoming a Responsive Science Teacher: Focusing on Student Thinking in Secondary Science*, published by National Science Teachers Association (NSTA) Press (2012).

Robert A. Logan

Robert A. Logan, PhD, is a member of the senior staff of the U.S. NLM and is a professor emeritus at the University of Missouri-Columbia School of Journalism. Dr Logan has published more than 45 articles in refereed journals and is the first author of two books and 11 book chapters. He is a member of the editorial boards of the *Journal of Mass Media Ethics, Science Communication*, and *Mass Communication and Society*. His research areas include health literacy, consumer health informatics evaluation, public understanding of science and medicine, theory and applications of Q methodology, and journalism ethics. Logan corepresents the NLM at the Institute of Medicine's Health Literacy Roundtable. He writes and narrates NLM's weekly "Director's Comments" podcast.

Michelynn McKnight

Michelynn McKnight is associate professor in the School of Library and Information Science, College of Human Sciences & Education, Louisiana State University, Baton Rouge, Louisiana, where she teaches graduate courses in health sciences information services and science and technology information sources. For more than 20 years she was the Director of the Health Sciences Library and Consumer Health Information Service for Norman Regional Health System in Norman, Oklahoma. A former member of the Medical Library Association Board of Directors and the U.S. NLM Biomedical Informatics and Review Committee, she is the author of many articles, book chapters, and books, including *The Agile Librarian's Guide to Thriving in Any Institution* (Westport, CT: Libraries Unlimited, 2010).

Linda Neuhauser

Linda Neuhauser, DrPH, MPH, is clinical professor of community health and human development at the University of California, Berkeley School of Public Health, USA. Her research, teaching, and practice are focused on translating research findings into improved health programs and policies. She uses participatory approaches to create programs and communication that are relevant to the specific needs of the intended audiences. She also heads the UC Berkeley Health Research for Action, center that works with diverse groups to codesign, implement, and evaluate health programs in the US and globally. Her numerous publications focus on participatory design, public health interventions, and many areas of health communication, including health literacy, risk communication, eHealth and mHealth strategies. She was previously a health officer in West and Central Africa with the US Agency for International Development.

Ellen N. Sayed

Ellen N. Sayed, MLS, M.Acc., AHIP, is currently the Director of the Distributed eLibrary at Weill Cornell Medical College in Qatar. Ms Sayed has more than 20 years of experience in academic medical librarianship, spanning a wide range of library services, including reference, resource sharing and outreach, instruction, and collection development. As a member of the Medical Library Association, Ms Sayed has made frequent presentations, and has authored a number of publications. Ms Sayed is currently serving on the IFLA Health and Biosciences Section Standing Committee. Ms Sayed is a member of the Academy of Health Information Professionals (AHIP) at the distinguished level. Her areas of interest include collection development, open access, library management, and innovative applications of technology to promote library services.

Nancy C. Seeger

Nancy C. Seeger is the health librarian in the Family Resource Center at Rainbow Babies and Children's Hospital/Case Medical Center in Cleveland, Ohio. Since 2007, she has had the privilege of helping countless health-care consumers navigate the maze of shared medical decision making. As part of the Family and Child Life Services Department, she works with an interdisciplinary team striving to meet the psychosocial and information needs of hospitalized children and their families. She also provides consumer health reference services to outpatient families, medical students, hospital staff, community visitors, and the general public. Prior to her current position, she worked as a deaf services reference librarian with the Cleveland Heights-University Heights Public Library, and as a reference and outreach specialist with the Foundation Center-Cleveland (a special library for the philanthropy/nonprofit sector). Librarianship is her second career. She also spent 12 years as an early childhood and special education teacher working primarily with deaf and hard-of-hearing students in Beachwood, Ohio.

Catherine Arnott Smith

Catherine Arnott Smith is associate professor, School of Library & Information Studies, University of Wisconsin-Madison, an affiliate associate professor in the School of Nursing at the same university, and a Discovery Fellow attached to the Living Environments Laboratory, a research lab focusing on virtual reality and visualization.

A former medical librarian at Northwestern University and Lincoln National Reinsurance Companies, Dr Smith was a NLM medical informatics trainee at the Center (now Department) for Biomedical Informatics, University of Pittsburgh, USA between 1997 and 2002. She has master's degrees in library and information science and American history/archives administration from the University of Michigan (both 1992); a master's degree in information science/biomedical informatics (University of Pittsburgh, 2000); and a doctorate in library and information studies/biomedical informatics (University of Pittsburgh, 2002). Her research interests and publications center on consumer interactions with clinical information systems through text, in environments from public libraries to personal health records, in the history of consumer health information provision in diverse settings by diverse people. She is pretty sure she is the only member of the American Medical Informatics Association whose first published work was a critical bibliography of the playwright Tennessee Williams.

Kay Hogan Smith

Kay Hogan Smith, MLS, MPH, CHES, is a professor and community services librarian at the University of Alabama at Birmingham (UAB) Lister Hill Library of the Health Sciences, where she has worked for 20 years. She has taught workshops on health literacy for librarians and health-care providers for 10 years, including a Medical Library Association-certified continuing education course on the subject. She frequently provides guest lectures for the UAB Geriatric Education Center, and collaborated with the Center in 2012 on a research project examining the relationship between health literacy levels of older adults and satisfaction with provider communication skills.

Marie-Thérèse Rudolf von Rohr

Marie-Thérèse Rudolf von Rohr is a PhD candidate in English linguistics at the University of Basel, Switzerland. She is a research member of the project *Typing yourself healthy—language and health online* (143286), funded by the Swiss National Science Foundation, which investigates e(lectronic)-health interaction in asynchronous, written computer-mediated communication. In connection to this project, she is currently working on her PhD thesis in which she investigates persuasive mechanisms in public health discourse online. She compares language use on different smoking cessation sources with the aim of shedding light on the interpersonal dimension of persuasion; that is, how it exploits relational as well as informational aspects of language.

Michael W. Wagner

Michael W. Wagner is assistant professor and Louis A. Maier Faculty Development Fellow in the School of Journalism and Mass Communication at the University of

Wisconsin-Madison. He holds an affiliated appointment in the Department of Political Science. His research focuses on questions related to political communication and has appeared in *Journalism Practice, Journalism and Communication Monographs, Annual Review of Political Science, Political Research Quarterly*, and many other academic journals and edited volumes. He is the coauthor of *Political Behavior of the American Electorate* from CQ/Sage Press.

Alan S. Weber

Alan S. Weber, PhD, is an associate professor of English who teaches the first-year writing seminar in humanities in the Pre-medical Program at Weill Cornell Medical College in Qatar (WCMC-Q). Dr Weber previously taught literature, writing, and the history of science and medicine at Cornell University, Ithaca, The Pennsylvania State University, and Elmira College. His research interests include language, history, and the social and cultural dimensions of science and medicine. He is the editor of *19th Century Science* (2000), and *Because It's There: A Celebration of Mountaineering Literature* (2001), and is the author of specialized publications on Shakespeare, women in medicine, and seventeenth century medicine. He also publishes frequently on Arabian (Persian) Gulf education, sociology, and science and technology studies.

Lisl Zach

Dr Lisl Zach is an associate teaching professor at Drexel University's College of Computing and Informatics. Her research interests include studying the information-seeking behaviors of a wide range of users and investigating ways of measuring and communicating the value of information services in diverse settings. She is also involved with specific questions related to health information literacy and providing health information to vulnerable populations. Dr Zach has published award-winning articles on the contributions of information services in hospitals and academic health science centers and on the ways in which various groups look for, evaluate, and use information. Dr Zach holds a PhD in information studies from the University of Maryland—College Park, an MBA from New York University, and an MSLS from the University of North Carolina—Chapel Hill.

Albert Zeyer

Albert Zeyer, Dr med., dipl. math., is professor in medical education at Berne University of Applied Sciences, Health Division, Switzerland, and senior lecturer in Science and Health Education at the University of Zurich, Institute of Education, Switzerland. He worked for many years as a mathematics and science teacher and as a medical doctor in various Swiss schools and hospitals. His research interests include Science|Environment|Health, cognitive style and motivation to learn science, and public understanding of science. Albert Zeyer is a strand co-chair and, together with coauthor Alla Keselman, a co-coordinator of the Special Interest Group Science|Environment|Health at ESERA (European Science Education Research Association) conference. Dr Zeyer is the author of recent peer-reviewed articles in the International Journal of Science Education, and Journal of Research in Science Teaching. He is also a coeditor (with Regula Kyburz-Graber) of the book entitled *Science|Environment|Health: Towards a renewed pedagogy for science education* (Springer) and of the Science|Environment|Health special issue of the *International Journal of Science Education*.

Editors' foreword

This volume began as a gleam in the eye of coeditor Smith, who was frustrated as a doctoral student by the Ad Hoc Committee on Health Literacy of the American Medical Association and its definition of health literacy: the "constellation of skills, including the ability to perform basic reading and numerical tasks required to function in the health care environment." As author Logan notes in this volume, this definition seems to exclude factors external to clinical care settings. Authors Lence and Capozza address the same distinction between "consumer" and "patient" in regards to digital environments: "consumers" when they are engaged in making decisions about obtaining health care (such as choosing a health plan) and "patients" when they are "interacting directly with health-care providers and services about personal health concerns." Consumers and patients are two different, albeit considerably overlapping, groups of people, neither of which ceases to exist when they move away from under the clinical gaze. In order to most effectively provide health information to them that has meaning in their everyday lives, we need to meet both consumers and patients where they live and understand what that clinical gaze does to the interaction and the information.

Smith and Keselman became collaborators on research in consumer health vocabulary, but have since expanded their portfolio to include work on exchange of health information in nonclinical settings, such as public libraries. Out of this collaboration was born this book. The call for chapters was sent to a diverse set of audiences: practitioners in librarianship and informatics as well as researchers and teachers in fields from library and information studies, to medical informatics, to linguistics, to journalism, to health communications.

The book is divided into three parts. The *Overview* section features chapters on topics which are foundational to understanding health information: health communication (Kreps and Neuhauser) and health literacy (Logan). The *Libraries* section is devoted to discussions of consumer health information. Smith's historical overview examines the professional literature of librarianship and the dominant themes in that literature regarding challenges of medical information provision to the public, regardless of library type. There follow chapters on ethical information provision (McKnight); the public library (Flaherty); consumer information services in the hospital library setting (Seeger); and the considerably underexplored community college library (Chernaik). Finally, the *Contexts* section explores different settings in which health information for the public is delivered: outreach for cancer prevention in the developing nation of Qatar (Sayed and Weber); older adults (Smith); health disparities populations in the community center (Dalrymple and Zach); the science classroom (Zeyer, Levin, and Keselman); online bulletin boards (Von Rohr); digital environments (Lence and Capozza); and the media (Wagner).

The editors hope that by publishing this volume they have prompted more discussion and more study of this important topic by interested consumers and patients everywhere.

Catherine Arnott Smith, PhD
Alla Keselman, PhD

Acknowledgments

Capozza and Lence

The authors wish to thank their colleagues and team members at HealthInsight who contributed to the projects described in this chapter, and the funding agencies which made them possible: The Office of the National Coordinator for Health Information Technology, the Centers for Medicare and Medicaid Services, the Center for Technology and Aging, the National Eczema Association, and the Rasch Foundation.

Hogan Smith

The author gratefully acknowledges the expert review and input of Dr Patricia Sawyer, Director of the Gerontology Education Program at the University of Alabama at Birmingham Comprehensive Center for Healthy Aging, in finalizing the content for this chapter.

Sayed and Weber

The authors report no personal or financial conflicts of interest related to this research. No human or animal experimental subjects were used in this research. The authors wish to thank the Qatar Cancer Society for providing background information, brochures, and in-depth interviews necessary for completion of the research.

Smith

The author expresses her considerable gratitude to two other authors cited in her chapter, in chronological order. The first is Audrey Powers, editor of *The ethics and problems of medical reference services in public libraries*, a summary of the September 1979 Bay Area (California) Reference Center Workshop that opened Smith's eyes to the timeless quality of consumer health questions. The second is Ellen Hull Poisson, whose 1983 doctoral dissertation for Columbia University was a tremendously important resource informing not only this chapter, but Smith's own thinking about medical information and the public.

Von Rohr

I would like to thank the Swiss National Science Foundation for funding the research project Language and Health Online (143286), of which this study is a part. I am grateful to the editors Catherine Arnott Smith and Alla Keselman for their helpful comments. Thanks also to Miriam Locher and Franziska Thurnherr for their valuable comments on different drafts of this chapter. Finally, thanks to Mirjam Wilhelm for double coding a part of the data.

Overview

Overview

Designing health information programs to promote the health and well-being of vulnerable populations: the benefits of evidence-based strategic health communication

Gary L. Kreps[1], Linda Neuhauser[2]
[1]George Mason University, Virginia; [2]University of California, Berkeley

1.1 Introduction

Health communication is the "the central social process in the provision of health-care delivery and the promotion of public health" (Kreps, 1988, p. 238). The most important resource for promoting health is relevant health information that can guide people's health decisions and can motivate them to adopt healthy behaviors. And, relevant health information can only be accessed through the *process* of communication. Access to relevant, timely, accurate, and persuasive health information can enable people's active participation at home, in the health-care system and in community settings.

Yet, the communication process is complex and fragile! It often breaks down and can confuse and frustrate some groups, especially when it is not culturally sensitive, appropriate, engaging, and informative. Health-care providers and health information specialists often underestimate the difficulties inherent in sharing relevant health information with consumers, especially with consumers who come from diverse cultural backgrounds and have limited levels of health literacy. Members of vulnerable populations include those who are at greatest risk for negative health outcomes. Vulnerable populations often include people who are poor, have low levels of education, are immigrants, are elderly, have disabilities, and/or are members of minority groups. It is critical that people in such vulnerable populations have access to relevant, accurate, timely, up-to-date, understandable, culturally sensitive, actionable, and easy-to-use health information to make good health decisions about avoiding health risks, responding to health problems, and promoting their personal health.

Over a half-century of health communication research provides strong evidence that strategic health communication efforts can help reduce health risks, disease incidence, morbidity, and mortality, as well as improve quality of life for at-risk populations by enabling at-risk individuals to make informed decisions about avoiding health risks, identifying health problems at an early stage when they are most treatable, and getting the best care to address health issues (Kreps, 2003, 2012a, 2012b; Kreps & Sivaram, 2008; Kreps & Sparks, 2008; Neuhauser & Kreps, 2003, 2010).

However, research also shows that many efforts to communicate health fail to meet their goals, or have only modest effects (National Research Council, 2000). A key weakness is that traditional health communication is often overly generic and not adequately aligned with the abilities, preferences, and life situations of specific audiences (Emmons, 2000; Neuhauser & Kreps, 2010). Traditional health communication weaknesses are greatly magnified for vulnerable populations who may require significant adaptation for communication content, format, and delivery. Unfortunately, health messages for at-risk groups are often presented in ways that are overly technical, complex, and foreign from the experience of many of the people who might benefit the most from the health information presented. As a result, those most in need of health communication may receive the least benefit from it.

Creating and disseminating relevant information about health issues to vulnerable populations is challenging and—as documented by research outcomes—often ineffective. Health information specialists, such as health educators, health promotion experts, campaign designers, health librarians, journalists, science writers, and others, who prepare educational materials for consumers, face tremendous obstacles to design and deliver effective health information to these groups. These obstacles include a range of complex societal, cultural, educational, physical/sensory, and linguistic and literacy-based health communication barriers. Therefore, it is imperative to design and deliver health messages for these populations that are easy-to-access, clear, interesting, and personally relevant to the audiences for which the messages are designed.

This chapter will examine (1) challenges to develop and deliver effective health information to vulnerable populations; (2) community-based, participatory strategies to improve the design and dissemination of health communication that better meets the needs of these populations; and (3) research methods to monitor and evaluate the effects of health communication for at-risk populations.

1.2 Barriers

Serious sociocultural communication barriers exist that can complicate health communication efforts targeted to vulnerable populations (Kreps & Kunimoto, 1994). These barriers are often related to significant differences in the backgrounds and experiences between those who are creating and disseminating health information and the at-risk audiences they want to influence (Kreuter & McClure, 2004). Key sociocultural divides that can complicate health communication efforts include differences in health beliefs, education levels, health literacy levels, access to communication channels, access and function needs (disabilities) to obtain and use information, digital literacy levels, as well as English language proficiency (Kreps, 2005, 2012b; Neuhauser et al., 2007; Neuhauser & Kreps, 2008). These cultural differences typically make it difficult for health information specialists to communicate relevant health information to at-risk populations because they do not have well-developed shared frames of reference (Kreps, 2005; Kreps & Sparks, 2008; Thomas, Fine, & Ibrahim, 2004). Strategic health communication programs should be designed to overcome these intercultural communication barriers to health education by providing vulnerable populations with personally relevant, timely, accurate, and motivating information that supports their unique individual needs and helps them to achieve their specific health goals.

Low levels of health literacy is potentially one of the most pressing cultural barriers that should be addressed when developing strategic health communication programs for vulnerable populations (Gazmararian, Williams, Peel, & Baker, 2003; Neuhauser & Paul, 2011). The US Institute of Medicine has defined health literacy as "the degree to which individuals have the capacity to obtain, process and understand basic health information and services needed to make appropriate health decisions" (Nielsen-Bohlman, Panzer & Kindig, 2004). The World Health Organization (WHO) has defined it as "the cognitive and social skills which determine the motivation and ability of individuals to gain access to, understand and use information in ways which promote and maintain good health" (Nutbeam, 2008; WHO, 1998). The concept of health literacy covers a broad range of communication-related abilities such as reading, comprehension, speaking, and numeracy (the ability to use and understand numbers in everyday life).

Population-based data from the US National Assessment of Health Literacy (NAAL) estimate that about half of American adults have low health literacy skills and are likely to have difficulty understanding and acting on health information (Kutner, Greenberg, Jin, Paulsen, & White, 2006). In addition, only about 12% would have the "proficient" level skills required to make informed decisions about complex information related to managing diseases and choosing health insurance. Studies of health literacy in the European Union show similar results: about half of adults are estimated to have limited health literacy (HLS-EUC Consortium, 2012). Health literacy skills tend to be lower among people with lower education, lower income, and those who are members of a minority group and/or are age 65 or older (Nielsen-Bohlman et al., 2004). Of these groups, older adults were found to have the lowest health literacy levels in the NAAL study.

The average American adult is estimated to read between the seventh and ninth grade levels (Doak, Doak, & Meade, 1996; Kirsch, Jungeblut, Jenkins, & Kolstad, 1993; National Work Group on Literacy and Health, 1998). The literacy levels of people who are deaf are of special concern. Although little evidence exists, members of the deaf community are estimated to read between the third and sixth grade levels (LaVigne & Vernon, 2003). Because so many people have difficulty understanding health information, it is recommended that readability of health information be aligned with people's reading skills. Unfortunately, over 1000 studies of the readability of print health materials have shown that they significantly exceeded the estimated reading skills of the audiences for whom they were developed (Rudd, Anderson, Oppenheimer, & Nath, 2007). Most studies have reported that readability of print health information materials is above the 10th grade, and college and graduate school levels are commonly found in more complex health information, including that which relates to health risks and taking medications.

The notable mismatch between the literacy demand of health information materials and people's health literacy skills puts many people at risk of misunderstanding health information. For example, the NAAL survey (Kutner et al., 2006) estimated that only about half of American adults would be able to understand enough from medication labels to know how to take their medication at the right time and at the right dose. Another study of 600 consent forms for medical procedures found that readability was at the college level (Hopper, TenHave, Tully, & Hall, 1998)—indicating that only about 5% of these forms could be understood by the average adult in the US people with limited health literacy are less likely to use preventive services and are more likely to have uncontrolled chronic conditions, take medications incorrectly, be hospitalized, report poor health status, and use costly health services. It is critically important to consider health literacy when developing communications for the lay public.

Research is also defining factors that affect the reading ease and usability of health information for general lay audiences, and especially for more vulnerable populations. Although there is no one set of such characteristics, they are often called "health literacy principles," "clear communication," or "plain language" design criteria. Such criteria include font size and type, formatting, list length, interactivity, Web navigation, and many other factors. US government reports, such as the Centers for Medicare & Medicaid Services *Toolkit for Making Written Material Clear and Effective* (2012), the U.S. Department of Health and Human Services' *Quick Guide to Health Literacy* (n.d.), the Centers for Disease Control and Prevention's *Simply Put* (2009), and others describe these principles and how to apply them to health communication materials.

People with limited English proficiency (LEP) are another important vulnerable group to consider when designing health communication strategies. In the US, in 2011, about 9% of people who spoke languages other than English at home were estimated to have LEP (Ryan, 2013). Because the majority of health communications are in English only, many people who have LEP face major barriers in accessing and understanding important health information. People with LEP are at especially high risk for medication errors, repeated hospitalizations, poor cancer screening rates, and other problems (Wilson, Chen, Grumbach, Wang, & Fernandez, 2005). A further issue is that LEP populations disproportionately include people with limited health literacy skills (especially recent immigrants with low education levels); LEP can contribute to and exacerbate the health consequences of low health literacy (Sentell & Braun, 2012; Sentell, Braun, Davis, & Davis, 2013). People with multiple vulnerabilities are among those with the greatest need for health information, and such communication is complicated to develop and deliver.

People with low socioeconomic status (SES) are also an important vulnerable group to consider when planning health communication. It can be difficult to capture the attention of low-SES at-risk groups with health messages because they are often dealing with many other pressing issues. This hinders audience exposure to health messages and is often a major roadblock to successfully disseminating health information (Fishbein & Hornik, 2008; Hornik, 2002; Slater, 2004). As a result, health information specialists view for audience attention within a complex multichannel information environment, and at-risk populations may not consider traditional health communication messages as interesting or exciting as those from competing sources of information on television, radio, the Internet, and other channels, especially since at-risk audiences often possess relatively low literacy levels (Randolph & Viswanath, 2004). A further issue is that low-SES and other at-risk populations tend to be less receptive to information about health threats, which they may perceive as uncomfortable or frightening, especially when they are already trying to manage many other life issues (Icard, Bourjolly, & Siddiqui, 2003). For such information to be effective, it is especially important that low-SES groups understand the specific importance of the messages to benefit them personally or someone they care about. In addition, low-SES populations may have multiple vulnerabilities, such as limited health literacy and low English proficiency—all of which pose additive challenges to effective health communication.

Finally, access and functional abilities (and disabilities) are important factors to keep in mind. As noted above, deaf populations typically have very low English reading skills. Deaf audiences may require communications in American Sign Language or in other special linguistic and/or technological formats—such as mobile video (Neuhauser, Ivey et al., 2013). People with visual impairments may need to receive health communication in larger font sizes, Braille or spoken through a computer screen reader or audiotape. People with mobility issues may have trouble manipulating communication in certain formats and may need adaptive communication aides to access information and to navigate within health-care environments. The US Census Bureau estimated that 18.7% of American noninstitutionalized adults had a disability in 2010, and that 12.6% had a severe disability (Brault, 2012).

Given the many barriers that vulnerable groups face in accessing, being engaged with, understanding, and effectively acting on health information, it is not surprising that traditional generic health communication strategies have often shown disappointing results. To address these challenges, there is increasing attention to the responsibilities and skills of communicators.

1.3 Lessons learned: improving health communication for vulnerable populations

During the past 30 years, researchers and practitioners have paid increasing attention to examining issues about communicating with at-risk audiences and developing better communication strategies. A key lesson learned is that health communicators should be careful to use terminology and concepts related to health that can be easily understood by consumers with low levels of health literacy. Such consumers are likely to have difficulty understanding the science behind health promotion and how to follow complicated disease prevention recommendations. Messages need to be framed to match consumers' level of understanding and their unique backgrounds. This is especially important when communicating with consumers who may be recent immigrants and who may also have low English proficiency problems, minority group members who may have particular health issues and concerns, and older adults and others who may have limited levels of health literacy.

The overall learning from many decades of health communication research is that health information specialists must develop "strategic"—rather than "generic" messages—to address health literacy and many other communication challenges. Such messages are designed to explain health information clearly to targeted audience members by using familiar language and examples that can promote a fuller understanding about how to use health information to enhance their health (Parker & Kreps, 2005). Strategic health communication efforts can help reduce confusion concerning complex health issues and help at-risk populations make informed decisions about the best ways to avoid health risks and address health problems. Well-informed consumers are prepared to participate more fully in disease prevention and health promotion efforts, and are empowered to adopt health promoting behaviors, such as appropriate forms of exercise, nutrition, and risk avoidance, as well as to engage in early detection and screening tests for serious health problems. Providing consumers with relevant, timely, and understandable health information can also encourage them cooperate with prescribed therapeutic regimens (Kreps et al., 2011).

To effectively address the aforementioned communication challenges for at-risk consumers, health information specialists should consistently develop clear, easy-to-understand, linguistically and culturally appropriate communication strategies to break through the confusion these consumers often confront when dealing with health issues. Further, health information specialists should develop engaging and persuasive messages that motivate members of vulnerable populations to adopt healthy behavior recommendations. We now examine strategies for addressing the challenges to communicate relevant health information to vulnerable populations and recommend best practices for designing meaningful messages and effectively using relevant media to promote people's health.

1.4 Strategies to develop strategic communication

Researchers and practitioners are increasingly using the following evidence-based approaches to develop strategic health communication for at-risk groups. These approaches help meet the challenge by moving away from traditional, generic "expert" messages to those that are oriented to consumer needs and preferences (Kreps, 1996).

1.4.1 Audience analysis

Strategic health communication efforts are informed through careful analysis of the backgrounds, health orientations, and communication competencies of the individuals who are in need of more customized health information, particularly members of vulnerable populations. Health information specialists should identify important background factors, such as strongly held health beliefs, level of understanding of key health issues, participation in health care, lifestyle practices, willingness to accept new health information, readiness to adopt recommended health strategies, familiarity and level of use of different media and communication channels, level of participation in communication networks, and audience members' judgments about the credibility of different information sources. This information can be used as a good foundation to design communication programs that are appropriate and effective with targeted audience members.

Specific kinds of audience analysis data that can help guide strategic health communication efforts include demographic information about targeted audience members (such as their age, gender, race, ethnicity, education, income, etc.); audience members' current and past health-related behaviors; communication characteristics (such as media use patterns, media preferences, literacy levels, and language preferences); knowledge, attitudes, values, and emotions related to the health topics; cultural habits and preferences; effective motivational factors; and potential barriers to accepting information and changing health behaviors. Collectively, these data can provide important insights into targeted audience members' perceptions and beliefs about health issues, identify any perceived concerns about targeted health issues, and help health information specialists to develop evidence-based communication strategies for reaching, educating, and influencing them.

1.4.2 Defining criteria for strategic message design

As described above, careful audience analysis is essential to identify the salient consumer characteristics that can be used to guide message design (Kreps, 2002, 2013). At an early stage, health communicators can use this analysis to answer important questions related to communication factors about members of targeted audiences. For example, what are the typical message exchange and information sharing processes employed by targeted groups of consumers? Who do these consumers typically talk to and acquire health-related information from? Who do they trust? How do they receive and provide social support? What are their predispositions for interpreting health messages? What are most influential factors to persuade them to engage with and respond positively to health messages? Which communication channels do they prefer to use? What are the best ways to provide these consumers with feedback about their health behaviors that can promote and reinforce health behavior changes? What are the most influential communication strategies for developing cooperative and trusted relationships with members of targeted groups?

With this background information, health communicators can begin to define criteria for creating communication that is responsive to audience communication patterns, needs, and interests, and that is easily accessible and culturally appropriate. Further, communicators can plan communication features so that messages are adaptive to changing social situations, motivational and reinforcing, as well as engaging, interesting, and interactive.

1.4.3 User-centered design

In addition to conducting audience analyses and planning specific communication objectives, participatory, or user-centered design is a powerful way to continuously test and revise communication strategies. User-centered design has been defined as "an approach to the assessment, design and development of technological and organizational systems that places a premium on the active involvement of...potential or current users of the system in design and decision-making processes" (Computer Professionals for Social Responsibility, 2005). User-centered design approaches originated in the 1970s in the fields of architecture, engineering, computer science, and other sociotechnical fields, and are now being adopted in health and social sciences—in particular in health communication (Neuhauser, 2001; Neuhauser, Kreps, Morrison et al., 2013). This approach consists of interconnected "feedback loops" of defining issues and developing and testing solutions. These are also called "build and evaluate loops" (Markus, Majchrzak, & Gasser, 2002).

In health communication, user-centered design methods consist of a variety of approaches to iteratively engage the intended audience members in defining issues and cocreating messages and distribution strategies. Methods can include focus groups, in-depth interviews, usability testing, observations, surveys, and other ways of gathering input from the targeted audiences throughout the development and evaluation of the communication. For example, *usability testing* refers to a broad range of structured methods to engage users in designing communication or other products (Nielsen, 2000). A common form of usability testing consists of in-person cognitive interviews and observations of individual audience members to test and revise draft communication messages or resources. Multiple rounds of such usability tests and resulting revisions help elicit problems and ensure that the final messages appeal to key beliefs, attitudes, and values of targeted audience members, and use familiar and accepted language, engaging images and examples to illustrate key points.

This kind of participatory testing can be used after the audience analysis is conducted, communication objectives are set, and draft messages/resources are developed. User-centered design can be helpful to uncover subtle, but important audience needs and characteristics that may have been missed during the initial audience analysis. In this case, even though patients identified their perceived needs at the outset of the planning phase (audience analysis), participatory techniques continued to identify ever-finer preferences and suggestions for improvement over time (Neuhauser, Kreps, Morrison et al., 2013). In another project, usability testing proved essential to develop effective communication about health plan choices for people with multiple communication challenges. This population consisted of low-SES, low-LEP consumers who also had disabilities. Sequential rounds of usability testing and revisions were necessary to create effective resources for this population (Neuhauser, Rothschild, Graham, Ivey, & Konishi, 2009).

Pretesting sample health education messages with representatives of targeted audiences is important to make sure the messages hit the intended mark with these audiences before implementing health communication intervention programs. Formative evaluation data gathered through message pretesting are a form of user-centered design in which health messages are shaped and further refined by actively gathering feedback from representatives of the actual audiences intended to benefit from the intervention. Pretesting can also enhance audience members' receptivity to and cooperation with health promotion efforts. Involving consumers, their family members, key members of their social networks, and community representatives can increase social support and encourage people to pay attention to and accept the information, and to use it to make positive health changes (Kinzie, Cohn, Julian, & Knaus, 2002; Minkler, 2000; Minkler & Wallerstein, 2008).

1.4.4 Complementary message strategies

Another lesson learned from communication research is that health information is more effective when presented through multiple reinforcing messages delivered through different complementary communication channels at several points over time. This multiple complementary message strategy is informed by the communication principles of redundancy and reinforcement to enhance message exposure and impact (Donohew, Lorch, & Palmgreen, 1998). These principles acknowledge the importance of repeating health messages because long-term exposure is often necessary for people to understand new concepts, especially for at-risk populations. Multimedia approaches help accommodate people's varying preferences for receiving information, and help keep people's attention by presenting information in different, "fresh" ways. In addition, research is increasingly demonstrating the power of using narrative (storytelling) and vivid imagery in illustrations and photos to educate and motivate audiences who have limited health literacy. These strategies are also useful for audiences who have problems with numeracy (understanding numerically presented information, such as statistics and numerical risk estimates) (Maibach & Parrott, 1995).

1.4.5 Tailored communication

Another effective approach to design customized health messages that meet the unique needs, backgrounds, and preferences of specific individuals is tailored communication (Neuhauser & Kreps, 2008). To tailor communication, key characteristics of an individual's background such as their name, age, cultural affiliations, or health status are gathered and then used to personalize messages sent to that person.

The following example is a typical scenario: a health plan sends a reminder letter to Sharon Smith who is 60 years old and has not had a mammogram in the past year. Sharon is African-American and has a sister with breast cancer. The letter Sharon receives says "Dear Ms. Smith, Our records show that you are due for your mammogram. It is especially important for African American women with a family history of breast cancer to have regular mammograms. Please call...." Tailored communication systems usually employ interactive computer systems to gather relevant background information from consumers concerning key demographic and communication variables. These data can be gathered from existing databases (such as at a health plan) or by surveying the individual about their demographics, psychographics, and health beliefs/behaviors.

Once key background information is obtained, the information is used to select specific messages that match the unique background features of the individuals, from those stored in a library of health education messages. In this way, information about the individual health risks and cultural characteristics of a specific consumer—for example, a Latino male who has diabetes—will automatically be selected and used to provide appropriate health content to that consumer. As the consumer continues to interact with the tailored health information system, the system continuously gathers and stores new information about the individual that can be used to further tailor messages in the future.

1.4.6 Communication channels

Identifying the best communication channels is another important challenge in developing effective health communication for at-risk populations. The best communication channels are those that are close, familiar, and easily accessible for targeted audience members (Maibach, Kreps, & Bonaguro, 1993). For example, it would be a serious error to develop an online health education Web site for consumers who do not have access to computers or who are not sophisticated computer users. Further, it is important to select communication channels that are dramatic and memorable (Maibach, Kreps, & Bonaguro, 1993). Another consideration is that the communication channels should be accessible over time so that the audiences can have multiple exposures to the messages and retain information to review later. Interactive channels that enable consumers to ask questions and receive clarifications about complex health information are especially effective. And, as noted above, multimedia approaches can present complementary, repeated messages that reinforce health messages—and their influence—over time.

1.4.7 Credibility

Another lesson learned from communication research is that it is important to decide what the best, most trusted sources are for delivering key health messages to audiences—especially for at-risk audiences who often are less trusting of new information (Maibach & Parrott, 1995). Decisions need to be made about whether it is best to utilize familiar sources of information, expert sources, or peers. Sometimes, several kinds of sources can be used together. For example, in a successful statewide parenting education program in California, parents received videos that featured advice from well-known parenting experts, celebrities, and typical (peer) parents (Neuhauser, Rothschild, & Rodriquez, 2007). Message-testing research (pretesting) that examines the impact of different communication sources (using different media channels) on targeted audience members can help communicators make good decisions about the best representatives to employ to deliver health information.

1.5 Evaluating health communication

Evaluation research should be built into all phases of health promotion efforts (Kreps, 2013). Although traditional evaluation designs for health programs have tended to emphasize baseline and outcome stages, the emerging trend is to have much more emphasis on formative evaluation and on iterative changes to the communication intervention (Neuhauser, Kreps, & Syme, 2013). This emerging model draws from theory and methods of the design sciences, user-centered and participatory design and action research (Minkler & Wallerstein, 2008; Neuhauser, Kreps, & Syme, 2013). In those kinds of evaluations, the process of building and gradually testing the communication intervention is just as important as determining the effects. As noted earlier, Markus et al. (2002) described this approach as "build and evaluate loops." This focus on iterative, user-centered evaluation processes is especially important for health communication interventions for vulnerable populations in which creating effective communication depends on uncovering and addressing multiple, and often complex, audience needs and preferences. In other words, better user-centered design of messages requires more user-centered evaluation activities.

As described above, health communication efforts should begin with careful needs analysis and audience analysis to identify the best goals, targets, and strategies for the interventions. This audience analysis is a type of formative, user-centered design. Although these early "planning stages" are not always considered part of traditional evaluation designs, in our view, this critical first phase should be considered a formal component of the evaluation and be carefully examined and reported. In the next evaluation phase, the messages designed and channels identified for delivering health promotion messages should be carefully pretested with representatives from targeted groups to make sure they communicate effectively to these groups and meet the objectives of the intervention. This message-testing phase is essential to iteratively refine and improve message strategies. Various kinds of usability testing can also be used to test consumer access to, comfort with, and ability to effectively use communication channels and tools. By carefully documenting the results of each of these formative evaluation activities, communication developers will be well informed about potential weaknesses and strengths of the communication *before* the outcome evaluation is completed and project resources are exhausted.

After the health communication planning and initial message-testing phases, the communication intervention is ready to implement and assess for outcomes. At this point, it is important to establish clear baseline measures of the targeted consumers' understanding, attitudes, and behaviors before the program is launched (Kreps, 2013). These baseline measures can be used as a starting point for tracking the influences of communication efforts. Feedback mechanisms, such as consumer surveys, observations, focus groups, hotlines, help-desks, comment cards, and others can be used to monitor and evaluate consumer exposure and reaction to health messages. In keeping with user-centered evaluation designs, these methods can also be used to refine the health communication, if necessary. The methods are effective strategies to track the influences of communication efforts on consumer beliefs, behaviors, and even their physiological outcomes. Further, it is important to conduct cost–benefit analyses to determine whether the health communication program is cost effective.

1.6 Practice implications

What policies and best practices are needed to guide effective communication efforts about health promotion and their evaluation? Most importantly, communication interventions to educate vulnerable populations need to be strategic and evidence based. Decades of research indicate that such health promotion efforts are highly complex and without careful planning and data collection, can show disappointing results. Fortunately, communication researchers and practitioners can draw on a growing body of research about the value of iterative, user-centered strategies to design and refine health communication programs. Approaches such as audience analysis, message testing, and continuous feedback loops from the targeted audiences are critical to help health information specialists adopt culturally sensitive communication practices to reach and influence vulnerable populations to make informed decisions about adopting prevention recommendations, changing their negative health habits, and choosing healthy lifestyles. Highly sophisticated strategies, such as computer-tailored communication, now make it possible to align communication messages with detailed consumer needs and preferences. Community participatory communication interventions are a valuable strategy to integrate consumers' perspectives into health education efforts and build stronger community commitment to health communication interventions. Including stakeholders, such as providers, policy makers, and others in the participatory design can further strengthen the communication approach and foster longer-term sustainability of the program.

Success in developing health communication that is better aligned with the needs of at-risk groups will also require major changes to traditional evaluation designs. Pre- and postevaluation focused primarily on outcomes is not sufficient to provide the detailed data essential to plan and refine strategic communication programs. Therefore, it is important for consumers to not only collaborate with health information specialists in designing culturally sensitive health communication programs, but also to help implement, evaluate, and sustain these programs.

For these reasons, it is important to provide health communication training for both health information specialists and consumers to enhance the quality of communication efforts.

The development and implementation of strategic health communication interventions holds great promise for promoting the health of vulnerable populations. Investing in the development of strategic, evidence-based, and collaborative health communication efforts can help establish a sound public health infrastructure for disseminating effective health information.

References

Brault, M. W. (July 2012). *Americans with disabilities: 2010. (Household economic studies)*. Retrieved from http://www.census.gov/prod/2012pubs/p70-131.pdf.

Centers for Disease Control and Prevention. (2009). *Simply put*. Retrieved from http://www.cdc.gov/healthliteracy/pdf/Simply_Put.pdf.

Centers for Medicare & Medicaid Services. (2012). *Toolkit for making written material clear and effective*. Retrieved from http://www.cms.gov/WrittenMaterialsToolkit.

Computer Professionals for Social Responsibility. (2005). *Participatory design*. Retrieved from http://cpsr.org/issues/pdf.

Doak, L., Doak, C., & Meade, C. (1996). Strategies to improve cancer education materials. *Oncology Nursing Forum, 23*, 1305–1312.

Donohew, L., Lorch, E. P., & Palmgreen, P. (1998). Applications of a theoretic model of information exposure to health interventions. *Human Communication Research, 24*, 454–468.

Emmons, K. M. (2000). Behavioral and social science contributions to the health of adults in the United States. In *National Research Council, promoting health: Intervention strategies from social and behavioral research* (pp. 254–321). Washington, DC: The National Academies Press.

Fishbein, M., & Hornik, R. (2008). Measuring media exposure: an introduction to the special issue. *Communication Methods and Measures, 2*(1), 1–5.

Gazmararian, J. A., Williams, M. V., Peel, J., & Baker, D. W. (2003). Health literacy and knowledge of chronic disease. *Patient Education and Counseling, 51,* 267–275.

HLS-EUC Consortium. (2012). *Comparative report of health literacy in Eight EU Member states: The European Health Literacy Survey (HLS-EUC).* Retrieved from http://www.healthliteracy.ie/wp-content/uploads/2012/09/HLS-EU_report_Final_April_2012.pdf.

Hopper, K. D., TenHave, T. R., Tully, D. A., & Hall, T. E. (1998). The readability of currently used surgical/procedure consent forms in the United States. *Surgery, 123,* 496–503.

Hornik, R. C. (2002). Exposure: theory and evidence about all the ways it matters. *Social Marketing Quarterly, 8*(3), 30–37.

Icard, L. D., Bourjolly, J. N., & Siddiqui, N. (2003). Designing social marketing strategies to increase African Americans' access to health promotion programs. *Health and Social Work, 28,* 214–223.

Kinzie, M. B., Cohn, W. F., Julian, M. F., & Knaus, W. A. (2002). A user-centered model for web site design: needs assessment, user interface design, and rapid prototyping. *Journal of the American Medical Informatics Association, 9,* 320–330.

Kirsch, I., Jungeblut, A., Jenkins, L., & Kolstad, A. (1993). *Adult literacy in America: A first look at the results of the national adult literacy survey [NCES 1993-275].* U.S. Department of Education, Office of Educational Research and Improvement, Institute of Education Sciences. Retrieved from http://nces.ed.gov/pubs93/93275.pdf.

Kreps, G. L. (1996). Promoting a consumer orientation to health care and health promotion. *Journal of Health Psychology, 1,* 41–48.

Kreps, G. L. (1988). The pervasive role of information in health and health care: Implications for health communication policy. In J. Anderson (Ed.), *Communication yearbook 11* (pp. 238–276). Newbury Park, CA: Sage.

Kreps, G. L. (2002). Enhancing access to relevant health information. In R. Carveth, S. B. Kretchmer, & D. Schuler (Eds.), *Shaping the network society: Patterns for participation, action and change: May 16-May 19, 2002, Seattle, Washington: Proceedings.* Seattle, WA: Computer Professionals for Social Responsibility.

Kreps, G. L. (2003). The impact of communication on cancer risk, incidence, morbidity, mortality, and quality of life. *Health Communication, 15*(2), 161–169.

Kreps, G. L. (2005). Disseminating relevant information to underserved audiences: implications from the Digital Divide Pilot Projects. *Journal of the Medical Library Association, 93*(4), 65–70.

Kreps, G. L. (2012a). Strategic use of communication to market cancer prevention and control to vulnerable populations. [Special issue] *Comunicação e Sociedade,* 11–22.

Kreps, G. L. (2012b). Strategic communication for cancer prevention and control: reaching and influencing vulnerable audiences. In A. Georgakilas (Ed.), *Cancer prevention* (pp. 375–388). Vienna: Intech.

Kreps, G. L. (2013). Evaluating health communication interventions. In D. K. Kim, A. Singhal, & G. L. Kreps (Eds.), *Health communication: Strategies for developing global health programs* (pp. 352–367). New York: Peter Lang.

Kreps, G. L., & Kunimoto, E. (1994). *Effective communication in multicultural health care settings.* Newbury Park, CA: Sage.

Kreps, G. L., & Sivaram, R. (2008). The central role of strategic health communication in enhancing breast cancer outcomes across the continuum of care in limited-resource countries. *Cancer, 113*(S8), 2331–2337.

Kreps, G. L., & Sparks, L. (2008). Meeting the health literacy needs of vulnerable populations. *Patient Education and Counseling, 71*(3), 328–332.

Kreps, G. L., Villagran, M. M., Zhao, X., McHorney, C., Ledford, C., Weathers, M., et al. (2011). Development and validation of motivational messages to improve prescription medication adherence for patients with chronic health problems. *Patient Education and Counseling*, *83*, 365–371.

Kreuter, M. W., & McClure, S. M. (2004). The role of culture in health communication. *Annual Reviews of Public Health*, *25*, 439–455.

Kutner, M., Greenberg, E., Jin, Y., Paulsen, C., & White, S. (2006). *The health literacy of America's adults: Results from the 2003 National Assessment of Adult Literacy.* Retrieved from http://nces.ed.gov/pubs2006/2006483_1.pdf.

LaVigne, M., & Vernon, M. (2003). An interpreter isn't enough: deafness, language and due process. *Wisconsin Law Review*, *5*, 843–935.

Maibach, E. W., Kreps, G. L., & Bonaguro, E. W. (1993). Developing strategic communication campaigns for HIV/AIDS prevention. In S. Ratzan (Ed.), *AIDS: Effective health communication for the 90s* (pp. 15–35). Washington, DC: Taylor and Francis.

Maibach, E. W., & Parrott, R. (Eds.). (1995). *Designing health messages: Approaches from communication theory and public health practice.* Thousand Oaks, CA: Sage.

Markus, M. L., Majchrzak, A., & Gasser, L. A. (2002). Design theory for systems that support emergent knowledge processes. *Management Information Systems Quarterly*, *26*, 179–212.

Minkler, M. (2000). Using participatory action research to build healthy communities. *Public Health Reports*, *115*(2–3), 91–197.

Minkler, M., & Wallerstein, N. (Eds.). (2008). *Community based participatory research for health: Process to outcomes* (2nd ed.). San Francisco: Jossey-Bass.

National Research Council. (2000). *Promoting health: Intervention strategies from social and behavioral research.* Washington, DC: The National Academies Press.

National Work Group on Literacy and Health. (1998). Communicating with patients who have limited literacy skills: report of the National Work Group on Literacy and Health. *Journal of Family Practice*, *46*, 168–176.

Neuhauser, L. (2001). Participatory design for better interactive health communication: a statewide model in the U.S.A. *Electronic Journal of Communication/La Revue Electronique de Communication*, *11*(3–4). Retrieved from http://www.cios.org/EJCPUBLIC/011/3/01134.HTML.

Neuhauser, L., Constantine, W. L., Constantine, N. A., Sokal-Gutierrez, K., Obarski, S. K., Clayton, L., et al. (2007). Promoting prenatal and early childhood health: evaluation of a statewide materials-based intervention for parents. *American Journal of Public Health*, *97*(10), 813–819.

Neuhauser, L., Ivey, S. L., Huang, D., Engelman, A., Tseng, W., Dahrouge, D., et al. (2013). Availability and readability of emergency preparedness materials for deaf and hard of hearing and older adult populations: issues and assessments. *PLoS One*, *8*(2), e55614. http://dx.doi.org/10.1371/journal.pone.0055614.

Neuhauser, L., & Kreps, G. L. (2003). Rethinking communication in the e-health era. *Journal of Health Psychology*, *8*(1), 7–22.

Neuhauser, L., & Kreps, G. (2008). Online cancer communication interventions: meeting the literacy, linguistic, and cultural needs of diverse audiences. *Patient Education and Counseling*, *71*(3), 365–377.

Neuhauser, L., & Kreps, G. (2010). eHealth communication and behavior change: promise and performance. *Journal of Social Semiotics*, *20*(1), 9–27.

Neuhauser, L., Kreps, G. L., Morrison, K., Athanasoulis, M., Kirienko, N., & Van Brunt, D. (2013). Using design science and artificial intelligence to improve health communication: ChronologyMD case example. *Patient Education and Counseling*, *92*(2), 211–217. http://dx.doi.org/10.1016/j.pec.2013.04.006.

Neuhauser, L., Kreps, G. L., & Syme, S. L. (2013). Community participatory design of health communication programs: methods and case examples from Australia, China, Switzerland and the United States. In D. K. Kim, A. Singhal, & G. L. Kreps (Eds.), *Global health communication strategies in the 21st century: Design, implementation and evaluation.* New York: Peter Lang.

Neuhauser, L., & Paul, K. (2011). Readability, comprehension and usability. In B. Fischhoff, N. T. Brewert, & J. S. Downs (Eds.), *Communicating risks and benefits: An evidence-based user's guide.* Silver Spring, MD: U.S. Department of Health and Human Services.

Neuhauser, L., Rothschild, B., Graham, C., Ivey, S., & Konishi, S. (2009). Participatory design of mass health communication in three languages for seniors and people with disabilities on Medicaid. *American Journal of Public Health, 99,* 2188–2195.

Neuhauser, L., Rothschild, R., & Rodriquez, F. M. (2007). MyPyramid.gov: assessment of literacy, cultural and linguistic factors in the USDA food pyramid website. *Journal of Nutrition Education and Behavior, 39*(4), 219–225.

Nielsen, J. (2000). *Designing web usability.* Indianapolis: New Riders Publishing.

Nielsen-Bohlman, L., Panzer, A., & Kindig, D. A. (Eds.). (2004). *Health literacy: A prescription to end confusion.* Washington, DC: The National Academies Press. Retrieved from http://www.iom.edu/~/media/Files/Report%20Files/2004/Health-Literacy-A-Prescription-to-End-Confusion/healthliteracyfinal.pdf.

Nutbeam, D. (2008). The evolving concept of health literacy. *Social Science and Medicine, 67,* 2072–2078.

Parker, R., & Kreps, G. L. (2005). Library outreach: overcoming health literacy challenges. *Journal of the Medical Library Association, 93*(4), 78–82.

Randolph, W., & Viswanath, K. (2004). Lessons learned from public health mass media campaigns: marketing health in a crowded media world. *Annual Review of Public Health, 25,* 419–437.

Rudd, R. E., Anderson, J. E., Oppenheimer, S., & Nath, C. (2007). Health literacy: An update of public health and medical literature. *Review of adult learning and literacy, 7,* 175–204.

Ryan, C. (2013). *Language use in the United States: 2011 (American community survey reports).* Washington, DC: U.S. Census Bureau. Retrieved from http://www.census.gov/prod/2013pubs/acs-22.pdf.

Sentell, T., & Braun, K. L. (2012). Low health literacy, limited English proficiency, and health status in Asians, Latinos, and other racial/ethnic groups in California. *Journal of Health Communication, 17,* 82–99.

Sentell, T., Braun, K. L., Davis, J., & Davis, T. (2013). Colorectal cancer screening, low health literacy and limited English proficiency among Asians and Whites in California. *Journal of Health Communication, 18,* 242–255.

Slater, M. D. (2004). Operationalizing and analyzing exposure: the foundation of media effects research. *Journalism and Mass Communication Quarterly, 81*(1), 168–183.

Thomas, S. B., Fine, M. J., & Ibrahim, S. A. (2004). Health disparities: the importance of culture and health communication. *American Journal of Public Health, 94,* 2050.

U.S. Department of Health and Human Services. (n.d.). *Quick guide to health literacy fact sheet: Health literacy and health outcomes.* Retrieved from http://www.health.gov/communication/literacy/quickguide/factsliteracy.htm.

Wilson, E., Chen, A. H., Grumbach, K., Wang, F., & Fernandez, A. (2005). Effects of limited English proficiency and physician language on health care comprehension. *Journal of General Internal Medicine, 20*(9), 800–806.

World Health Organization. (1998). *Health promotion glossary.* http://www.healthliteracypromotion.com/upload/hp_glossary_en.pdf. Accessed July 2014.

Health literacy research's growth, challenges, and frontiers

2

Robert A. Logan
U.S. National Library of Medicine, Maryland

2.1 Introduction

This chapter discusses how health literacy research has evolved. It is designed to provide a brief history, review some of health literacy's conceptual underpinnings, note the range of current research, and address some current research gaps, as well as encourage future research.

This chapter introduces the rapid expansion of health literacy research into diverse areas of clinical medicine, the health-care delivery system, nonclinical settings, and health communication areas. This chapter also explores some of the current challenges in health literacy research and provides a brief conclusion.

Although this chapter emphasizes health literacy research and developments in the US, some international health literacy work is included. The chapter's intent is not to review or summarize the health literacy research literature, or provide a systematic review of research findings—as the latter was published recently (Agency for Healthcare Research and Quality, 2011; Berkman et al., 2011).

The topics within the chapter's other four sections are four milestones (and other developments) in health literacy research's growth; health literacy's evolving definition and conceptual underpinnings, the spectrum of health literacy research, health literacy research's current needs and frontiers, as well as a conclusion.

Before moving to these topics, Smith's (2009) and Nutbeam's (2000, 2008) contextualization of the role and scope of health literacy provide this chapter's conceptual inspiration. Smith (2009, 2011) and Smith and Moore (2011, 2012) suggest suboptimal life skills progression, unfavorable individual and family clinical outcomes (such as depression), substandard health status, risk behaviors, lower quality of life, higher likelihood of family dysfunction, less effective maternal care, less efficient use of health services, and deficient public health may be associated with low maternal health literacy. Conversely, Smith (2009, 2011) and Smith and Moore (2011, 2012) suggest more optimal life skills progression, improved individual and family clinical outcomes (such as depression management), improved health status, healthier behaviors, more efficient use of health services, a higher quality of life, enhanced family interactions, more effective maternal care, and enhanced public health may be associated with higher maternal health literacy. Under Smith's (2009) and Nutbeam's (2000, 2008) integrative and multidimensional framework, health literacy is a core individual and social determinant of individual and community health and may be associated with the development (or reduction) of individual, family, and public health disparities. In addition, maternal health literacy is suggested by Smith to be a pragmatic intervention strategy that fosters a life skill which enhances individual, family, community, clinical, and public health outcomes (Carroll, Smith, & Thomson, 2015; Smith & Moore, 2012; Wollesen & Peifer, 2006). Smith's (2009) and Nutbeam's (2000, 2008) conceptual inspiration is informed (and sometimes challenged) by three decades of health literacy scholarship, which is introduced in the next section.

2.2 Four milestones in health literacy research

Parker and Ratzen (2010) described health literacy as a new and evolving research and practice discipline within clinical medicine and public health. These authors provided a timeline of health literacy's diffusion and acceptance in clinical medicine, health-care administration, and public health in the US within the first decade of the twenty-first century. Parker and Ratzan's timeline covers 23 developments (emphasizing US governmental, health-care delivery system, and public health administrative initiatives), which include the initial definition of health literacy in 2000 (discussed in the next section), as well as the results from the US National Assessment of Adult Literacy (NAAL) and the 2006 Surgeon General's Workshop (discussed below).

In contrast with Parker and Ratzen's (2010) focus on administrative developments, this section of this chapter describes health literacy's diffusion in terms of four research milestones. The first two led to health literacy's acceptance as a research construct, and the third and fourth fostered health literacy's use as a constructive intervention within the health-care delivery community. In addition, there is discussion of the reasons that health-care organizations, public health practitioners, insurers, pharmaceutical companies, health foundations, and government health agencies recently became interested in health literacy interventions. Similar to Parker and Ratzen (2010), the interest in health literacy research is presented as a convergence of diverse elements among health-care delivery organizations, government, as well as investigators. This section also covers a 2011 systematic review that provided a challenging overview of the health literacy field's progress and promise (Agency for Healthcare Research and Quality, 2011; Berkman et al., 2011).

2.2.1 Milestone 1: Health literacy and adult literacy

Adult literacy often is conceived as a combination of reading skills, document comprehension, writing skills, as well as functional abilities to understand pragmatic information, such as directions, instructions, numerical descriptions, and maps (Institute of Medicine, 2004). While *health literacy* research encompasses the latter issues, the field focuses on a different set of abilities, skills, tasks, and conceptual underpinnings within clinical care, public health, health-care management, patient/consumer, home, and other health information-seeking contexts that are discussed below (Institute of Medicine of the National Academies, 2004). For example, Nutbeam (2000) described health literacy skills within three categories: functional/technical skills (ability to read and understand numbers), interactive/social skills (listening, speaking), and critical thinking skills (the ability to integrate information within new life situations and challenges).

Just as health communication developed as a separate discipline within the broader fields of interpersonal and mass communication, and consumer health informatics as a research branch within biomedical informatics, health literacy has become a distinct discipline within the broader fields of literacy, health education, and health communication (Institute of Medicine of the National Academies, 2004, 2006; Keselman, Logan, Smith, LeRoy, & Zeng-Treitler, 2008). Health literacy's research interests dovetail with many other disciplines including not only consumer health informatics, but numeracy, cultural competence, plain language, eHealth, mHealth, patient activation, patient health self-management, health information seeking, shared decision making, health prevention, adult literacy, risk assessment, and the public understanding of science (Coulter, 2011; Hibbard, Mahoney, Stockard, & Tusler, 2005; Keselman et al., 2008; Logan, 2014; see the Consumer Health Informatics Research Resource at http://chirr.nlm.nih.gov/health-literacy.php). The National Library of Medicine PubMed bibliographic gateway to biomedical literature provides a gateway to research published in major medical and public health research journals. PubMed's topic-specific query page on health literacy encompasses the aforementioned disciplines and continues to expand as relevant research emerges from related subdisciplines (National Library of Medicine, 2015).

The initial milestone fostering health literacy's development as a separate conceptual and research construct may have been the Doaks' observations that health literacy and adult literacy seemed to reflect different skill sets. The Doaks' observations, based on their experiences as health communicators and educators, suggested that there might not be an association between a person's level of adult literacy and his or her health literacy—the understanding of health information, medical terms, and health information seeking (Doak & Doak, 1987; Doak, Doak, Friedell, & Meade, 1998; Doak, Doak, & Root, 1996; Doak, Doak, & Meade, 1996).

Other researchers noted a positive association among levels of adult literacy with educational attainment and health knowledge proficiency. However, the Doaks explained that their health educational experiences suggested a person's age, income, educational attainment, and adult literacy levels sometimes did not predict a person's understanding of medicine and medical terms, as well as consumer interest in seeking information about health and medicine (Doak & Doak, 1987; Doak et al., 1998; C. C. Doak et al., 1996; L. G. Doak et al., 1996). Overall, the Doaks challenged researchers to assess the comparative degree that health literacy—not adult literacy, cultural competence, or English proficiency—predicted patient and public health outcomes, as well as the degree to which health literacy was distinguishable from adult literacy. The Doaks also provided an array of suggestions about improving the effectiveness of patient educational initiatives via the assessment of reading difficulty and cultural appropriateness of patient materials. This work provided a foundation for some contemporary health literacy practices.

2.2.2 Milestone 2: The NAAL

A second milestone in health literacy's development as a separate research construct occurred with the confirmation of prior scholarship via the research methods—as well as the results—of the 2003 US NAAL, published in 2006 (Kutner, Greenberg, Jin, & Paulsen, 2006; White, 2008). In the first and only nationally generalizable sample that assessed diverse adult literacy skills level in the US, the NAAL used separate measures of adult reading skills and health literacy (White, 2008). This methodological decision reinforced and operationalized prior research, which suggested levels of health literacy and adult reading skills were different research issues, or represented two research constructs. The separation of adult reading skills and health literacy into different constructs provided impetus for the idea—influenced by the Doaks and suggested by others—that health literacy-oriented research interventions should be assessed via tailored health literacy instruments (Institute of Medicine of the National Academies, 2004).

In addition, the NAAL findings suggested that health literacy and adult reading skills, as well as other literacy levels of US adults, were inconsistent (Kutner et al., 2006; National Center for Education Statistics; White, 2008). The levels of health literacy proficiency, or the ability to understand medical terms and information, were somewhat dissimilar from some parallel measures of adult reading skills. So, besides the methodological separation of health and adult reading skills, the NAAL findings provided generalizable evidence that adult literacy and health literacy might be conceptually distinct. Both the methodological separation and eventual research findings suggested the importance of separating health literacy and adult literacy into self-contained research constructs, which helped reinforce prior qualitative and quantitative findings.

In the past several years, similar findings in the US and internationally seem to have settled the question about prioritizing health literacy constructs to measure audience understanding of health and medical terms and information (Berkman, Davis, & McCormack, 2010; Institute of the Medicine of the National Academies, 2013; McCormack, Haun, Sorensen, & Valerio, 2013; Pleasant, 2014; Royal College of General Practitioners, 2014). There has been expansion of literature that assesses health literacy instruments, that is, quantitative, applied measures of health literacy, and raises questions about which instrument to use. Certainly, this suggests a tacit consensus that health literacy constructs should be utilized to measure the public understanding of health and medicine. For example, a description of a few contemporary health literacy instruments is outlined within the "health literacy" section of the National Library of Medicine's Consumer Health Informatics Research Resource (CHIRR) (http://www.nlm.nih.chirr.gov). Exemplar instruments include the Rapid Estimate of Adult Literacy in Medicine (REALM), the Test of Functional Health Literacy in Adults (TOFHLA), and the Newest Vital Sign (NVS). Haun, Valerio, McCormack, Sorensen, and Paasche-Orlow (2014) recently identified and compared 51 health literacy research instruments while O'Neill, Goncalves, Ricci-Cabello, and Ziebland (2014) compared a subset of 35 self-administered health literacy instruments. Haun et al. (2014) as well as O'Neill et al. (2014) describe some strengths and weaknesses of some commonly used health literacy instruments, while both articles stress the need for further assessment of the psychometric rigor underlying health literacy instruments. While we will return to this issue later in this chapter, it must be emphasized that analyses of the quantitative (or qualitative) merits of health literacy instruments probably would not occur without some prior establishment of health literacy as a self-contained research construct.

2.2.3 Milestone 3: Interventions

The 2003 NAAL findings provided the first of two more milestones that influenced the adoption of health literacy interventions within clinical settings, health-care organizations, and related venues. In a widely cited finding, the NAAL reported that only 12% of Americans were proficient in understanding medical terms or information (Kutner et al., 2006). That is, 88% of Americans either had below basic, basic, or an intermediate understanding of basic medical terms and information. The suboptimal levels of health literacy proficiency in the US provided an empirically generalizable snapshot of the underwhelming public understanding of medical terms and health information. International research findings suggest that similarly low levels of health literacy exist in other nations (Institute of Medicine of the National Academies, 2013; Royal College of General Practitioners, 2014; Sorensen et al., 2012).

As Parker and Ratzan (2010) noted, the NAAL findings added to the mounting evidence that the US health literacy was a foundational clinical practice and public health concern. Within a short period after its release, many US organizations, such as the Institute of Medicine (IOM), the American Medical Association (AMA), the US Surgeon General, and agencies within the U.S. Department of Health and Human Services cited the NAAL data and similar findings to reinforce calls to action to improve public understanding, as well as foster more interest in health literacy research and initiatives (Koh et al., 2012; Paasche-Orlow, Wilson, & McCormack, 2010; Parker & Ratzen, 2010). Hence, the exposure of the nation's low health literacy status provided some momentum to intervene, and encouraged organizations on the national, state, and local levels to support diverse health literacy initiatives and research.

Prior to the official release of the NAAL findings, a 2006 conference hosted by the US Surgeon General noted a sense of urgency in addressing the nation's health literacy gaps (Office of the Surgeon General, 2006). This conference presented diverse research that assessed health literacy efforts to address clinical challenges. Examples include improving provider/patient communication and communication between health-care organizations and patients. In addition, scholarship was presented that focused on health literacy initiatives within broader public health settings. The conference's combination of research interests helped establish a subtle, but important foundation that health literacy research should focus on clinical medicine and public health as well as a range of public health and patient-oriented interventions. Underlying these developments was a tacit interest in the degree that health literacy represented an individual and social determinant of the nation's health.

Similar research interests and conceptual underpinnings also were evident prior or parallel to the release of the NAAL findings at the US Health Literacy Annual Research Conferences that began in 2008. There was also activity through the public workshops held by the US IOM's Health Literacy Roundtable, the Institute for Healthcare Advancement, and other conferences in North America and around the world.

Once a narrow discipline, health literacy research blossomed in the twenty-first century as the urgency and extent of the problems surrounding the public understanding of health and medicine became apparent. This research fostered concurrent interest to respond among health-care providers, health care and provider organizations, public health practitioners, insurers, pharmaceutical companies, health foundations, and government health agencies (Institute of Medicine of the National Academies, 2004, 2006; Parker & Ratzen, 2010).

2.2.4 Milestone 4: Outcomes

The US Surgeon General's 2006 conference additionally advanced a second milestone that influenced the adoption of health literacy initiatives and interventions among health care and other organizations. During the conference, Baker (2006) reported a series of findings that suggested health literacy, measured as an independent construct, might be a robust predictor of patient health, such as mortality and self-reported health status, as well as a few health-care administrative outcomes, such as hospitalization rates. These findings (among others) suggested health literacy might be an empirically robust predictor even after controlling for social-demographic variables, such as education and income levels, that previously were positively associated with favorable health outcomes (Baker, 2006). The latter findings provided some empirical support for the Doaks' earlier observations that health literacy was a foundational and predictive demographic variable. The findings also advanced the idea that health-care organizations might turn to health literacy initiatives as a pragmatic tactic to address an array of patient, community, public health, and institutional challenges (Doak & Doak, 1987; Doak et al., 1998; C. C. Doak et al., 1996; L. G. Doak et al., 1996).

Since the Surgeon General's 2006 conference, an array of research has suggested health literacy interventions are associated with some favorable patient care and health system outcomes. Koh et al. (2012) and Bailey, Oramasionwu, and Wolf (2013) note, for example, clinical benefits including reduced mortality, improved patient adherence to medical instructions, and enhancement of overall patient safety. The National Network of Libraries of Medicine summarized findings showing that health literacy interventions therapeutically assist patients with cancer, diabetes, asthma, and hypertension (Almader-Douglas, 2013). Koh et al. (2012) also described health administrative benefits linked to health literacy interventions: improved diabetes patient self-management skills, more use of preventive services, and a reduction in hospitalization and re-hospitalization rates, which, in aggregate, lower medical costs.

Overall, these and many other findings advanced health literacy interventions *as a tactic* to influence patient health and health delivery system outcomes. This is because these interventions seem to be well-focused and less expensive than the larger-scale social interventions often promoted as strategies to enhance patient and community health. For instance, a Robert Wood Johnson Foundation initiative announced in 2014 (Robert Wood Johnson Foundation, 2014) advocates comprehensive interventions to build a healthier nation. These interventions include improving educational opportunities for vulnerable children and enhancing the quality of life in low-income areas. These recommendations are undergirded by research that suggests individual and community health outcomes are associated with improvements in key sociodemographic indicators, such as income and educational levels (Center on the Developing Child at Harvard University, 2010).

In contrast, some health literacy research suggests comparatively modest initiatives. Examples are strategies designed to boost patient and community understanding of health and specific conditions; foster more health self-management skills; improve quality of patient interaction at key points of health-care interaction (such as before hospital discharge); and generate health information seeking (often using new, less costly digital technologies). These initiatives *also* foster therapeutic individual and public health outcomes. As a result, health literacy interventions comparatively provide a portfolio of pragmatic and affordable strategies for health-care organizations to consider—especially if they are interested in enhancing social welfare in addition to improving individual and overall health-care organizational outcomes.

Overall, health literacy research received a boost from these findings. They suggested that health literacy interventions may impact desirable health and organizational outcomes. Health literacy initiatives emerged as a comparatively prudent and constructive investment to improve individual and community health, as well as address quality improvements and expensive challenges within the health-care delivery system.

Certainly, the US Joint Commission's inclusion of health literacy within their accreditation standards in 2011 (National Committee for Quality Assurance, n.d.) reinforced that health literacy initiatives and interventions represented pragmatic strategies to address clinical care and broader administrative health-related issues.

More recently, innovations in the US Patient Protection and Affordable Care Act, such as Patient-Centered Medical Homes and Accountable Care Organizations, also encouraged a new range of programs and institutional accountability that embraced health literacy initiatives and evaluation research (Koh, Baur, Brach, Harris, & Rowden, 2013; Koh et al., 2012; Parker & Ratzen, 2010).

In addition to these research, administrative, and health policy initiatives, the potential of health literacy interventions and health education/communication activities simultaneously was enhanced by changes in mass media and health information technology. The sudden growth of eHealth and mHealth provided a comparatively less expensive, interactive technological infrastructure for health literacy interventions sponsored by health-care and public health organizations (Kreps, 2012; Kreps & Neuhauser, 2010; Logan, 2014; Neuhauser & Kreps, 2003).

In summary, the interest in health literacy initiatives and research was accelerated by a convergence of elements. These included health information/communication technological changes; research findings about health literacy's desirable impact on patient and health administrative outcomes; the comparative advantages of health literacy as a tactic of clinical, organizational, and social intervention; population research that suggested widespread, low health literacy levels; and increasing government and private health-care organizational interest (Kreps & Neuhauser, 2010; Logan, 2014; Neuhauser & Kreps, 2003; Parker & Ratzen, 2010).

On the other hand, a 2011 systematic review of health literacy research challenged some of the previous inferences about health literacy interventions' impact on clinical and health-care organizations. This served to provide a more tempered overview of the field's accomplishments and promise (Agency for Healthcare Research and Quality, 2011; Berkman et al., 2011). The systematic review found health literacy interventions did not consistently result in improved clinical or health-care delivery system outcomes. The systematic review also did not find health literacy was a more robust, or predictive variable to influence clinical and administrative outcomes compared to other demographic measures, such as income and educational levels, and health status (Agency for Healthcare Research and Quality, 2011; Berkman et al., 2011).

The systematic review did suggest that health literacy interventions remained one among several coordinated strategies to improve clinical and health-care delivery system outcomes. Hence, health literacy's impact as a comparative individual and social determinant of health was advanced as a significant topic for future assessment.

Additionally, the review suggested (Agency for Healthcare Research and Quality, 2011; Berkman et al., 2011) that the health literacy field needed more basic research about the reliability and validity of prevailing health literacy constructs, as well as more consistency in defining health literacy. The adoption of a common health literacy definition and an underlying conceptual model, as well as more rigorous evaluations of research instruments, are the topics of the next section.

However, the good news for future researchers is the emerging yet inconsistent evidence about health literacy's potential use and therapeutic value in clinical medicine, organizational health-care delivery, and public health. The inconsistency presents important research opportunities. While the field's growth and expansion has been brisk, thanks to the convergence of elements noted above, the basic research questions about the therapeutic or beneficial impact of health literacy initiatives and interventions require future assessment opening doors to new frontiers.

2.3 Health literacy's evolving definition and conceptual underpinnings

This section outlines the range of recent efforts to define health literacy, as well as the possible impacts of the field's lack of a consensus definition.

The definition known as the "Calgary Charter" was adopted in 2008 by the participants at a meeting in Canada, organized to provide a new definition of health literacy. This definition recognized health literacy's role as an individual and social determinant of health (Coleman et al., n.d.) The Calgary Charter defined health literacy as such:

Health literacy allows the public and personnel working in all health-related contexts to find, understand, evaluate, communicate, and use information.

Health literacy is the use of a wide range of skills that improve the ability of people to act on information in order to live healthier lives.

These skills include reading, writing, listening, speaking, numeracy, and critical analysis, as well as communication and interaction skills (Coleman et al., n.d.)

CHIRR reports on the evolution of health literacy's definition prior to the articulation of the Calgary Charter. It concludes that the field was formed from diverse emphases that started with the abilities of individuals to adjust to clinical settings, which was augmented by definitions that increasingly included the sociocultural dimensions of health literacy. For example, the Ad Hoc Committee on Health Literacy of the American Medical Association (1999) defined health literacy as the "constellation of skills, including the ability to perform basic reading and numerical tasks required to function in the health-care environment." This definition included skills related to everyday health functions such as the "ability to read and comprehend prescription bottles, appointment slips, and other essential health-related materials" (Ad Hoc Committee, 1999). Although the latter definition captured some key elements of health care, the IOM noted that it seemed to exclude factors external to clinical care settings (Institute of Medicine of the National Academies, 2004). Smith (2009) also criticized the AMA's definition as an individual "deficit" model of health literacy.

In 2000, the IOM's developed a new definition. This viewed health literacy more broadly as an intermediate outcome that interacts with forces from culture and society, as well as individual characteristics, such as educational levels and health status.

[T]he degree to which individuals have the capacity to obtain, process, and understand basic health information and services needed to make appropriate health decisions (Ratzan & Parker, 2000; Selden, Zorn, Ratzan, & Parker, 2000).

The latter definition was revised slightly by several authors who emphasized health literacy should help individuals make *informed* rather than *appropriate* decisions. The revised definition was the following.

The degree to which individuals can obtain, process, understand, and communicate about health-related information needed to make informed health decisions (Berkman et al., 2011, 2010; McCormack et al., 2010).

On an international level, the World Health Organization (WHO) had previously expanded the definition of health literacy to incorporate the behavioral motivations and social skills that empowered individuals to take control of their health (Nutbeam, 1998). The WHO also maintained health literacy should not only help recipients make informed clinical decisions, but assist them with prevention and health self-maintenance skills. Nutbeam (2008) noted that the WHO's definition envisioned health literacy as a more multidimensional process that empowered individuals to take control of their health. The WHO's definition reported in Nutbeam (1998) was the following.

[T]he cognitive and social skills which determine the motivation and ability of individuals to gain access to, understand, and use information in ways which promote and maintain good health.

CHIRR reports that the latter perspectives recognized the importance of individuals' motivations and abilities to *obtain, understand,* and *use* information to make informed health decisions and interact successfully with the health-care system, rather than simply possess literacy skills. The U.S. Department of Health and Human Services' Office of Disease Prevention and Health Promotion conceptually added that health literacy was influenced by communication skills, some knowledge about health topics, as well as culture (e.g., language, norms), the demands of the health care and public health systems, and situational/contextual considerations.

Other conceptions advocated that health literacy should be defined in a way that encompassed an individual's abilities to navigate the health-care system (including filling out complex forms and locating providers and services), share personal information with providers, engage in self-care and chronic-disease management, and understand quantitative concepts (numeracy skills) such as probability and risk. Numeracy skills also might include specific competencies, such as calculating cholesterol and blood sugar levels, measuring medications, and understanding nutrition labels (CHIRR, n.d.).

Sorensen et al. (2012) noted some other definitions, perspectives, and conceptual models focused on health literacy in terms of disease treatment, disease prevention, or health promotion. Some of these health literacy definitions and conceptual underpinnings tended to reinforce the role of health literacy in health and disease prevention, as opposed to disease treatment, and in enhancing a person's self-perceived quality of life.

Sorensen et al. (2012) also provided a sophisticated schema to compare an array of existing health literacy definitions and underlying conceptual models.

Overall, the Calgary Charter's early health literacy definition represented an effort to be more conceptually inclusive and provide a consensus definition that was mindful of many of the field's diverse conceptual frameworks. While it remains to be seen whether the Calgary Charter—or one of the definitions discussed above—will be widely adopted by scholars internationally, it is important for a new researcher in the field to understand that the selection of a health literacy definition communicates a conceptual direction. It is important to select a health literacy definition that is consistent with the research questions or hypotheses one seeks to address, as well as a conceptual focus one hopes the field will adopt.

Pleasant (2013, 2014) agreed with Sorensen et al. (2012) that the subtle differences among health literacy definitions and underlying conceptual models suggested the field's conceptual diversity. But Pleasant argued that the lack of a consensus about health literacy's definition—coupled with the presence of diverse underlying conceptual models—inadvertently challenged the field's long-range scholarly credibility and gravitas. Pleasant noted that many of the field's diverse definitions and conceptual models fostered the use of diverse and sometimes incompatible research instruments (Haun et al., 2014; Pleasant, 2013, 2014; Pleasant, McKinney, & Rikard, 2011). Haun et al. (2014) recently described the diversity of the 51 health literacy measures they identified as "instrument proliferation" (p. 301). Similar to the conclusions of the Agency for Healthcare Research and Quality's systematic review (2011), Pleasant (2014) explained that current diverse and often incompatible research instruments make it difficult to aggregate research findings, as well as compare results that might yield more systematic insights about conceptual models and instruments.

Pleasant (2013, 2014) and McCormack et al. (2013) also found that many health literacy quantitative research instruments have not been undergirded by basic research that demonstrates their psychometric rigor. Altin, Finke, Kautz-Freimuth, & Stock's systematic review (2014) found only 17 articles focused on the development and validation of 17 health literacy instruments. Pleasant (2013, 2014) suggests that these issues inhibit investigators' abilities to make informed choices about the optimal health literacy instrument and underlying conceptual model.

Despite health literacy research's recent growth, Pleasant (2013, 2014) warned the field's credibility might be compromised by its current diversity and disagreements about underlying definitions, conceptual models, instruments, and methods. As a remedy, Altin, Finke, Kautz-Freimuth, and Stock (2014), Haun et al. (2014), McCormack et al. (2013), O'Neill et al. (2014), and Pleasant (2013) noted the need for conceptual transparency, as well as more basic research and validation studies about health literacy's diverse instruments.

These researchers suggest a road map for a branch of future health literacy research that focuses on finding a consensus health literacy definition, as well as efforts to boost the psychometric rigor underlying health literacy's constructs, instruments, and research methods. More positively, the latter needs create a range of opportunities for research that could be foundational to the field's evolution as a more controlled and systematic science.

2.4 The range and vitality of health literacy research

The vitality of health literacy research is perhaps best illustrated by outlining the spectrum of areas in clinical medicine, health-care organizations, and other areas where health literacy research is conducted. This section briefly lists a few topics within broader research areas.

The CHIRR (n.d.) list of broad health literacy research topics includes health literacy research concepts, methods, and tools; social demographics; clinical outcomes; health behaviors; health-care providers; health-care delivery system; health educational and prevention interventions; health educational and prevention interventions (K-12, higher, and adult education); and health policy.

A second illustration of the diversity of health literacy research is found in the National Library of Medicine's Medical Subject Headings (MeSH) (http://www.ncbi.nlm.nih.gov/mesh). MeSH is used to index the literature published within PubMed MEDLINE. As a demonstration of the range of health literacy research distributed within MeSH headings, we selected controlled vocabulary terms in three categories: acute disease ("heart diseases"), chronic disease ("diabetes mellitus"), and health-care organization ("patient discharge"). Then, we chose three recent health literacy publications within each category.

For example, three recent, selected health literacy research citations indexed about "heart diseases" are the following:

- Chen et al.'s (2013) findings that health literacy promotes self-care among patients with heart failure.
- Dracup et al.'s (2014) assessment of health literacy levels among heart failure patients.
- Adeseun, Bonney, and Rosas' (2012) findings that health literacy is associated with blood pressure risk factors.

Three recent, selected health literacy research citations under the "diabetes mellitus" MeSH heading include these studies:

- Mbaezue et al.'s (2010) findings that health literacy impacts patient self-monitoring of blood glucose among patients in an inner city hospital.
- Mulvaney, Lilley, Cavanaugh, Pittel, and Rothman's (2013) validation of a diabetes numeracy test with adolescents with Type 1 diabetes.
- Osborn et al.'s (2011) findings that health literacy explains racial disparities in diabetes medication adherence.

Three recent, selected health literacy research citations under the "patient discharge" MeSH heading include these studies:

- Smith, Brice, and Lee's (2012) findings of associations among functional health literacy and adherence to emergency department discharge instructions in Spanish-speaking patients.
- Mitchell, Sadikova, Jack, and Passche-Orlow's (2012) findings that health literacy levels impact patient postdischarge hospital utilization after 30 days.
- Cawthon, Walia, Osborn, Niesner, Schnipper, and Kripalani's (2012) observations about how health literacy improves patient care transition.

A review of MeSH terms found in combination with health literacy research show that this work has been investigated in disease domains as various as chemically induced, eye, immune system, musculoskeletal, occupational, nervous system, endocrine system, parasitic, respiratory, skin and connective tissue, virus, hemic, and lymphatic diseases, as well as cancer. Health literacy research has incorporated diagnoses including bacterial infection, congenital heredity, neonatal disease and abnormalities, female urogenital disease and pregnancy complications, and wounds and injuries.

Healthy literacy research has also encompassed health-care delivery or patient service domains including adolescent health services; child care; community health services; dental health services; health services for the aged; nursing services; nursing care; mental health services; preventive health services; rehabilitation services; and emergency medical services.

Five other health-care delivery, or patient service, concepts studied in the context of health literacy include health services misuse; medical errors; life support care; withholding of treatment; and terminal care.

The vitality and diversity of the health literacy field also is evident by visiting the aforementioned health literacy topic query within PubMed (National Library of Medicine, 2015). The health literacy topic query provides curated and comprehensive access to health literacy citations, abstracts, and some full text articles published in major biomedical and public health journals.

While the range of topics in health literacy research is impressive, it is by no means exhaustive, as we will explore in the next section.

2.5 Health literacy research's current needs and frontiers

This section suggests areas where more health literacy research is needed and notes some broader considerations about the integration of health literacy research with allied disciplines. Some of the field's significant research gaps and resulting needs have been noted by Rudd (2013) and Koh et al. (2013) recently. We must understand the gaps in our knowledge in order to encourage health literacy researchers to enter these areas—and shape new frontiers.

Aldoory, Ryan, and Rouhain's literature review (2014) recently found inattention to health literacy research within an enduring health communication topic—the challenges of patients and research participants to understand informed consent forms and instructions. These researchers found few extant studies about informed consent comprehension among adults with low health literacy. They additionally reported a dearth of research within community-based settings about informed consent comprehension. In contrast to the expansive research outlined in the previous section, Aldoory et al.'s (2014) findings underscore the need for more health literacy research within an acknowledged area of poor patient understanding and health-care delivery system challenges—an area where insights from health literacy research might be expected to occur (Institute of Medicine of the National Academies, 2014).

Similarly, Dewalt and Hink (2009) found that evidence-based assessments of health literacy interventions were less frequent than health literacy initiatives. Abrams, Klass, and Dreyer (2009) found a dearth of evidence-based assessments of health literacy interventions, especially within community-based interventions tailored to low-literacy populations.

Turning to other areas, Rudd (2013) recently outlined health literacy research gaps within an array of clinical and health-care organizational domains. Overall, Rudd (2013) noted health literacy research "must now provide insights for needed change to make information more accessible, tools more useable, information exchanges more productive, and navigation of institutions easier and more dignified" (p. 6).

Within the topic of provider/patient communication, Rudd (2013) found more health literacy research needed for questions about the reading tasks required of patients; understanding patient reading levels; enhancing the clarity of written and spoken communication; removing clinical jargon between health-care organizations and culturally diverse audiences; and the impact of a health-care professional's communication skills on health literacy and patient health outcomes.

Specific areas of health-care organizational/patient communication practices need further investigation in connection to health literacy, including the physical and social environments within health-care settings as barriers/catalysts to improve patient/consumer health literacy; and the creation of a more cheerful, empathetic environment in which health care is delivered in medical offices, hospitals, clinics, and other health-care organizations.

More health literacy research is needed in specific areas affecting health-care organizations, such as the clarity of written and spoken communication; assessments of the comparative degree of difficulty of written and posted health information routinely distributed by medical centers; the barriers that discourage people from health information seeking (such as problematic Web sites, phone interactions, poor maps, and signage); and the mismatch between the literacy demands of health materials (written or online) and the literacy skills of adults with a secondary school education. There are gaps in understanding about unnecessarily dense and complex forms (e.g., hospital entry, consent, and discharge forms), clear signage, and enhancing the capacity of visitors to navigate building entrances, passageways, and destination points within hospitals, clinics, and other health-care organizations.

Finally, health literacy needs to be investigated in connection to the rigorous pilot testing of materials intended for consumers and patients, the measurement and ranking of the communication skills of clinical and health-care professionals; and theory-driven research is needed based on assessments of consumer and patient interventions.

These areas in aggregate present an array of opportunities for future scholars.

Similarly, Koh et al. (2013) identified a range of research needs including how health literacy (in conjunction with health communication and health information technology) contribute to more shared decision making between patients and providers, and how health literacy strategies create personalized self-management tools and resources for patients. More health literacy research is needed regarding the creation of accurate, accessible, and actionable health information that is targeted or tailored for specialized audiences, such as persons who are medically underserved.

Koh et al. (2013) added there is a pressing need for research about foundational health-care administrative issues, such as how intraorganizational communication affects the processes and outcomes for health-care organizations that seek to be more health literate; what interpersonal and/or organizational strategies help reduce the burden on individuals to coordinate their own medical care; and how health information systems best engage patients and caregivers by providing easily understandable, personalized medical record data as well as other clinical information.

For US-based researchers, Koh et al. (2013) noted that there is an array of additional health literacy needs and opportunities to evaluate how health literacy interventions and initiatives advance the accountable care organizations and patient-centered medical homes initiatives introduced by the US Patient Protection and Affordable Care Act.

Altin et al. (2014), Pleasant (2013, 2014), and McCormack et al. (2013) noted a need for more systematic and comparative evaluations of health literacy instruments. Sorensen et al. (2012) identified a parallel need for more comparative evaluations of health literacy's underlying conceptual frameworks. In addition, the Agency for Healthcare Research Quality's health literacy systematic review also suggested that health literacy research needs more conceptual consistency, more research about its underlying measures, and a new emphasis on best research practices to generate data sets that provide a better foundation for future systematic analyses (Agency for Healthcare Research and Quality, 2011; Berkman et al., 2011).

In terms of frontiers, the author recently noted the need for research that bridges the conceptual overlaps and gaps among health literacy research and allied fields, such as health communication and consumer health informatics. As health information technology evolves and is used to monitor clinical biometrics, inform targeted audiences about health, and modify health-related behaviors, there is a pressing need for cross-disciplinary research that embraces consumer health informatics, health communication, and health literacy. There also may be a need to embrace research in related disciplines, such as the public understanding of science (Logan, 2014).

Health literacy research is needed within diverse areas of clinical care, public health, and health-care administration, as is research that bridges multidisciplinary boundaries. The array of aforementioned topics suggests there are abundant options for current and future health literacy researchers as well as potential new frontiers.

2.6 Conclusions

The paradox of health literacy research is that its evidence-based gaps become more evident as the field matures. While health literacy research's growth is expansive, even a bit breathtaking, scholars are constantly reminded that there are foundational areas of clinical, public health, and health administration where health literacy research has yet to occur (Aldoory, 2014); basic research needs to be conducted (McCormack et al., 2013; Pleasant, 2014) and inconsistent results point to the need for further research (Agency for Healthcare Research and Quality, 2011; Berkman et al., 2011).

For the moment, a synopsis of health literacy's impact is best summarized as more moderate than the comprehensive and empowering influence of health literacy suggested within the chapter's introduction (Smith, 2009, 2011; Smith & Moore, 2011, 2012). Instead of describing health literacy—or the specific instance of maternal health literacy—as a robust independent research variable, associated with therapeutic individual, social, cultural, clinical, and health-care organizational outcomes, the Agency for Healthcare Research and Quality's (2011) recent systematic review and the Health Literacy Roundtable of the Institute of Medicine of the National Academies (2004, 2006) suggest health literacy is an intermediate variable that interacts with other demographic, clinical, health-care organizational, and sociocultural influences. Although health literacy initiatives are cited as a constructive intervention to improve individual as well as social health outcomes, the extent to which health literacy *by itself* influences clinical health, health-care delivery, and public health outcomes remains to be demonstrated (Institute of Medicine of the National Academies, 2004, 2006).

Still, the latter suggests health literacy is an individual and social determinant of health. It opens the door to research to assess how health literacy compares to other individual, social, cultural, clinical, health-care organizational, community, and family influences. This provides a range of research opportunities and challenges for health literacy scholarship.

Finally, a future challenge to health literacy research is the lack of an international organization that integrates health literacy into a self-contained, academic discipline. Health literacy practice and research lack a member-based, international, academic society that hosts periodic meetings, publishes refereed journals, provides news about research and practice to members, encourages practitioner–scholar, practitioner–practitioner, scholar–scholar dialogue, establishes leadership, provides a clearinghouse for practitioner services, such as the Institute for Healthcare Advancement's health literacy listserv™, and organizes the field into a more coherent whole (Institute for Healthcare Advancement, n.d.). While some of these needs may be served by existing international conferences and organizations, there is no focal point for interaction and diffusion among the world's health literacy researchers and practitioners. In addition, there is no central organization to define health literacy professionalism, which lets the field's future be defined externally—by clinical medicine, public health, and other disciplines its scholars and practitioners represent. Certainly, the development of an international health literacy disciplinary infrastructure to elevate research and practice standards, similar to professional societies in clinical, medical administration, public health, and social science disciplines, remains an area where future researchers can contribute to the field's evolution, growth, gravitas, and success.

References

Abrams, M. A., Klass, P., & Dreyer, B. P. (2009). Health literacy and children: recommendations for action. *Pediatrics, 124*(Suppl. 3), S327–S331.
Ad Hoc Committee on Health Literacy for the Council on Scientific Affairs, American Medical Association. (1999). Health literacy: report of the Council of Scientific Affairs. *Journal of the American Medical Association, 281,* 552–557.
Agency for Healthcare Research and Quality. (2011). *Health literacy interventions and outcomes: An updated systematic review [Evidence report/technology assessment number 199].* Rockville, MD: Agency for Healthcare Research and Quality. Retrieved from http://www.ahrq.gov/downloads/pub/evidence/pdf/literacy/literacyup.pdf.
Aldoory, L., Ryan, K. E. B., & Rouhain, A. M. (2014). *Best practices and new models of health literacy for informed consent: Review of the impact of informed consent regulations on health literate communications.* In Informed consent and health literacy: A workshop Washington, DC: Institute of Medicine of the National Academies Health Literacy Roundtable. Retrieved from https://www.iom.edu/Activities/PublicHealth/HealthLiteracy/2014-JUL-28.aspx.
Almader-Douglas, D. (June 2013). *Culture in the context of health literacy update.* Retrieved from http://nnlm.gov/outreach/consumer/hlthlit.html.
Altin, S. A., Finke, I., Kautz-Freimuth, S., & Stock, S. (2014). The evolution of health literacy assessment tools: a systematic review. *BMC Public Health, 14,* 1207.
Andeseun, G. A., Bonney, C. C., & Rosas, S. E. (2012). Health literacy associated with blood pressure but not other cardiovascular disease risk factors among dialysis patients. *American Journal of Hypertension, 25*(3), 348–353.
Bailey, S. C., Oramasionwu, C. U., & Wolf, M. S. (2013). Rethinking adherence: a health literacy-informed model of medication self-management. *Journal of Health Communication, 18,* 20–23.
Baker, D. W. (2006). The associations between health literacy and health outcomes: self-reported health, hospitalization, and mortality. In Office of the Surgeon General (Ed.), *Proceedings of the Surgeon General's Workshop on improving health literacy.* Rockville, MD: Office of the Surgeon General (US). Retrieved from http://www.ncbi.nlm.nih.gov/books/NBK44260/#proc-healthlit.panel1.s14.
Berkman, N. D., Davis, T. C., & McCormack, L. (2010). Health literacy: what is it? *Journal of Health Communication, 15*(S2), 9–19.
Berkman, N. D., Sheridan, S. L., Donahue, K. E., Halpern, D. J., Viera, A., Crotty, K., et al. (2011). *Health literacy interventions and outcomes: An updated systematic review. Evidence report/technology assessment No. 199 [AHRQ publication number 11–E006].* Rockville, MD: Agency for Healthcare Research and Quality.
Carroll, L. N., Smith, S. A., & Thomson, N. R. (2015). Parents as teachers health literacy demonstration project: integrating an empowerment model of health literacy promotion into home-based parent education. *Health Promotion and Practice, 16*(2), 282–290.

Cawthon, C., Walia, S., Osborn, C. Y., Niesner, K. J., Schnipper, J. L., & Kripalani, S. (2012). Improving care transitions: the patient perspective. *Journal of Health Communication*, *17*(Suppl. 3), 312–324.

Center on the Developing Child at Harvard University. (2010). *The foundations of lifelong health are built in early childhood.* Retrieved from http://developing-child.harvard.edu/index.php/resources/reports_and_working_papers/foundations-of-lifelong-health/.

Chen, A. M., Yehle, K. S., Albert, N. M., Ferraro, K. F., Mason, H. L., Murawski, M. M., et al. (2013). Health literacy influences heart failure knowledge attainment but not self-efficacy for self-care or adherence to self-care over time. *Nursing Research and Practice*, *2013*, 353290. Published online 2013 Jul 24. doi: 10.1155/2013/353290.

Coleman, C., Kurtz-Rossi, S., McKinney, J., Pleasant, A., Rootman, I. & Shohet, L. (n.d.). *The Calgary charter on health literacy: Rationale and core principles for the development of health literacy curricula.* Retrieved from http://www.centreforliteracy.qc.ca/sites/default/files/CFL_Calgary_Charter_2011.pdf.

Consumer Health Informatics Research Resource. (n.d.). *Health literacy.* Retrieved from http://chirr.nlm.nih.gov/health-literacy.php.

Coulter, A. (2011). *Engaging patients in healthcare.* Berkshire, England: Open University Press.

Dewalt, D. A., & Hink, A. (2009). Health literacy and child health outcomes: a systematic review of the literature. *Pediatrics*, *124*, S265–S274.

Doak, L. G., & Doak, C. C. (1987). Lowering the silent barriers for patients with low literacy skills. *Promoting Health*, *8*(4), 6–8.

Doak, C. C., Doak, L. G., Friedell, G. H., & Meade, C. D. (1998). Improving communication for cancer patients with low literacy skills: strategies for clinicians. *CA: A Cancer Journal for Clinicians*, *48*(3), 151–162.

Doak, C. C., Doak, L. G., & Root, J. H. (1996). *Teaching patients with low-literacy skills* (2nd ed.). Philadelphia, PA: Lippincott.

Doak, L. G., Doak, C. C., & Meade, C. D. (1996). Strategies to improve cancer education materials. *Oncology Nursing Forum*, *23*(8), 1305–1312.

Dracup, K., Moser, D. K., Pelter, M. M., Nesbitt, T., Southard, J., Paul, S. M., et al. (2014). Rural patient's knowledge about heart failure. *Journal of Cardiovascular Nursing*, *5*, 423–426.

Haun, J. N., Valerio, M. A., McCormack, L. A., Sorensen, K., & Paasche-Orlow, M. K. (2014). Health literacy measurements: an inventory and descriptive summary of 51 instruments. *Journal of Health Communication*, *19*(Suppl. 2), 302–333.

Hibbard, J. H., Mahoney, E. R., Stockard, J., & Tusler, M. (2005). Development and testing of a short form of the patient activation measure. *Health Services Research*, *40*(6 Pt. 1), 1918–1930.

Institute for Healthcare Advancement. (n.d.). *Health literacy discussion list home page.* Retrieved from http://listserv.ihahealthliteracy.org/scripts/wa.exe?INDEX

Institute of Medicine of the National Academies. (2004). *Health literacy: A prescription to end confusion.* Washington, DC: The National Academies Press.

Institute of Medicine of the National Academies. (2006). *Preventing medication errors: Quality Chasm series.* Washington, DC: The National Academies Press.

Institute of Medicine of the National Academies. (2013). *Health literacy: Improving, health, health systems, and health policy around the world.* Washington, DC: The National Academies Press.

Institute of Medicine of the National Academies. (2014). *Informed consent and health literacy: A workshop*. Washington, DC. Institute of Medicine Health Literacy Roundtable. Retrieved from https://www.iom.edu/Activities/PublicHealth/HealthLiteracy/2014-JUL-28.aspx.

Keselman, A., Logan, R. A., Smith, C. A., LeRoy, G., & Zeng-Treitler, Q. (2008). Developing informatics tools and strategies for consumer-centered health communication. *Journal of the American Medical Informatics Association, 15*(4), 475–483.

Koh, H. K., Baur, C., Brach, C., Harris, L. M., & Rowden, J. N. (2013). Towards a systems approach to health literacy research. *Journal of Health Communications: International Perspectives, 18*(1), 1–5.

Koh, H. K., Berwick, D. M., Clancy, C. M., Baur, C., Brach, C., Harris, L. M., et al. (2012). New federal policy initiatives to boost health literacy can help the nation move beyond the cycle of costly 'crisis care.' *Health Affairs, 31*, 434–443.

Kreps, G. L. (2012). The maturation of health communication inquiry: directions for future development and growth. *Journal of Health Communication, 17*, 495–497.

Kreps, G. L., & Neuhauser, L. (2010). New directions in eHealth communications: opportunities and challenges. *Patient Education and Counseling, 78*, 329–336.

Kutner, M., Greenberg, E., Jin, Y., & Paulsen, C. (2006). *The health literacy of America's adults: Results from the 2003 National Assessment of Adult Literacy (NCES2006–483)*. Washington, DC: U.S. Department of Education, National Center for Education Statistics. Retrieved from http://nces.ed.gov/pubs2006/2006483.pdf.

Logan, R. A. (2014). Health campaign research: enduring challenges and new developments. In M. Bucci, & B. Trench (Eds.), *Routledge handbook of public communication of science and technology* (2nd ed.) (pp. 198–213). New York: Routledge.

Mbaezue, N., Mayberry, R., Gazmararian, J., Quarshie, A., Ivonye, C., & Heisler, M. (2010). The impact of health literacy on self-monitoring of blood glucose in patients with diabetes receiving care in an inner-city hospital. *Journal of the National Medical Association, 102*(10), 5–9.

McCormack, L., Bann, C., Squiers, L., Berkman, N. D., Squire, C., Schillinger, D., et al. (2010). Measuring health literacy: a pilot study of a new skills-based instrument. *Journal of Health Communication, 15*(S2), 51–71.

McCormack, L., Haun, J., Sorensen, K., & Valerio, M. (2013). Recommendations for advancing health literacy measurement. *Journal of Health Communication, 18*, 9–14.

Michell, S. E., Sadikova, E., Jack, B. W., & Paasche-Orlow, M. K. (2012). Health literacy and 30-day postdischarge hospital utilization. *Journal of Health Communication, 17*(Suppl. 3), 325–328.

Mulvaney, S. A., Lilley, J. S., Cavanaugh, K. L., Pittel, E. J., & Rothman, R. L. (2013). Validation of the diabetes numeracy test with adolescents with type 1 diabetes. *Journal of Health Communication, 18*(7), 795–804.

National Center for Education Statistics. Publications and products. Retrieved from http://nces.ed.gov/pubsearch/getpubcats.asp?sid=032.

National Committee for Quality Assurance. (n.d.). *Accreditation programs*. Retrieved from http://www.ncqa.org/Programs/Accreditation.aspx.

National Library of Medicine. (February 15, 2015). *MEDLINE/PubMed search and health literacy information resources*. Retrieved from http://www.nlm.nih.gov/services/queries/health_literacy.html.

Neuhauser, L., & Kreps, G. L. (2003). Rethinking communication in the E-health era. *Journal of Health Psychology, 8*(1), 7–22.

Nutbeam, D. (1998). Health promotion glossary. *Health Promotion International, 13*, 349–364.
Nutbeam, D. (2000). Health literacy as a public health goal: a challenge for contemporary health education and communication strategies into the 21st century. *Health Promotion International, 15*, 259–267.
Nutbeam, D. (2008). The evolving concept of health literacy. *Social Science & Medicine, 67*, 2072–2078.
Office of the Surgeon General. (2006). *Proceedings of the Surgeon General's Workshop on improving health literacy.* Rockville, MD: Office of the Surgeon General (US).
Osborn, C. Y., Cavanaugh, K., Wallston, K. A., Kripalani, S., Elasy, T. A., Rothman, R. L., et al. (2011). Health literacy explains racial disparities in diabetes medication adherence. *Journal of Health Communication, 16*(Suppl.3), 268–278.
O'Neill, B., Goncalves, D., Ricci-Cabello, I., & Ziebland, S. (2014). An overview of self-administered health literacy instruments. *PLoS One, 9*(12), e109110.
Paasche-Orlow, M. K., Wilson, E. A. H., & McCormack, L. (2010). The evolving field of health literacy research. *Journal of Health Communication, 15*, 5–8.
Parker, R. M., & Ratzan, S. C. (2010). Health literacy: a second decade of distinctions for Americans. *Journal of Health Communication: International Perspectives, 15*(Suppl. 2), 20–33.
Pleasant, A. (2013). *Health literacy measurement. Lecture 3 of 5: Better health: Evaluating health communication symposium.* National Library of Medicine.
Pleasant, A. (2014). Advancing health literacy measurement: a pathway to better health and health system performance. *Journal of Health Communication, 19*(12), 1481–1496.
Pleasant, A., McKinney, J., & Rikard, R. V. (2011). Health literacy measurement: a proposed research agenda. *Journal of Health Communication, 16*(S3), 11–21.
Ratzan, S. C., & Parker, R. M. (2000). Introduction. In C. R. Selden, M. Zorn, S. C. Ratzan, & R. M. Parker (Eds.), *National library of medicine current bibliographies in medicine: Health literacy [NLM Pub. No. CBM 2000-1].* Bethesda, MD: National Institutes of Health, U.S. Department of Health and Human Services.
Robert Wood Johnson Foundation. (2014). *Commission to build a healthy America.* Retrieved from http://www.rwjf.org/en/about-rwjf/newsroom/features-and-articles/Commission.html.
Royal College of General Practitioners. (2014). *Health literacy: Report from an RCGP-led health literacy workshop.* London, UK: Royal College of General Practitioners.
Rudd, R. E. (2013). Needed action in health literacy. *Journal of Health Psychology, 18*(8), 1004–1010.
Selden, C. R., Zorn, M., Ratzan, S. C., & Parker, R. M. (2000). *National library of medicine current bibliographies in medicine: Health literacy [NLM Pub. No. CBM 2000-1].* Bethesda, MD: National Institutes of Health, U.S. Department of Health and Human Services.
Smith, S. A. (2009). *Promoting health literacy: Concept, measurement & intervention.* Cincinnati, OH: Union Institute & University, Dissertation Abstracts International, 70, 9.
Smith, S. A. (2011). Health literacy and social service delivery. In S. A. Estrine, H. G. Arthur, R. T. Hettenbach, & M. G. Messina (Eds.), *New directions in behavioral health: Service delivery strategies for vulnerable populations.* New York: Springer Publishing.
Smith, P. C., Brice, J. H., & Lee, J. (2012). The relationship between functional health literacy and adherence to emergency department discharge instructions among Spanish-speaking patients. *Journal of the National Medical Association, 104*(11–12), 521–527.
Smith, S. A., & Moore, E. (2011). Health literacy and depression in the context of home visitation. *Journal of Maternal and Child Health, 16*(7), 1500–1508.

Smith, S. A., & Moore, E. (2012). Health literacy and depression in the context of home visitation. *Journal of Maternal Child Health, 16*(7), 1500–1508.

Sorensen, K., Broucke, S. V., Fullam, J., Doyle, G., Pelikan, J., Slonska, A., & HLS-EU Consortium Health Literacy Project European, et al. (2012). Health literacy and public health: a systematic review and integration of definitions and models. *BMC Public Health, 12*(80). Retrieved from http://www.biomedcentral.com/1471-2458/12/80.

White, S. (2008). *Assessing the nation's health literacy: Key concepts and findings of the National Assessment of Adult Literacy (NAAL)*. Chicago, IL: American Medical Association Foundation.

Wollesen, L., & Peifer, K. (2006). *Life skills progression: An outcome and intervention planning instrument for use with families at risk*. Baltimore, MD: Brookes.

Libraries

Medical information for the consumer before the World Wide Web

Catherine Arnott Smith
School of Library and Information Studies, University of Wisconsin-Madison

3.1 Introduction: "Closed to the Public"

> *I dislike the term 'lay public', for it gives me the unwanted feeling of a holier-than-thou attitude... It seems to me to smack so much of separation of the sheep from the goats.*
>
> Beehler (1955)

In 1934, Mildred Farrow had an article published in the *Bulletin of the Medical Library Association* (1934). Little is left to posterity about Ms Farrow. Born Mildred Stearns, in Maine, she grew up, married, and spent her life in Northern California. A member of the Medical Library Association (MLA)'s Publications Committee between 1933 and 1935, she contributed a total of three articles to the literature of librarianship, of which *Closed to the Public* was the first published. In it, Ms Farrow describes the problems at her medical library, that of the San Diego Medical Society. Ms Farrow's opinions and the portrait she paints are striking exemplars of the challenges of health information provision to the public before, during, and after the era in which she published.

Ms Farrow begins by making her position clear: the public is a problem.

> Looking back over the many years that the San Diego Medical Library was open to the public, I can but marvel why, as a part of the Medical Society, this problem was not solved long ago (p. 225).

The public was also a cost to the Medical Society:

> The medical profession was paying for the upkeep of their library, paying the salary of the librarian, to say nothing of the many incidentals of expense that accumulate from day to day in any medical library (p. 225).

However, the public was extremely interested in what the medical library had to offer. Its goals included not only self-treatment, but cost savings—although Ms Farrow feels that the public's ability to understand what it found on the shelves was doubtful:

> They read on subjects of which they had not the slightest knowledge. They copied prescriptions for themselves, their friends, their relatives and immediate families.

> Many informed me that they had saved hundreds of dollars by being allowed the use of the Medical Library. One woman said that her husband was a druggist and that she had copied prescriptions for him for a period of years and was desperate when informed that the library was no longer at her disposal. One man said that he had no faith in doctors – that he found all the treatment necessary in the books in our library. When he was informed that the books in this library were written by the medical profession – well, he had not thought of that! Anyway, he had saved money (p. 225).

In addition to consuming the library's resources, the public consumed the librarian:

> Another woman was a regular 'customer' for many years. Every few days she came with a list of subjects in which she was interested. Many a morning she kept me busy while the reading table would be piled high with material as a result of her many wants. When she had finished copying and reading, she would 'borrow' the telephone to call up her friends and relatives, read her copied notes, give out prescriptions, describe symptoms and tell her listeners what to do (p. 225).

All this despite the librarian's attempts to refer the public to a health-care professional who could better meet its needs:

> I assured her that I felt sure that her family physician could give her far better advice than she could possibly obtain in our library. 'But,' said the dear lady, 'I will have to pay a doctor' (p. 226).

What was the public doing in the medical library? In addition to saving money and treating itself, the public was second-guessing its physicians despite the lack of credentials to understand what it was reading:

> Still another angle were the patients who came to check up on their physicians' treatment. Any doctor knows the danger of this. What layman can understand the comparative treatments? It meant nothing to them that the doctor had made a study of their particular case, their particular type of disease. No, this book says so and so and the other book says something different and the patient becomes dissatisfied. They lose confidence in their physician and they wander from office to office and this is where the various cults and isms come in (p. 226).

Ms Farrow makes it very clear that she considers medical literature itself a danger to the unwary; so much so, that she eventually petitioned her library administration for a change:

> After enduring this over a period of years and spending the time for which the Medical Society was paying me… I felt that a stop should be made to the promiscuous use of the medical information that the library contained… The laity were literally treading on thin ice by handing out information and advice on subjects about which they knew little or nothing, whereas the medical profession had acquired this knowledge by years of hard study and the expenditure of thousands of dollars (p. 226).

After Ms Farrow's petition to the Society's Council, the rules were changed and library admission restricted to members of the Society and a few other cases:

> The only exception being visiting physicians, nurses, college and high school students accompanied by a note from the dean or teacher indicating the subject on which the student has been asked to write (p. 226).

Ms Farrow brought the problem of the public to the attention of the medical library profession in its flagship journal, the *Bulletin*, because she considered provision of access to medical information "a dangerous practice—for the layman as well as the medical profession" (p. 226). What do we know of Mildred Stearns Farrow that would help us understand her reaction? As noted, the historical record is scant; she was never memorialized by the MLA after her death in 1957 (Fort Rosecrans National Cemetery, 1957); the San Diego County Medical Library no longer exists; in fact, current employees of the Society express surprise today that they ever had a library, let alone a librarian (K. Lewis, personal communication, January 13, 2015). But the Society has a Web site. Today, the Society's Web site directs consumers and patients to 109 "community resources," from 2-1-1 San Diego to the YWCA in San Diego County, five "helpful Web sites" such as the State of California's Office of the Patient Advocate and "Health Companion: Your Personal Health Resource." The Society has its own Foundation "with the mission of addressing unmet San Diego healthcare needs of all patients and physicians through innovation, education, and service" (San Diego County Medical Society, 2013). The reader needs absolutely no reminding that times have changed.

We do know two things about Ms Farrow which may have affected her viewpoints. First, she was married to an Army surgeon, Edgar James Farrow, from 1902 (Marriage Notices, 1902) until his death in 1947 (Fort Rosecrans National Cemetery, 1957). Second, her career in libraries was a long one, even predating her marriage. The 1900 San Francisco census (United States Bureau of the Census, 1900) shows Mildred Stearns, "librarian," living with her parents. Does Farrow's published allusion to "many years" of work mean that she was San Diego County's sole medical librarian for more than 30 years, before and at least during her marriage? Was she a member of the large untrained class of library working women, or did she have formal training? It is impossible to say. But while Ms Farrow's career remains obscure, her viewpoints in 1934 were common, and neatly encapsulate one professional attitude of medical librarians. For the Ms Farrows of the world, medical information is dangerous in the wrong hands. And these dangers reveal themselves in four principal dimensions, the depths of which make up the rest of this chapter:

- *Libraries as organizations and institutions* (optimal library types; access policies)
- *Librarians* (interpersonal challenges; appropriate and inappropriate roles)
- *The patron* (motivations for access; demands on staff time; expectations; self-diagnosis)
- *Content* (audience appropriateness; maintenance of quality collections; "bad" versus harmful books; terminology)

This chapter concentrates on a particularly interesting era: the pre-Web world, defined here as "before 1994," a world in which public access to medical information was considerably limited by gatekeepers, among them the professionals we know as librarians. The functions of these gatekeepers, and the public's interactions with them, varied according to library type and library policy. In public libraries, print and audiovisual materials had to be acquired for patrons to use them, but what personal contact occurred between librarian and library patron remained within the control of the patron. In medical libraries, conversely, gatekeepers controlled access not only to materials, but also to the collections and even the buildings within which those materials were placed. In addition, librarian gatekeepers set policies for assistance using medical information. In these libraries, policies were articulated by the library as an organization. To understand why medical information can be challenging to provide in nonclinical settings, such as libraries, the literature of the gatekeepers can be revealing.

Two good historical reviews have been published which focus either on particular library types or particular professional literature. Perryman (2006) examines the history of information provision to patients in hospital libraries; Rubenstein (2012) looks at the activity of librarians in the context of our evolving understanding of public health, viewed through the lens of two specific, important, and long-running generalist and specialist journals (2012). For this chapter, the author reviewed professional literature about *any* library type, published *anywhere*, between 1879 and 1993, that (1) directly addressed professional challenges of medical information provision to the public in libraries and (2) was indexed in either *Index Medicus*—the print precursor of today's MEDLINE database, produced by the National Library of Medicine, or the key databases indexing the professional literature of librarianship: Library Literature and Information Science and LISTA (Ebsco), and Library and Information Science Abstracts (ProQuest).

3.2 Background: the beginnings of consumer health information

The principal professional association in the United States focusing on medical information and its provision is the MLA. The Consumer and Patient Health Information Services section of MLA—hereafter CAPHIS—has made the definition of consumer health information clear for years. CAPHIS notes the important distinction between consumer health information *provision* and patient *education*:

> *Consumer health information (CHI) is information on health and medical topics provided in response to requests from the general public, including patients and their families. In addition to information on the symptoms, diagnosis and treatment of disease, CHI encompasses information on health promotion, preventive medicine, the determinants of health and accessing the health care system.*
>
> *Patient education is a planned activity, initiated by a health professional, whose aim is to impart knowledge, attitudes and skills with the specific goal of changing behavior, increasing compliance with therapy and, thereby, improving health.*
>
> *CHI and patient education overlap in practice, since patient behavior may change as a result of receiving health information materials. Patient education and CHI*

> *often differ in terms of the setting in which the process occurs, rather than in terms of the subject matter.*
>
> CAPHIS Task Force (1996, p. 238)

A note on terminology is necessary at this point. "Consumer health information" is a phrase that appears to have become popular in the United States through the various consumer empowerment movements of the late 1960s. It encodes both its politics and its designated audience—not physicians, not nurses, not health-care workers of any kind; but consumers, the general public, including but not limited to patients—in its name. The phrase does not appear in *Library Literature*—the longest-running index to the literature of professional librarianship—before 1978, with the publication of Goodchild (1978), but is there four times that year, three times used to index articles about the same health information service. The Google NGram Viewer (https://books.google.com/ngrams) permits searching of the Google Books corpus. Case-insensitive searching for the phrase "consumer health information" shows no record of the phrase in the corpus prior to 1967. Since the focus of this chapter is on professional literature published between 1879 and 1994, to avoid inducing confusion in quoting the older work, the author uses the phrase "medical information" from here on.

CAPHIS distinguishes medical information provision to the general public from patient education of the general public—a large group, as will be discussed later, that includes but is not exclusive to patients. The fact that the distinction needs to be made is evidence of two important truths. First: one does not have to occupy the role of "patient" to seek medical information. Second: the role of the librarian/information profession in relation to medical information provision does not *depend* on the information seeker's being a patient. In sum: patients are not prerequisite to the process.

The "general public," then, is an amorphous class of individuals. It is comprised not only of patients, but of their families, their friends, their caregivers, and those who have no present illness but seek information anyway for different reasons: homework; curiosity; and self-education. People have long been drawn to medical information. One clue that the public's appetite is timeless is the fact that every generation of librarians to publish about it considers it a trend. This recognition has been going on for a very long time. In 1921, physician John Farlow told the MLA: "An ever increasing number of the community is taking a larger and larger interest in medicine and health…we cannot help noticing a very lively interest on the part of the public in subjects which, not so very long ago, were supposed to be the monopoly of the medical profession" (p. 3). Stein and Lucioli (1958) wrote about public libraries "inundated by the recent flood of popular books in the health field" (p. 2110). Morrison (1976) wrote casually, 2 years before the phrase "consumer health information" appeared in her professional reading, that "we do know that many people are seeking health information these days" (p. 5). Cormier cites "increased demand" (1978, p. 2051) in the same year that Ellen Gartenfeld, a twentieth-century pioneer in expanding access to medical information, complained that even health sciences libraries "are now being asked to add to their collections materials that they do not know how to evaluate" (p. 1912). And in 1993, the year before the World Wide Web changed everything, Dahlen could say with confidence: "Large public libraries actively disseminate health information to patrons" (p. 165).

Discussion of libraries providing medical information to the public also appears frequently in the literature for decades before the Web, framed as apologetic, defense, or both. The usage of public libraries as venues for medical information was made quite explicit by the American Medical Association's House of Delegates in June 1909, at their annual meeting in Atlantic City. The American Medical Association's House of Delegates adopted a resolution:

> *Requesting the women physicians of the American Medical Association to take the initiative… to act through women's clubs, mothers' associations and other similar bodies for the dissemination of accurate information touching these subjects among the people.*
>
> Fishbein (1947, p. 999)

Physician surgeon Dr Rosalie Slaughter Morton introduced this 1909 resolution—what Morton herself later called the "first organized movement in history" (Morton, 1937, p. 165)—for general prevention of disease through specific education of the public. The work of this committee included not only production of health bibliographies, 8 pages of listings of topics in health and hygiene, but the printing and placement of these bibliographies in public libraries and distribution of bibliographies at the annual public librarian's meeting. The library of Dayton, Ohio appears to have been the only library to have actually followed through on the plan; librarians asked state representatives of the AMA Committee, working through the county medical society, to supervise books on health subjects acquired for the public library: "the best on civic, home, personal and social hygiene…classified for men, women, parents, boys, and girls." (Committee for Public Health Education Among Women, American Medical Association, 1912, p. 62).

Frankenberger (1936) similarly called for the American Social Hygiene Association to "foster a wider and more direct education in social hygiene material" (264) working through the mechanism of libraries, adding that a useful side benefit would be to keep the library staff themselves educated and informed about social hygiene. He further alluded to the prophylactic effect of reading about sexual health on "venereal disease morbidity" (264).

3.2.1 Who uses medical information?

One of the earliest records of medical libraries in action is the article by physician–librarian C.D. Spivak, published in his own journal, *Medical Libraries*, in January 1899. Spivak listed 120 medical libraries in the United States, of which 45 were public libraries. So the question "who uses" was asked almost from the beginning. Physicians are always assumed users. To a lesser extent, patients, too, are assumed, more obliquely described as "sick people" under the care of physicians (Brother Ignatius, 1941; Wallis, 1949).

But writers referred generally to "the public" or "the lay public" from the beginning (Cormier, 1978; Frankenberger, 1936; Garrison, 1921; Getchell, 1898; Loomis, 1926;

Richardson & McCombs, 1929; Shores, 1954; Wire, 1902; Wyer, 1930). The public were present in libraries from the beginning, too. Getchell, writing about Worcester's Medical Library, reported that the public library gave a room to the city's Medical Society in 1893:

> *The Medical Society extended to the public* still freer *[italics mine – CAS] use of its books which can now be taken from the building by any holder of a public library card under certain mild restrictions.*
> Getchell (1898, p. 35)

The implications of Getchell's remark is that the public had access to books before 1893, but could not take them home. A letter to the editor of *Medical Libraries* in 1898 came from a physician who was also on the Indianapolis Board of School Commissioners and wanted to create a public medical library. He hints similarly:

> *It would be more advantageous to both the students and the profession were the various small collections of medical literature united into one and made accessible to the public at large.*
> Sloan (1898, p. 45)

Miles (1983) reports that in the early years of the Surgeon General's Library (now the National Library of Medicine; during the period from 1895 to 1913), the Library received queries from laymen as well as physicians, although Miles unfortunately did not make clear which questions came from whom. It is certainly difficult to imagine a doctor asking a librarian "Were germs transmitted by postage stamps?" although they might well have asked "What was known about the morbid fear of thunderstorms?" (p. 203). But in the end, no librarian can really discern a profession simply from a patron's question:

> *Public librarians are involved in 'patient education' every day, but we don't call it that because we never know when our 'patrons' are 'patients.'*
> Morrison (1976, p. 5)

Or as M.S. Averill put it bluntly in the consumer health magazine *Medical Self-Care*: "If you are not a health worker, say so. It helps the librarian understand your needs" (Averill, 1980, p. 39).

The public displayed gender differences. Carolyn Ulrich, of the New York Public Library, was careful to point to both "men and women" in the public (1920), but Yellott and Barrier, in an early study, could report that 70% of medical information seekers surveyed were women, 30% men, and that "over a third" of the women were "of childbearing age" (1983). More interesting, and more illustrative of the potential reach and depth of medical information usage, are those writers who were more specific as to the public's job descriptions. Ballard (1927) describes a theatrical patron in search of medical information with an air of "all in a day's work":

> *'I come from the 'Enemy'', said a man as he walked into my office on afternoon, and I was rather surprised and perturbed for a moment until he finished by saying 'Now playing at the 'Hollis Theatre' ...' He was the leading man of a stock company (p. 18).*

Estelle Brodman, writing about users of health sciences libraries, used a well-established taxonomy of medical information users: "scientists, engineers, and technicians" (1974, p. 69). Mildred Langner, in the same special issue of *Library Trends*, categorized "the laity" as either professional or nonprofessional (1974, p. 15). Students were always mentioned as users of medical information, starting in high school (Chambers, 1955; Loomis, 1926; Monahan, 1955; Van Hoesen, 1948) through junior college (Brodman, 1974); dietetics school (Loomis, 1926; Monahan, 1955); and college (Chambers, 1955; Monahan, 1955). Nonstudent, nonmedical professionals also recur as medical information users: lawyers are mentioned multiple times (Chambers, 1955; Hawkins, 1963; Langner, 1974; Monahan, 1955), as well as social workers (Langner, 1974; Radmacher, 1963); journalists (Chambers, 1955; Langner, 1974; Wickes, 1934); novelists, editors, and writers (Chambers, 1955; Cole, 1963; Radmacher, 1963); geologists (Chambers, 1955); engineers (Chambers, 1955; Langner, 1974); metallurgists, research workers, movie-radio-television producers, businessmen, undertakers and morticians, businessmen, taxidermists, architects, translators, book-dealers and book-sellers, musicians and music teachers, advertising men, illustrators, policemen, athletes, beauticians, and taxi drivers (Chambers, 1955); clergymen (Langner, 1974); and employees of government agencies, insurance companies, freight agencies, industrial plants, the FBI, and narcotics bureaus (Monahan, 1954).

Barely mentioned in this historical laundry list are three categories of user later considered very significant in the consumer health literature of the twenty-first century: "housewives" (acknowledged exactly once, by Chambers, 1955); caregivers, defined in passing by Langner in 1974 ("those who wish to study some particular diseases because they themselves *or some member of their families are sufferers* [italics mine – CAS]" p. 15); and, oddly, librarians, presumably librarians using the collections of libraries not their own (Chambers, 1955).

However, Florence Van Hoesen's public library reference study (1948, but data collected in 1937–1938) broke question-askers down into occupational groups and found housewives the fourth largest category. In the Science and Technology subject area, Van Hoesen's findings are notable for the presence of *nonprofessional* job classifications. While "Science and Technology" is not completely synonymous with "medical questions," we know that these questions were asked; for example, "a skilled tradesman wanted 'a feeding formula for infants'" (p. 83). Table 3.1 shows the variety of occupational groups represented in Van Hoesen's data, in descending order by frequency asked at main libraries.

3.2.2 What

Clearly, there were very few people living who could not be imagined by someone as potential users of medical information. Thus, the scope of their possible interests was equally potentially wide. The oldest primary source of reference questions the author has been able to identify is the reference log retained by The Public Library of Cincinnati and Hamilton County, OH, which records telephone questions posed to reference librarians in the library's "Useful Arts Room" between December 1915 and August

Table 3.1 **Total questions asked by members of various occupational groups, Science and Technology category (Van Hoesen, 1948)**

Group	Number of questions[a]	
	Main library	Branch libraries
Professional	171	10
Skilled workers	84	30
Students	82	156
Housewives	51	25
Clerks, stenographers	23	10
Shopkeepers, salesmen	21	8
Unskilled workers	17	7
Unknown occupation	17	3
Farmers	2	2

[a]Asked in 1 week, winter, 1937, in 15 libraries in 6 library systems; 4 branches and 1 main library were surveyed for one additional week in winter 1938.

1920. While librarians did not log personal information about the callers, it is not hard to discern consumers, patients, and their caregivers in these examples:

Has library a book telling how to cook for sick people? (January 29, 1916)
Name some noted physicians who discovered important things? (October 14, 1916)
Has library trained nurse? (January 17, 1917)
What is Ward K at the general hospital? (July 2, 1917)
Recipe for old cough remedy or pepper syrup (January 3, 1918)
Name of a book on window trimming, one on diet and one on the origin of the gypsies (February 14, 1918)
Does library have the contagious lists received daily for last year? (March 8, 1918)
Medium height for girl of 22 (August 29, 1918)
Has library Dr Gunn's family physician? (January 25, 1919)
Spell migraine (November 14, 1919)
Has library Dr James Kelly's Highways of health? (January 22, 1920)
Has the Library a book by Robinson on sex hygiene for girls? (June 1, 1920)

There is an understandable gap in the record for a month during the worldwide flu pandemic of 1918. The library was closed for October through November of that year. However, once it was open again, librarians recorded this question: "What to use to fumigate books after influenza?" (January 7, 1919).

Writers typically refer to medical subjects as "most popular" with the public—or not. Ulrich (1920) identified her most frequent requests as books "on prospective parenthood," followed by marriage preparation and then sex hygiene for adolescents; one hopes that the patrons sought these books in the reverse order. Loomis (1926) may be hinting at this particular subject matter when she argues that "many agencies are popularizing the knowledge of medicine, and we should welcome every chance to supply the laity with a good, wholesome brand of this literature" (p. 35). Shores, in the second edition of his reference text for library science students, called for an increase in collecting works on "socialized and preventive medicine" based on anticipated demand (1939, pp. 331–332). Van Hoesen's (1948) study revealed that in branch libraries, within the larger category of Science and Technology, "medical terms and topics" was the most frequent subject asked about. Van Hoesen's examples of questions in this group featured diets for heart disease, recovery from infantile paralysis, hay fever, and life expectancy. Eakin, Jackson, and Hannigan (1980) surveyed public and medical library staff in that city and found that more health-related questions from the general public were asked at the Houston Public Library Science and Technology section than at the medical library: 361 (62%) versus 221 (38%) for the concurrent four-week period (p. 222). Five percent of public library reference questions, in the Science and Technology branch, were health related. Demand from the general public was apparently present in both library settings. In fact, the typical public user in both places was looking for information on "a specific disease, procedure, drug, or diet" (p. 222), proportions which have varied little in the research literature of consumer health ever since.

The interest of the public was driven in part by the media. Ulrich, for example, saw a direct connection between social hygiene broadcasts and requests for articles about VD (1920). Naturally, the public's own individual health status was always a motivation, but Bay, in 1924, reported somewhat defensively that it was not the *only* motivation: "It is not only that persons come with symptoms of blood-pressure, but that they wish to inform themselves about conditions and matters which really *can* be judged by average intelligence" [italics original – CAS] (p. 13). Bay continued: "Problems in obstetrics are of daily occurrence…" The patient-driven information need was apparently common enough at reference desks that Wallis, in 1949, felt it necessary to make a distinction between those interested in outright self-doctoring—acting as one's own physician in the absence of, or in opposition to, the physician—and "the many who merely want to find out for themselves the fundamental causes for their ailments" (p. 252). Wallis admits that some of the general public are looking for medical information in order to convince a doctor that they are sick; others want to verify their doctors' credentials. Beehler (1955) also pointed to the general public's role as critical consumer: "He no longer is satisfied with being given a pretty pink pill for whatever ails him; he wants to know the whys and wherefores" (p. 241).

3.2.3 How much

It was recognized decades ago that medical questions were a persistent type of specialized information need. Van Hoesen's (1948) was one of the first systematic studies of reference services in public libraries about any subject matter, including medicine. Van Hoesen was for years a faculty member at the library school at Syracuse University. Her dissertation, at the University of Chicago, was completed in 1948, but her reference data collection took place in 1937–1938, one of a handful of master's or doctoral theses centering on reference questions completed at Chicago during the 1930s; the reader is referred to the excellent bibliography in Berelson (1949) for more. Ms Van Hoesen collected data from 15 main libraries and branches in 6 different city systems—Tampa; Houston; Cincinnati; Washington, DC; Los Angeles; and Boston—during 1 week in the winter of 1937; she surveyed Houston and Boston for a second week the following winter. This all yielded 3596 questions, which Van Hoesen then categorized using a modification of the Dewey Decimal Classification (DDC). The Science and Technology class in which medicine is grouped was in the top three categories, ranked by popularity, in 15 of the 16 libraries investigated—the one exception being Los Angeles, where it was edged out by literature. And in branch, but not central, libraries, "medical terms and topics" was the most popular of the Science and Technology topics, followed by agriculture.

"Medical terms and topics" was the fifth most popular subtopic of *all* questions in one week, regardless of DDC category, asked in central libraries (the topics in order were: Biography: 205 questions; Education, 115; Geography, 74; Laws, 61; Medical Terms and Topics, 52 questions). In the branch libraries, it was sixth most popular: Biography, 188; Geography, 102; History, US, 58; History, 48; Poems, Songs, Hymns, 44; Medical Terms and Topics, 28 questions (Van Hoesen, 1948, p. 53).

The director of the "Halsted"[1] Hospital's new Consumer Health library described by Poisson (1983) reported that a public library director stated in May 1982 that there had always been a "fairly large number of questions" on health, which was one motivation for the construction and organization of their public library consumer health network (Poisson, 1983, p. 109). Marshall, Sewards, and Dilworth (1991) conducted what is considered to be the first systematic study of consumer health information needs in Canadian public libraries. These authors found that 8% of reference questions involved health information. Dewdney, Marshall, and Tiamiyu's study (1991) yielded the oft-cited-since proportion of 10%.

Finally, with the World Wide Web just around the corner, Sullivan, Schoppman, and Redman's reference study at the University of Michigan's public academic medical center library (1991) identified the category of user most in need of research services. It was the category dubbed "unaffiliated public," which asked 32% of all reference questions during the study period—and more *research-related* questions than any other type of user—including health professionals. This was a library type for which research-related questions are appropriate, because the primary population served by an academic medical center library such as Michigan's is clinical and student researchers. But because a public university is tax-supported, and open to the nonclinical public, the general public has admission, and will bring its questions to the desk. The discussions below relate to the challenges this poses for information providers in different library settings.

---

[1] Poisson assigned pseudonyms to all of the libraries in which she interviewed in order to ensure confidentiality. Every reference to one of Poisson's libraries will put the name of the library in quotes to make clear that this is not the library's real name.

3.3 Libraries

> *For generations physicians and other health sciences practitioners have considered their work so esoteric and so liable to be used wrongly or in nefarious ways that they built around themselves and their tools and records a wall of privacy.*
>
> Brodman (1974, p. 63)

3.3.1 Types

In 1983, Poisson summed up old and ongoing professional tensions around library types and consumer health information quite succinctly:

> *Medical librarians frequently feel that their training and expertise is and should be primarily directed toward serving clinical, teaching, and research needs of health care professionals. They feel ill-equipped, both in training and resources, to serve the public or patients who do not have medical training, who may not have a precise idea what they are looking for, who have information needs that may extend beyond published resources, and who may be personally and emotionally involved in the issue. The organizations which house medical libraries may not recognize public or patient health education as high priorities... or those organizations may not see their own libraries as optimal places for public or patient health information access.* (1983, p. 11).

What libraries could and should provide medical information is, in the United States, a question literally as old as the MLA itself; as Schell comments (1980), the very first meeting minutes of the Association provide support: "The object of this Association is to encourage the improvement and increase Public Medical Libraries" (p. 930). What changed over time was the definition of "the public."

The early physician-librarian named Spivak (1902) summarized the situation in this way: in 1902, there were 120 subject-specific library medical departments in the United States. Of these, 11 were housed in public libraries, and one in a school library. This was the year in which Philadelphia physician George Gould made an impassioned plea for the "union of medical and public libraries" in the *Philadelphia Medical Journal*. Citing statistics compiled by Spivak, Gould wrote of the current situation:

> *This shows in what a deplorable condition of barbarism is medical literature... In many cases the 'library' is a small collection of old rubbish of which the resident physicians know absolutely nothing, sometimes even do not know that it exists.* (1898, p. 238).

The MLA had been formed—by four doctors, including Gould, and four librarians—out of the same movement in 1898 (Groen, 1996). But between 1898 and 1902, Spivak wrote (1902) there had been a great increase in the number of public library medical departments. And a few years later the question of public access to medical information in libraries began to recur in the professional literature. The question was, "Who is the public? What does the public read…95% of the reading of the 'public' consists of novels, the bane of every librarian of the land" (Spivak, 1902, p. 20).

Initially, "the public" meant "doctors who otherwise would be dependent on their own small collections." At a special symposium on medical departments in public libraries, public library directors from cities across the country consistently pointed to the value of having medical reference texts available for consultation by physicians, who were therefore able to assist their patients better. "User of medical text" was so synonymous with "physician" that medical departments in public libraries were argued against precisely because the physician user was seen as too specialized and unrepresentative of the community. For example, doctor, medical and legal librarian George E. Wire had professional experience in Chicago, at both the Chicago Public Library and the Newberry Library, and insisted:

> *No public library had the right to the money of the people for books benefiting only one class… Clergymen, lawyers, physicians, if not engineers and high-grade machinists, should provide their libraries, just as some of them do their own tools or instruments…*
>
> Wire (1902, p. 9)

By 1907 things were deteriorating for medical departments in public libraries. The director of the Grand Rapids, Michigan library, Samuel Ranck, sent out a letter surveying librarian colleagues about their practices and policies for medical books, and heard from a number of public libraries doing cooperative collecting, lending, and borrowing with medical libraries.

However, Ranck still worried about collections. He wrote that in smaller cities like his, "the number of physicians who are interested in technical literature, outside of the current periodicals, is so small that independent library associations often have languished…" (p. 210). For that reason, these collections ended up at the public library, which, according to Ranck, couldn't take care of them. Ranck considered these collections of poor quality, in part due to "the lack of sufficient interest on the part of the average physician in the study of the literature of his profession" (1907, p. 211). Most physician users seem to read only the current issues: "They believe that all a practicing physician needs to read is the periodical" (p. 212) and current periodicals at that.

How about the nonphysician public? Ranck quotes "a well-known librarian" as saying "A public library is no place for medical books" (1907, p. 212). According to Ranck, "[M]ost people are agreed that in a library which is used by the public generally there must be some restriction in the use of such books." He gives the example of an anatomy text, the *Human body* by Martin, which "should not be given freely to a child" (p. 212). Furthermore, opinions seemed to vary based on the professional identity of the person in charge of the medical department. Doctors in charge put all the medical books in locked cases. "A librarian of a public library thinks books are for the use of the public. The medical librarian thinks they are for the use of physicians and students only" (p. 219). Interestingly, Ranck makes a distinction between the usefulness of periodicals—the same material that his physician users seemed to want—and that of books in the medical collection. "Some of our best medical periodicals contain a considerable number of articles that are of interest and profit to any man of ordinary intelligence" (p. 213). Frustratingly, Ranck gives the medical classic, Osler's *The Principles and Practice of Medicine*, as an example of a good book, but he gives no analogous examples of periodicals appropriate for laypeople.

The question of the most appropriate library type to provide medical information to physicians seems to have been considered settled in the United States before World War I. As late as 1928 Harry Lydenberg, of New York Public Library, was still arguing that the public library had a role for medical collections but made it clear that this should happen only when there *were* no medical libraries willing to take up the task:

> *If the medical special library is developed or can be developed the public library may well stand aside.*
>
> Lydenberg (1928, p. 20)

By 1955, Beehler could define the "lay public" was "that group of persons who have no affiliation whatever with the medical profession" (p. 242), which clearly had implications for what library they would use. Defined by who they were not, is it any surprise that the population of consumers remained nebulous? Wannarka, in her historical review (1968), reported that medical collections in public libraries were never more than a minority—they "reached an apex in 1916 when twenty-eight such collections made up 16.47 percent of the total number of medical libraries in the nation" (p. 8). Belleh and Van de Luft (2001) reflected that these collections ultimately dwindled because public libraries were unable to meet the demands of the original targeted audience—physicians—in either reference services or collection strength.

In the pre-Web literature of professional librarianship, even strong advocates of public access to medical information expressed concerns about public libraries acting as points of access. King (1955) stated, quite conscious of the irony: "It is a big problem, and I do not believe it will be solved by saying that the public library is the place for the lay public" (p. 244). That public librarians still saw a demand for medical information among their clientele, however, is very clear from the persistence in the library literature about attention to the challenges, the issues, the ethical dilemmas. The consumer empowerment movements of the 1970s appear to have reinvigorated longstanding tensions over the best location for medical information. It was in 1976, for example, that the Library of the Health Sciences at the University of Illinois at the Medical Center, Chicago, sponsored a one-day workshop for public librarians. This workshop was clearly aimed at enhancing public librarians' health information resources, for the principal output of the workshop was an annotated bibliography. (For example, about Samuel Andleman's "new home medical encyclopedia," published in 1973, the annotator writes: "Easily read, with fair coverage of very basic health-related topics. Not well illustrated and not very thorough, this work mostly consists of definitions" (Walker & Hirschfeld, 1976, p. 459).) Gartenfeld implicitly commented on the state of collections a few years later, writing that one problem with public libraries was that they "rarely have the materials that the user wants to see" (1978, p. 1912). Conversely, consumer advocate Averill stated confidently that public libraries "are often the best place to begin a search for health information" because the collection and staff were designed to serve the public (1980, p. 38). As Poisson articulated in 1983, it was a tension between two different library types with two different kinds of expertise: in the medical library, expertise in handling specialized information; in the public library, expertise in handling the public.

Schell reported statistics that clearly illustrated the problem of location and public access to medical information (1980). In 1979, a large survey focused on access was conducted among librarian members of the MLA. Eighty-two percent of librarians at publicly funded medical school libraries reported being open to the public. However, at 68% of those libraries, access to the collection was the only service provided. Of privately supported libraries in the same study, 60% served the general public, but only 23% of them gave complete reference service. Schell concluded that given these conditions of limited or completely restricted access, much more interlibrary cooperation across library types, as well as standards for delivery of medical information to the public, would be necessary to meet the apparent need.

Despite these decades of professional debate, medical information remains a stubbornly multitype resource. As of January 2015, the U.S. National Network of Libraries of Medicine lists 1060 member libraries offering consumer health information services to the public. Of these, 533 are hospital libraries; 216 are academic libraries; 232 are public libraries; and 79 are "other" library types, for example, special libraries serving sports medicine institutes, or cancer foundation libraries.

3.3.2 Access to collections

For some medical librarians, the notion of access was encoded in the very definition of the "lay public." In the medical library setting, Wickes (1934) voiced a typical attitude when she distinguished between patrons who had an "a priori claim" on library service, as "[those who] contribute to the library's support," for example, those patrons with university connections or who pay for memberships (p. 138). Similarly, Crowe, writing about a medical society's library inside a public library in the 1940s—a relatively late date for this kind of institution—commented that "books circulate only to primary clientele, chief source of financial support," and "persons to whom they grant permission to use their cards" (p. 223). However, Crowe is careful to note that the general public can use the room at all times, and that "unlimited reference service is given to the layman as well as the physician" (p. 223). Public versus private support is a longstanding and important distinction present in the access debates. Ferguson (1976) said in the very first issue of his consumer health journal that "The Yale Medical Library, like those of most private medical schools, is closed to the lay public…State medical schools tend to be a little more open. Anyone know of any medical libraries that welcome laypeople?" (p. 20). In 1980, Averill cautioned that "most medical libraries are tax-supported, which means their collections are accessible by law. But having access may not mean a right to borrow. Unless you are a health worker, borrowing privileges may be hard to come by" (p. 39).

Beehler (1955) described a spectrum of opinions consisting of two extremes: medical libraries denying the public access, and medical libraries that "[do] not call anyone a layman in the sense that he is not allowed to use their facilities" (p. 241).

Ergo, laymen are people who are not allowed in. Poisson (1983) comments that public access to medical information is "an inroad to privileged, professional knowledge" (p. 7). Libraries of different types set up different kinds of barriers and filters to enable access to medical materials by the persons the library considers to be able to use them most effectively.

Some barriers were physical. Wire, in 1902, dealt with his patrons' presumed prurient interest in "the plates"—that is, anatomical illustrations—by locking the good stuff up: "All sorts of lies were told and all sorts of dodges resorted to, and we had to keep something like 200 v[olumes] under lock and key" (p. 12). Even as late as 1949, Wallis described security restrictions on medical collections designed "to keep the more daringly illustrated books, and some kinds of sex books, out of the hands of this type of reader" (p. 252). Anatomy books, and "the more extreme ones dealing with sex," were housed in Special Collections, where they could be policed by staff. Librarian Wallis reports that in his librarian wife's army hospital library she locked up the medical *dictionaries*, but for a different reason: "so that some G.I.s could not learn new symptoms on the eve of being sent back to duty" (p. 253). Monahan (1955) quoted a psychiatrist who was in favor of separating the "lay" from the "technical" books and then making the lay materials more convenient to search, precisely so that laypeople could not get at the technical material. Hawkins (1963) suggested that librarians use their professional tools—knowledge representation and intellectual access—as another kind of barrier:

> *Libraries serving a large lay public sometimes maintain nonstandard publications in a special collection or indicate in some way the unorthodox publications listed in the catalog (p. 476).*

Branch (1979) wrote about a telephone medical information service called Health-Line run by Columbus and Franklin County, Ohio. Health-Line used a cassette system to vend information on specific health topics. She described a situation in which nervous patrons did their own self-censoring using the library's cassette tape classification system. One has to wonder if this worked well for the librarians too:

> *Many people use the number of the tape they want, rather than the name, as though this insulated them further from contact with what they may perceive as socially undesirable curiosity about sex or drugs. In fact, many ask for "extension 728" as though the very existence of a cassette system were unknown to them. (In some cases, it may have been: some practical jokers must find the existence of Health-Line irresistible, and some calls, at least, may be the result of a telephoned message for Mr Smith to call 221-7700, extension 728).*

Radmacher, at the Skokie, Illinois Public Library, discussed the "dearth" of literature discussing "nonstandard" medical literature in public libraries. Like Hawkins, she does not define what "standard" medical literature in public libraries would be (1963, p. 463). But Radmacher, too, describes the filtering tactic of cataloging and shelving particular medical materials as "reference" as opposed to "circulating" (Radmacher, 1963, p. 465). In the end, the final solution is to make the entire medical collection, physical and intellectual, difficult to access: the "Welch" Hospital described by Poisson (1983) had security protocols in place which, librarians admitted, certainly could be a disincentive to the general public.

Other barriers erected by libraries are policies—surmountable by those who carry the proper credentials, typically permission in the form of a doctor's note. However, to have access to the doctor and thus get the note, one must have status as a patient, and this constrains our understanding of consumers in general. Titley, in the third edition of Annan and Felter's medical librarianship textbook for library school students (1970) encourages a policy of helpful isolation of patrons and resources. Cordon off the layperson with the question ("their hesitancy and confusion makes them easy to identify") (p. 356). Segregate the material they seek, by enforcing the physician-permission requirement from a physician: "*Written* [italics original – CAS] requests from the physician for the patient to read particular material should be honored" (p. 356). Jeuell, Franciso, and Port (1977) surveyed public and private health sciences libraries and reported that "many libraries" in both settings "do not give out information to patients about their illnesses without a physician's note" (p. 294). Librarian Lisa Dunkel, speaking at California's Bay Area Reference Center workshop on consumer health in 1979, reported that "Many hospitals let patients use their libraries only with a doctor's permission" (Powers, unpublished manuscript, p. 2).

Library staff, of course, represent the ultimate consumable resource from the library director's perspective. Yet another variant on access restrictions, then, is the limit that has to be placed on certain library services. This is a common problem of libraries facing strains on organizational resources, as always appears to have been the case. Poisson (1983) says that "the mandate of some medical libraries extends only to the provision of information services to the organization's professional staff to support health-care services, teaching, and research" (p. 10). Some librarians have always allowed the public in, but limit reference services. For example, Poisson quotes one library's list of questions that cannot be answered by phone. This is so long that the reader may wonder just what questions were left to ask:

> *Telephone reference is strictly limited to queries which can be answered in five minutes... [not] [information about] credentials of health care professionals, medical biography and history of medicine, referrals to health care professionals, drug information, questions on 'research in progress', definitions or descriptions of diseases and syndromes, questions on library tests or uses of medical equipment or medical/diagnostic procedures, healthcare procedures for travelers, and biomedical statistics.*
>
> Poisson (1983, pp. 38–39)

Similar reference guidelines for conserving reference services recur over and over in the literature. For example, the Bay Area Reference Center workshop presenters surveyed public librarians and found that "most respondents," reference guidelines included time limitations—"many felt that if it took more than a few minutes the patron should come to the library" (Powers, unpublished manuscript, p. 7).

Other service restrictions existed to protect the interlibrary lending network. Statistics on the general public's use of interlibrary loan services for medical materials are very rarely published, but the public library network described by Poisson (1983) reported that 11.3% of the Hospital Library member's lending in 1980–1981 was to public library borrowing members, implying public library patrons (p. 130).

In the end, libraries as organizations define themselves when they define their user communities. For the librarians who work in these organizations, there are consequences to access restrictions—and to the absence of access restrictions—as will be discussed in the next section.

3.4 Librarian

3.4.1 Interpersonal challenges

> [The librarian] related an encounter she had had with a woman who came to the library and asked for information on Marfan's Syndrome. The Librarian provided her with pictures, articles and texts on the syndrome, and the woman 'got pretty hysterical', and told the Librarian that her son had this condition. The Librarian felt that she should say something, so she put her arm around her and said 'I'm really sorry that you had to see this.'
>
> Poisson (1983, p. 100)

> The Librarian said that she could envision an irate physician storming into the Library, waving a pamphlet, and shouting, 'How dare you give this to my patient!' This has never happened...
>
> Poisson (1983, p. 74)

Some challenges inherent in medical information provision lie in the emotional effects of information provision on the provider. And some medical information interactions are just plain awkward. Gartenfeld's thoughtful and important piece in *Library Journal* (1978) directly addresses this professional nervousness, and the barrier it can impose to successful library service, by admitting frankly that emotions happen: "We as librarians must learn how to deal with the occasional emotionally upset patron and not use this as a justification for denying information to those who want or need it" (p. 1914). "Physician lawyer" Norman Charney, quoted in Epstein (unpublished manuscript), noted that it was important for librarians not to prejudge the effect of giving bad news. But Fecher (1985) agreed that "providing information about fatal illnesses can be particularly stressful" (paragraph 12).

Ellsworth (1975) describes a very early unobtrusive study of medical reference around a controversial topic—abortion. Her study is also an illustration of the effect of sensitive reference questions on unprepared librarians. Ellsworth's goal was to measure the public library's effectiveness in comparison to telephone hotline services designated for people in search of abortion information. This small study was conducted while Roe v Wade (410 U.S. 113, 1973), the classic U.S. abortion rights case resolved by the Supreme Court, was being adjudicated. Ellsworth herself visited 15 Washington, D.C. area public libraries between March and April 1972 and asked librarians at each branch: "I have a friend who needs an abortion; do you know where she can get one?" (p. 29). In her account, published in *The Unabashed Librarian*, she noted the unhelpful response of one librarian whose immediate supervisor was "unable to have any children"; the librarian herself admitted to "mixed feelings" (p. 29). This library was the only branch of the 15 libraries where Ellsworth was "not only told there was no information but was given a decidedly chilly reception as well" (p. 29).

"As librarians," said the hospital librarian whose experience with the unhappy patron opened this section of the chapter, "we are trained to become fairly insensitive" (Poisson, 1983, p. 100). Medical questions are quite often sensitive ones, and they place strains on a professional who may desire the appearance of noninvolvement. Paradoxically, it requires awareness of the patron's individual situation, as well as sensitivity, to handle difficult questions dispassionately—and the very nature of "highly sensitive casework" (Donohue, 1976, p. 83) was one reason that lines of demarcation had to be drawn between information and referral services maintained by other public agencies, and existing public library reference services. Branch's account of the Ohio Health-Line service (1979) noted that one advantage of the cassette-driven telephone service was the cloak of anonymity it provided callers: "It is the epitome of anonymous communication. Those timid about asking personal questions can call Health-Line in full assurance that the library staff will not know their identity" (p. 328) which may have had something to do with the use statistics: most Health-Line questions turned out to involve sex (masturbation, male and female sexual response in particular). The Bay Area Reference Center survey of consumer health information services, conducted in 1979, asked library staff: "How does your library handle 'touchy' questions?" and the responses were illuminating:

- Very carefully!
- With great tact.
- All are touchy and all are handled the same (Powers, unpublished manuscript, p. 7).

"Touchy" was not defined, but librarians seemed to know it when they saw it. One respondent offered a clue about what might be particularly "touchy"—the delivery of bad news:

> *If we recognize the illness as being rare and very serious, we prefer not to be the one to give the patron 'the news' unless their physician has directed them to the library.*
>
> Powers (unpublished manuscript, p. 7)

Touchy questions make librarians uncomfortable because they know patrons are also made uncomfortable. "When [librarians] ask for more detailed information, they are not prying," cautions Averill to consumer readers: "They are just trying to elicit an 'answerable question'" (1980, p. 93). "Many" librarians interviewed by Yellott and Barrier (1983) reported that they "initially felt uncomfortable probing for the underlying questions; they felt they were prying" (p. 34). These librarians were doubly tasked, because Yellott and Barrier asked them to note their patrons' emotional states during their reference interactions. Fortunately, 90% of the patrons were calm to begin with; 4% were calmer at the end of the interview than at the beginning; and only 6% felt worse, although the authors do not tell us why or if the librarians' behaviors had anything to do with this.

A number of authors have noted the central paradox of reference work, particularly reference work with sensitive topics, in which the patron must give something up for the patron to be assisted (Wood, 1991). Librarians perceive additional stresses: particular attention needs to be paid to the patron's privacy in situations of medical information need (Dewdney et al., 1991). Ellsworth (1975) notes the significance of telephone manners in her unobtrusive abortion information study. Here, she comments on the skill set needed by abortion hotline volunteers:

> *Volunteers [on the hotlines] must be able to play two roles simultaneously: they must be able to locate and relay accurate information (and thus play reference librarian) and they must be able to deliver the information in a non-judgmental manner, and most particularly, in a non-negative manner (and thus play peer counselor). Any slight hint that the hotline volunteer is making a negative value judgment about the caller's moral fiber is likely to bring about a quick 'hang up' from the caller (p. 30).*

Donohue and Kochen (1976) call this special sensitivity one important aspect of information and referral services that is "beyond the competence of the library" (83), requiring staffing and resources and, one presumes, customer services skills to be found in a separate community service altogether and not in a library reference environment.

But Averill, in an early issue of the early consumer health magazine *Medical Self-Care*, recommends librarian-shopping to her readers instead: "If one librarian seems un-cooperative, try another. Odds are you will find one sympathetic to your search" (1980, p. 38).

Librarians are afraid that they lack the experience that they believe to be essential for medical information handling. "They may not feel completely 'in command' of the health collection. Librarians are trained not to give medical *advice* [italics original – CAS] – and rightly so" (Averill, 1980, p. 38). Paterson (1988) reviewed 10 years of library literature on the topic of health information provision, and found a common complaint: "Librarians and other nonmedical people have no background for answering medical information questions and would not be up to date on the latest treatments" (p. 84). Dewdney et al. (1991) found her public librarian respondents less confident about the accuracy of their answers to medical questions than they did about nonmedical questions, although, oddly, the same respondents didn't believe that "lack of subject knowledge" was a common problem in this domain (p. 190). Perhaps they were expressing a belief that other librarians had to be better at this than they were themselves. Furthermore, librarians' nonclinical expertise meant that they lacked context. One of Poisson's "Vesalius" library subjects commented about the barriers that lack of context placed on what librarians could do for patrons:

> *[A] woman... had recently written for information on a critical illness which her doctor thought she might have. She wrote that he had been unable to give her much information on the disease, and asked the Library to send information on diagnosis, prognosis and treatment. The Librarian said that as librarians they do not want to be in a position of censoring information, but without knowing more about the person, her education, stage of the disease, other health-related factors, and personal circumstances, it would be unacceptable to send her*

> *a clinical and technical discussion on prognosis (which was very poor) and treatment of the disease.*
>
> Poisson (1983, p. 40)

Most significantly, librarians have long feared being held legally liable for providing medical information without a license (Poisson, 1983). The earliest acknowledgment this author has been able to find of the "lawsuit" threat and its accompanying anxiety appeared in a *Library Journal* editorial published on January 1, 1978, by John Berry, entitled "Medical information taboos." At a meeting of the New York regional chapter of the MLA, Berry reported being "assaulted" by librarians concerned about the risks they faced. "Some librarians," wrote Berry, "claimed that medical information to a general public would place them in legal jeopardy" (p. 7). While Berry acknowledged the fear, he warned against overreaction, quoting Ellen Gartenfeld, coordinator of the Massachusetts Community Health Information Network (CHIN): "As long as librarians don't pretend to be doctors, there's really no problem" (p. 7). Six months later, Gartenfeld published her own article about CHIN. Gartenfeld, like Berry, is clearly addressing and responding to an actual voiced professional worry when she writes:

> *The last and most frequently asked question has to do with the legal liability of librarians who provide medical information to the health care consumer. We have checked with librarians knowledgeable in this area and they assure us that professional librarians, providing information from recognized sources, cannot be held liable for misuse of that information. As long as we take responsibility for communicating that we are librarians and not health care professionals, we need not worry about being sued (p. 1914).*

An echo of the liability theme is found in the Bay Area Reference Workshop conducted in 1979 (Powers, unpublished manuscript). A sample public library reference policy was reproduced in its entirety for conference attendees. That policy states: "Librarians must be careful not to practice medicine" (p. 8). The reference librarian is directed to tell the patron: "You may want to check with a qualified medical person for advice" (p. 8), thereby reinforcing the message that the librarian is not a "qualified medical person." Vaillancourt and Bobka (1982) stress that "Anxieties…can be put to rest by following the standard careful reference procedures: To provide information, and information only (NOT interpretation)" [emphasis original – CAS]. By embedding the caution in the context of reference services and policies, Vaillancourt and Bobka are attempting to densensitize librarian readers to sensitivity.

Four years later, Poisson parenthetically comments: "Public librarians in particular are sometimes concerned about their legal liability in providing medical information" (1983, p. 11). In her description of the "Vesalius" medical society library, "recommendations of physicians or clinics or other kinds of medical advice are restricted to avoid possible legal implications" (p. 39). In the similarly cloaked "Billings Medical School Library," serving an academic medical center "in a relatively poor and largely black city," the Supervisory Reference Librarian interviewed by Poisson noted that—"by providing published material or reading directly from published sources – *without providing an interpretation* – [italics mine – CAS], the Library could not be held legally liable" (p. 59).

3.4.2 Roles

Apart from these anecdotes, reactions to anecdotes, and clearly real—albeit pseudonymized—reference guidelines based apparently on reports of other anecdotes, there is very little documented in the professional literature after 1983, and nothing found before 1978, about this apparently legally groundless fear of liability on the part of librarians. The author would like to make the case here that the central organizing principle of librarians' professional worries around medical information provision was fear of transgression: of violating the boundaries of the librarian's role. Frankenberger (1936), concerned with social hygiene, made this point early. He related the interesting case of the physician who shifted responsibility directly onto Frankenberger's library: "[The physician] asked us to suggest a book he could recommend. This responsibility of the physician should not be transferred to us" (Frankenberger, 1936, p. 263).

Why? The reason was expertise.

> *The patient had come to him seeking such advice and his guidance based on first hand knowledge was what she was desirous of obtaining.*
> *Frankenberger (1936, p. 263)*

Furthermore, library staff were not able to assume this role of expert and advice-giver because, as laypersons themselves, they could not step into the physician expert's role and interpret information:

> *[I]t is not possible [for them]… to read all the books in any class of literature nor without medical training would they be qualified to pass upon the soundness or correctness of the opinions or text matter contained in these works.*
> *Frankenberger (1936, p. 264)*

Chambers, speaking at a symposium devoted to "the lay public," echoed this same opinion 18 years later, citing the physician's, not the librarian's, perspective:

> *One physician summarized his views this way: the patient is the doctor's problem. If knowledge about his condition can and should be obtained through reading, it is the doctor's responsibility to guide him, and if the doctor is unfamiliar with the literature, it is* his *responsibility [italics original – CAS] to find out about it.*
> *Chambers (1955, p. 260)*

Not only were librarians not physicians, their restriction from interpreting content meant they were not educators, calling again to the distinction between medical information provision and patient education:

> *The function of a medical reference service is primarily to make information available, and educational function is secondary and limited … although the layman does not always realize this.*
> *Chambers (1955, p. 258)*

The Bay Area workshop proceedings (Powers, unpublished manuscript) refer to "a fundamental theme, mentioned on nearly every form" (returned by the workshop participants to the organizers) "...librarians must take special care with medical reference and *never* [italics original – CAS] give personal advice, interpretation, or diagnosis" (p. 6). Conference organizers warn against interpretation by librarians no less than six times in 13 pages of narrative, quoting examples of policies from Kaiser-Permanente Library, Oakland, CA; Fremont (CA)'s Main Library; and the San Joaquin Valley (CA) Library System.

Speaking to attendees at the Bay Area Reference Center workshop Dr Tom Ferguson—a pioneer in consumer health informatics—was observing the demarcation of librarian and health-care professional roles when he stated that while he did see librarians as a "type of health worker,"

> *[T]here are many opportunities for librarians to become more closely involved in the health education [italics mine – CAS] field while still remaining reference sources rather than practitioners...*
>
> Ferguson quoted in Powers (unpublished manuscript, p. 3)

The librarian is not a doctor; nor, yet, is the librarian a teacher; the librarian is a conduit.

The reader should understand that the librarian–patron relationship is directed by the librarian's role in the context of the library as an organization. That relationship is at the heart of the medical information provision domain. If the patron is unhappy, so too will the librarian be. In the following section, the patron as challenge is discussed in detail.

3.5 The patron

> *The fear that the prurient-minded youth will use the medical books in a public library to his undoing, is born of the absurd notion that if the boys don't half-learn things there they will, of course, get them straight elsewhere. As a matter of fact the true story, half understood though it be, of the medical book is a decided advance on the street-corner stories which every boy gets his fill of anyway.*
>
> Dana (1902, p. 13)

John and Jane Q. Public in search of medical information are manifest in caricature very early in the professional literature of librarianship. Dr Gould's petition for the "union of public and medical libraries," published in the *Philadelphia Medical Journal* (1898), lampooned the patron public as "Lord Demos," while at the same time admitting that Lord's power:

> *I confess as I look over the shelves of the ordinary public library, or watch its statistics, that I am a bit frightened and disgusted at the trend and outcome of our much vaunted public-school system.*
>
> Gould (1898, p. 238)

Democratic book preferences were bound to have an effect on library collections:

> *How soon will Lord Demos find his soul? And come to a consciousness of his own rights and privileges? At present he is mightily interested in his quacks, his faith-cures, pink pills, wizard-oils, and a thousand evil spawnings of magic–mongering and savagery. How soon will it be, before you have to order 100 copies of each new work on hygiene, economics, domestic cookery, and the care of babies? You smile, but Demos has done stranger things than this in his time.*
>
> Gould (1898, p. 238)

Lord Demos was just unscientific. The lawyer, librarian, and physician George Wire saw the public as possessed by the public library's ancient evil, Prurient Interests:

> *There is a certain class of readers who frequent these medical sections for the unclean purpose of reading books on certain subjects, or of looking at the plates. This crowd followed the medical books from the Chicago public library to the Newberry library. Needless to say they were refused the books, and we gradually got rid of them in that way; but in a tax supported institution they demand, as they did in the Chicago public library, to see certain books, and if refused are liable to make trouble... This is one of the most disagreeable features of the whole thing, and one which the physicians who ardently advocate the medical department know nothing of.*
>
> Wire (1902, p. 12)

Why are patrons challenging? As the mob or as individual human beings, they can exhibit challenging behaviors. Florence Wickes (1934), reference librarian at Stanford's Lane Medical Library, saw problem patrons as two types. There were problem professionals: doctor friends of doctor friends who believe they should be given things; lawyers, journalists, writers. Then there was the problem public: "giggling adolescents" after the medical images, the spiritual descendants of the louts who followed Dr George Wire around in Chicago; the "victims of senility seeking various thrills in a certain type of medical literature" (p. 139); high school students too immature for the material; and patients. Miss Wickes presented this typology, complete with illustrative quotations, at an MLA conference. One conference attendee made comments on Wickes' remarks which were published as afterwords to Wickes' article. This person seems to have missed the memo about librarians not giving out medical advice:

> *Met school teacher – asked for material on insomnia. I suggested that she lie flat on her back until she fell asleep, or go to the clinic or to the doctor's office on State Street. She asked about doctor's fee. Two weeks later met her to find she is sleeping soundly and no visit to the doctor because his fee was $25.00.*
>
> Wickes (1934, p. 142)

In 1963, patrons from the public were still problems, although Radmacher, the advocate for separating technical from nontechnical content, was more concerned with the public's motives and mental health than Florence Wickes had been in 1934. Books needed to be hidden, said Radmacher, "from the pornographically minded, the morbid curiosity seeker, the neurotic, the hypochondriac, and those susceptible to impressions gleaned from the medical literature" (Radmacher, 1965, p. 465). Lillias Alexander of Toronto's Sunnybrook Hospital Library was thinking of people like this, as well as the very old medical idea of "maternal impressions," when she wrote in the August 1943 *Ontario Library Review*:

> People, when ill, are very susceptible to be suggestion...A young mother read a book that mentioned a child with a cleft palate. It worried her into feeling that her child would be similarly affected (p. 489).

In the very first issue of his pioneering consumer health journal, *Medical Self-Care*, Tom Ferguson, MD, reported another concern about patron motives:

> One of the Yale medical librarians, who wants to stay anonymous, was telling me that medical librarians were discouraged from giving laypeople medical information for fear it might be a patient gathering evidence for a malpractice suit.
>
> Ferguson (1976, p. 20)

3.5.1 Competition for time

The public was a problem for medical libraries in part because it competed for the librarian's scarcest resource—his/her own time—with physicians, the designated primary service population of any medical library. "These all take up time in a day that does not lack incident," said Florence Wickes (1934, p. 139). Frankenberger, in a published response to Wickes, agrees: "We have to do things within reason, unless we have unlimited funds and staff" (p. 142). Medical librarian Cathy Schell, writing in *Library Journal* in 1979, spoke of "the undeniable necessity of providing first priority service to the physician. When there is not enough time to satisfy the physician's information demands, how can time be lent to others?" (p. 930). And the less familiarity a patron has with information sources, the more time the patron requires, another problem for resource management. Ellen Hull Poisson learned from staff at the pseudonymous "Vesalius" medical society library that time had to be spent simply "trying to guide people who are not experienced library users" (1983, p. 41).

3.5.2 Expectations

In addition to challenges posed by patron behaviors, there are psychosocial issues. Patrons have expectations of their libraries and their librarians.

For example, Yellott and Barrier's early study of public librarians found that patrons expected the librarians to provide medical opinion (1983). The same year, Poisson reported the same to be said by her medical society librarians: "The librarians frequently find that callers expect to be talking to a physician and to be able to obtain medical advice" (1983, p. 39). Since librarians are trained *not* to interpret or offer advice, and institutional policies stress the prohibition, these assumptions become tensions that shake the structure of the reference interview before it starts. Patrons have emotions: anger "[I]f refused, are liable to make trouble" (Wire, 1902, p. 12). There is also unhappiness. Fear of possible patron reactions is another of Berry's "medical information taboos" (1978) and about which readers are reassured by Gartenfeld (1978): "[T]he potential emotional strain caused by information about their illness" (Berry, 1978, p. 7). One physician said to Yellott and Barrier (1983) that he "wondered if inquirers sometimes become upset by the information the library provided" (p. 35).

Poisson's "Welch Hospital" Library story includes an instructive anecdote:

> The Assistant Librarian told of the visit of a hospital volunteer whose grandchild was critically ill, and of her request for information on the child's disease. The Assistant Librarian was concerned because she was not sure that the volunteer knew that the condition was usually fatal. She did provide the information, however, and told the grandmother that although it did not look hopeful, there might be more current therapies available and that she should speak with a physician.
>
> Poisson (1983, p. 81)

Paterson, in summing up 10 years of literature, wrote: "People want advice rather than just information" (1988, p. 84). Patrons, of course, may not be clear on what they want in the first place; this is why they are time-consuming: "Most don't know specifically what they want or what their options are" (Poisson, 1983, p. 41). Another phenomenon impacting medical information provision is what Poisson (1983) calls "self-censorship" by patrons who use euphemisms to obscure the real question. She quotes a public librarian, part of a public library community health information network, who

> [o]bserved that many patrons will not tell the librarian outright what they are looking for... they might use euphemisms, such as 'skin diseases', rather than 'venereal disease' (p. 128).

3.5.3 Self-diagnosis

The single most frightening patron-centered challenge, and the one most present in decades of library literature, is that posed by the patron who works on his or her own without assistance from the librarian, attempting to diagnose his or her own illness. Wallis states flatly that the public wants information on how to treat the public (1949), even while Hawkins calls them "a problem in ethics" (1965). Radmacher of Skokie, IL, reports that her library will not even acquire books about diagnosis and treatment precisely because:

> The layman who wants to diagnose his own ailment and the patient obsessed with his own disease is also the person who would perform his own surgery if a do-it-yourself handbook were available (1963, p. 464).

John Berry, quoting Ellen Gartenfeld, had to reassure librarian readers that "We're not providing 'do-it-yourself' medicine; we're supporting people in their interaction with their health professionals" (1978, p. 42). Vaillancourt and Bobka (1982) acknowledge the thin line between supporting "health enhancement" via information, and facilitating self-diagnosis via information. Public librarians surveyed by Yellott and Barrier (1983) expressed fears to the researchers that self-diagnosis was going on. In 4 of the 66 cases investigated in their study, this actually proved to be the case. Paterson's 10-year review in 1988 brought the worry out again as a theme: "Self-diagnosis may mean that a person will delay seeking medical care when needed" (1988, p. 84).

Broadway (1983) used an interesting strategy to investigate the medical information needs in her small Utah community: She asked health-care professionals about their needs and their perceptions of both their patients' needs and how well the public library could meet them. She conducted a focus group of community physicians and found the opinion there that

> Patients read "just enough to get themselves in trouble." One physician noted that patients who read about illnesses tend to self-diagnose and may not listen to doctors' instructions.

Broadway asserts, however, something borne out in this literature review: "The assumption that inadequately informed patients will be inclined to self-diagnose is probably based on a few negative experiences and not on logic or research" (p. 256). While it does appear to have been a worry on the part of physicians, few instances turned up in the published literature beyond those mentioned by Yellott and Barrier (1983).

For a patron to engage in self-diagnosis, prior to 1994, the librarian had to be intimately involved. She would have had to have provided him with access—intellectual, physical, both—to medical content that enabled the self-diagnosing. The final dimension discussed below presents the challenges inherent in medical content.

3.6 Content

> My father hunts in a doctor book,
> While sweat adorns his brow;
> My mother weeps, but will not look
> For the pin that hurts me now.
>
> The baby book (1915)

3.6.1 Terminology

The entry-level obstacle to understanding medical content is medical terminology, and the problematic status of terminology clearly relates directly to the fear of inducing confusion. This is a very old fear, as this poem published in the *Chicago Journal* (1895) makes clear:

> Oh, deep and learned doctors, can you not permit us pray,
> To have old-fashioned ailments in a good old-fashioned way?
> The language you employ's designed to take away one's breath;
> Your terms are quite enough to frighten timid folks to death…

Our good old grandmas never dreamed of dreadful things we see,
But pinned their blind and simple faith to herbs and boneset tea;
They never guessed that when their friends from earth were called away,
'Twas all because of microbes and the dread bacteria.

'Twas well they never ran across these later fearful germs,
Not ever had to look up on these brand new-fangled terms.
Those days a patient never guessed of what was just inside,
And that is why no more of them turned up their toes and died.

Lillard (1895, p. 109).

The medical terminology problem is first explicitly mentioned in the library literature in 1934 by Florence Wickes, of the Lane Medical Library. She describes a patient patron "floundering in terms so new and strange to him that he arrives at a wrong conclusion if he arrives at any at all" (1934, p. 139). Frankenberger (1936) calls attention to the terminology and resulting translation problem as a reason for physicians to read *popular* materials: "[T]he physician can then 'answer his patients' questions in terms which he can understand" (p. 263).

Wallis, 12 years later, argued that patients had a right to expect material information written in their own language, not jargon (1949); clearly, for Wallis, the idea that patients and doctors necessarily spoke different languages was already ingrained. Schell (1980) points out that access to medical information in any type of library, whether public or medical, cannot be the whole solution to the problem of access if the patron still faces the "roadblock" of technical jargon once she is handed the material. Jargon may make materials "incomprehensible to the person or even to the medical librarian, who is usually not a physician" (p. 930). The last point is seconded by Averill, who comments: "Be extra-precise about the names of medical conditions and drugs; a minor misspelling can send you and the librarian off on a wild goose chase" (1980, p. 39). Averill also points to nursing terminology as a solution to the medical terminology problem: "Nursing periodicals are often very useful to the non-doctor because they use less technical language than medical journals" (1980, p. 39). Poisson calls attention to "the ability of library patrons to comprehend and apply" medical information as a concern (1983, p. *lv*) and quotes the director of her pseudonymous public library on the problem of medical dictionaries:

I always felt guilty giving someone a medical dictionary even though someone would come in and ask for it... it wasn't going to be helpful. The definition just threw in three or four more words that no one understood (p. 109).

Paterson (1988) connected terminology with the physician's role, not the librarian's: "Lay people need nontechnical information with interpretation by a doctor" (p. 84). As medical terminology was a problem in print, so too did it become a problem for CD-ROMs; a library school student at Kent State University investigated two consumer-centered databases, Health Periodicals and Combined Health Information, for comparison with MEDLINE because MEDLINE "furnishes information that is too technical for the average person" (Gawdyda-Merolla, 1992, p. 3). Clearly the medium had no effect on the central message.

3.6.2 Appropriate for audience

Another problem of medical content in library collections comes from its assumed audience. It is not simply a question of "physician" versus "public" use, but of generalist versus specialist information needs. The general public, in fact, can be as specialized in its needs as can physicians. Louise King, reference librarian at the New York Academy of Medicine, summed the dilemma up as a problem directly related to both content and patron:

> *There are many problems which the public has to solve and the literature for which may not properly belong in a public library. I am now speaking of those who wish to read further into medical problems than most of the literature written only for the lay public will go.*
>
> *King (1955, p. 243)*

The need for materials at different levels of understanding makes for selection problems, as well as reference problems, in medical as in other special library settings (Wyer, 1930). These general sources may be insufficient for the user and often stimulate further questions, so that the librarian refers him/her to a physician or another library (Schell, 1980). Dewdney et al. (1991) wrote that the usability of sources was a problem for librarians, too, who can be looked at as a particular kind of consumer.

It is frustrating to be a librarian facing expression of a critical, deep information need, but know that you do not have the right resources to meet those needs. Sometimes no specialized material exists for use by the public (Fecher, 1985; Poisson, 1983). More often, material is simply not appropriate. Poisson cites an anecdote involving *Index Medicus*, the print ancestor of today's MEDLINE database produced by the National Library of Medicine in one form or another since 1879:

> *The Associate Librarian… spoke of a couple who had come to the library with a small child who was critically ill. The Associate Librarian had been able to locate material in texts and some articles from* Index Medicus, *and discussed the possibility of a computer search with them. It was apparent to her that "the* Index Medicus *articles (were) not what they wanted" and that they didn't need to know all the worst clinical possibilities. She felt that they really needed a support group or pamphlet material written specifically for parents of children with this condition. She said 'This happens all the time, and they don't need my help, they need someone else's help.'*
>
> *Poisson (1983, pp. 40–41)*

3.6.3 Aspects of quality collections

What are some other content-related challenges? There is the challenge of maintaining a quality collection. Wallis, as early as 1949, argued that laypeople should have, and had a right to expect, reliable, accurate materials on the shelf. Gartenfeld, 29 years after Wallis, called this "quality assurance" (1978, p. 1914). This causes anxiety for public librarians. The "Dickinson" Public Library reference librarian commented that "they had never been sure that they were purchasing 'the best material'" (Poisson, 1983, p. 107) and were concerned about "material which is inaccurate" (Poisson, 1983, p. 95). On the medical side, libraries open to the public can have "schizophrenic" collections due to the library's need to anticipate lay and research needs at the same time and on the same shelf (Poisson, 1983, p. 47). Sometimes, the items of highest interest are the items that go missing, which means that collections are uneven and out of date (Dahlen, 1993; Dewdney et al., 1991). This was a problem noted in the first edition of Alan Rees and Blanche Young's classic *Consumer Health Information Source Book*; the "high theft rate" of health information items also affects the availability of items for other patrons (Averill, 1980, p. 38).

3.6.4 "Bad" books

Medical information can also be controversial information. The equation of the two is a very old one in the literature, and a review of medical subjects considered "good" or "bad" between 1934 and 1989 is instructive. Late nineteenth- and early twentieth-century physicians and librarians were most concerned about books produced by the "pathies" (homeopathy, osteopathy, and the like) and the public enamored of them was "the poor, silly community, hypnotized by ignorance and befouled by quackery" (Gould, 1898, p. 239). George Wire was so worried about the potential conflict of advice on the shelves between, for example, allopaths and homeopaths, that he urged the library profession "not to rely on the advice of any physician or committee of physicians for selection of its medical books" (1902, p. 10). Even in 1963, Fleming could agree that "popular" was synonymous with "crackpot," and wasn't wild about the free availability of sexual information, even for student nurses:

> *I am sure that most of you have some sort of device by which you are making it impossible for the freshman student or the student nurse to come in and get Hirsch's* The Power to Love *and Stone's* Marriage Manual *and then come back later and get Dickinson's* Technique of Contraception *without some screening. The same device can be used with pseudoscientific materials (p. 474).*

Connor (1963) considered "pseudotechnical" popular books to be "truly dangerous" (p. 468). He gave as examples titles on cancer pathogenesis, cellular therapy, and counting calories. Hawkins (1963) cited "quack" literature (p. 476). The Director of the "Dickinson" Public Library also interviewed by Poisson pointed to "diet books" as another controversial area (p. 108), while the Melville director commented, as have hundreds of thousands of public librarians before and since, "If they are things that people are asking for, we provide them...We even have materials on Christian Science" (p. 113).

Medical and food quackery, not to mention food faddism, were to be avoided (Cole, 1963). But Radmacher (1963) at Skokie would not acquire even orthodox and standard medical textbooks for fear that they would give her patrons ideas. Cormier (1978) would rather see the real thing than the popular translation—"these books are a poor source for current and standard medical knowledge" (p. 2051). Vaillancourt and Bobka (1982) refer to controversial materials "in terms of reliability, or what some professionals call 'quackery'" (p. 46). Ironically, among popular works, the lack of scientific value perceived by librarians at "John Shaw Billings Library" made them a poor choice even for a consumer collection (Poisson, 1983). This same library ruled out collection of "borderline" medical material, defined as books that crossed boundaries into the subjects of "religion, magic, astrology, alchemy…and quackery" (Poisson, 1983, p. 44). Emery (1989) wrote, in one of the few pre-Web articles making explicit the connection between popularity and poor quality, about factual inaccuracies, "redundant" and "contradictory" information as problems of popular works (p. 5).

3.6.5 Harmful books

A second category of bad books is the type considered not just low-quality, but actively harmful to the reader. Different generations of librarians in different settings have defined "harm" differently. At times it seems to have meant "loss of innocence." Morrison (1976), writing about consumer health information in public libraries, notes: "We've come a long way from the time when we hid the medical dictionary and a few anatomy books behind the desk – lest the normally curious youngster learn the 'facts of life' from our resources" (p. 5). But at other times it meant "inducing confusion" and health-care professionals expressed fears about causing that confusion. For example, an anonymous "Halsted Hospital" physician interviewed by Poisson (1983)

> admitted that he had not referred his own patients to the Consumer Library, and added… In my practice they just don't bring up questions to research. I may have a different kind of practice… backs, disks. That's the kind of thing that if they read the literature I think they could become confused (p. 99).

In the "Vesalius" medical society library, one physician member of the Library Committee gave his opinion of the consumer health movement:

> One has to support the position that knowledge is good, therefore I am in favor of people knowing more although I have no very strong expectations.

However, Poisson notes:

> He was concerned with the potential for misunderstanding and use of the information, which he felt to be aggravated by the "condition of gangrene" currently prevalent in the U.S. educational system as a whole (p. 42).

Vaillancourt and Bobka's early guidelines (1982) give insight into what must have been received wisdom in the profession about potentially harmable patrons:

> The philosophy that information may be hurtful to some patients: psychiatric patients, individuals with cancer, those suffering from diseases with painful, bizarre or possibly embarrassing symptoms (pp. 42–43).

And Jackson (1985) quotes the President's Commission on Ethics in Medicine and Research:

> *Contrary to the fears of many that too much information frightens patients, most refusals of treatment stem from too little information and other lapses in communication (quoting Frederick, 1982, p. A-2).*

As noted in the discussion of librarian-centered challenges, above, librarians worried about causing emotional damage. Gartenfeld reminds us that the potential "harm that may be caused" (1978, p. 1914) should not excuse the librarian from providing information at all. She acknowledges that the consumer will not always be happy with what they learn—and that this does not justify "denying information to those who want or need it" (1978, p. 1914). Paterson lists fear and confusion as a common theme in the literature of the past 10 years (1988), summing it up as: "Not everyone can nor should they be able to handle health information and such knowledge can frighten" (p. 84). In addition, "patrons are often unable to reconcile several possible medical solutions" (p. 84).

3.7 Conclusions

The author almost resists, but cannot quite, the urge to conclude this historical review with the obvious truisms. To say there is nothing new under the sun is to miss the point: medical information is apparently very, very important to the general public and has been for a very long time; this was not an information need that was created, although it is fair to say that it was an information need considerably accelerated, by the World Wide Web. This chapter characterized four dimensions of challenge inherent in medical information provision in libraries. Perhaps they are not dimensions, but an Escher-style staircase in which librarians and patrons meet themselves going up and down. Medical information begins at the level of the physical instantiation of the library building, when it is considered for acquisition using library funds and for housing in specialized circumstances based on library policies. Having been acquired and housed, medical information is available to the professionals who staff the library, who find that their provision of it to patrons places them in new and perhaps inappropriate roles—counselor, interpreter, navigator—in relation to those patrons. The patron who receives medical information may have her motives come into question; her expectations about what library staff can do for her, and her use of professional staff time, may be new and different. Finally, the material that is provided to patrons may require interpretation and staff time to be at all useful and appropriate to specific patrons' needs; care must be taken to buy quality and avoid "bad" materials and ensure that the collection stays up-to-date; if library funds are to be spent on medical information, what kind of patrons should be allowed access to it, and how can that access best be controlled in the building? No matter what the medium—pages or bytes—in which medical information is exchanged for the use of the general public, these are timeless questions which will continue to arise as long as information has intermediaries.

References

Alexander, L. (1943). Patients' reading. [Ontario library review, August 1943. Quoted in]. *Library Journal, 69,* 489.
Averill, M. S. (1980, Summer). The library: passport for health information. *Medical Self-Care,* 38–39.
Ballard, J. F. (1927). Information, reference and bibliographic service. *Bulletin of the Medical Library Association, 17*(1), 18–27.
Bay, J. C. (1924). Sources of reference work. *Bulletin of the Medical Library Association, 14*(1), 10–15.
Beehler, I. A. (1955). Who is the "lay public"? *Bulletin of the Medical Library Association, 43*(2), 241–242.
Belleh, G. S., & van de Luft, E. (2001). Financing North American medical libraries in the nineteenth century. *Bulletin of the Medical Library Association, 89*(4), 386–394.
Berelson, B. (1949). *The library's public: A report of the public library inquiry.* New York: Columbia University Press.
Berry, J. (January 1, 1978). Medical information taboos. *Library Journal,* 7.
Branch, S. (1979). Health-line: a new reference service. *RQ, 18*(40), 327–330.
Broadway, M. D. (1983). Medical and health information needs in a small community. *Public Libraries, 32*(5), 253–256.
Brodman, E. (1974). Users of health sciences libraries. *Library Trends, 23*(1), 63–72.
Brother Ignatius. (1941). Bibliotherapy: books as cure-alls. *Hospital Progress, 22,* 325–327.
Chambers, J. L. (1955). Services to the lay public. V: what services should we give the lay public? *Bulletin of the Medical Library Association, 43*(2), 257–262.
Cole, G. (1963). Questionable medical literature and the library: a symposium: the hospital library. *Bulletin of the Medical Library Association, 51,* 480–482.
Committee for Public Health Education Among Women, American Medical Association. (1912). *Report for the year 1911–1912.* Chicago: Author.
Connor, J. M. (1963). Questionable medical literature and the library: a symposium: the medical society library. *Bulletin of the Medical Library Association, 51,* 467–471.
Consumer and Patient Health Information Services [CAPHIS] Task Force, Medical Library Association. (1996). The librarian's role in the provision of consumer health information and patient education. *Bulletin of the Medical Library Association, 84*(2), 238–239.
Cormier, J. (1978). Practicing librarian: medical texts for public libraries. *Library Journal, 103,* 2051.
Crowe, M. R. (1944). A medical society library in a public library. *Bulletin of the Medical Library, 32*(2), 221–229.
Dahlen, K. (1993). The status of health information delivery in the United States: the role of libraries in the complex health care environment. *Library Trends, 42*(1), 152–179.
Dana, J. C. (1902). Medical departments in public libraries. *Medical Libraries, 5*(3), 13.
Dewdney, P., Marshall, J. G., & Tiamiyu, M. (1991). A comparison of legal and health information services in public libraries. *RQ, 31*(2), 185–197.
Donohue, J. C., & Kochen, M. (1976). *Information for the community.* Chicago: American Library Association.
Donohue, J. C. (1976). The public information center project. In M. Kochen, & J. C. Donohue (Eds.), *Information for the community* (pp. 79–93). Chicago: American Library Association.
Eakin, D., Jackson, S. J., & Hannigan, G. G. (1980). Consumer health information: libraries as partners. *Bulletin of the Medical Library Association, 68*(2), 220–229.

Ellsworth, S. M. (1975). The public library as an information disseminator. *The U*N*A*B*A*S*H*E*D Librarian, 14*, 29–30.

Emery, M. W. (1989). The responsibility of librarians for collecting and making AIDS materials available. In *Annual conference of the Utah library association, St. George, UT, February 15–17, 1989*. ED 322 922.

Epstein, H. A. (Unpublished manuscript). Patient/consumer information. In P. J. Wakely & R. S. May (Eds.), *Basic library management for health sciences librarians* (2nd ed.). Midwest Health Sciences Libraries Network.

Farlow, J. W. (1921). The relation of the large medical library to the community. *Bulletin of the Medical Library Association, 11*(1–4), 2–4. Retrieved from http://www.ncbi.nlm.nih.gov/pmc/articles/PMC234874/.

Farrow, M. S. (1934). Closed to the public. *Bulletin of the Medical Library Association, 22*(4), 225–227. Retrieved from http://www.ncbi.nlm.nih.gov/pmc/articles/PMC234235/.

Fecher, E. (1985). Consumer health information: a prognosis. *Wilson Library Bulletin, 59*(6), 389–391.

Ferguson, T. (1976, Summer). Medical libraries. *Medical Self-Care, 20*.

Fishbein, M. (1947). *A history of the American Medical Association, 1847 to 1947*. Philadelphia: W.B. Saunders.

Fort Rosecrans National Cemetery. (May 15, 1957). *Report of interment [Mildred S. Farrow]*. Provo, UT: U.S. National Cemetery Interment Control Forms, 1928–1962 [database on-line], Retrieved from Ancestry.com.

Frankenberger, C. (1936). Medical libraries, the medical profession and social hygiene literature. *Journal of Social Hygiene, 22*, 262–264.

Frederick, D. (October 26, 1982). *Be more open with patients, doctors told*. Pittsburgh Press, A-2.

Garrison, F. H. (1921). Community interests from the viewpoint of the medical library. *Long Island Medical Journal, 15*, 105–113.

Gartenfeld, E. (1978). The community health information network: a model for hospital and public library cooperation. *Library Journal, 103*(17), 1911–1914.

Gawdyda-Merolla, L.E. (1992). *Medical reference: A comparative analysis of two consumer health databases with MEDLINE*. Unpublished master's thesis, Kent State University, Kent, Ohio.

Getchell, A. C. (1898). The Worcester district medical library. [Sketches of medical libraries III.]. *Medical Libraries, 1*(5–6), 33–40.

Goodchild, E. Y. (1978). CHIPS: consumer health information program and services—in Los Angeles. *California Librarian, 39*, 19–24.

Gould, G. (July 30, 1898). Union of medical and public libraries. *Philadelphia Medical Journal, 1*.

Groen, F. K. (1996). Three who made an association: I. Sir William Osler, 1849–1919 II. George Milbry Gould, 1848–1922 III. Margaret Ridley Charlton, 1858–1931 and the founding of the medical library association, Philadelphia, 1898. *Bulletin of the Medical Library Association, 84*(3), 311–319.

Hawkins, M. (1963). Questionable medical literature and the library: a symposium: the National Library of Medicine. *Bulletin of the Medical Library Association, 51*, 475–479.

Jackson, M. (1985). Mental health information access? Yes, but not in my neighborhood. *Catholic Library World, 56*(7), 287–290.

Jeuell, C. A., Franciso, C. B., & Port, J. S. (1977). Brief survey of public information services at privately supported medical school libraries: comparison with publicly supported medical school libraries. *Bulletin of the Medical Library Association, 65*(2), 292–295.

King, L. (1955). Service to the lay public. II. Shall we purchase lay material and if so to what extent? *Bulletin of the Medical Library Association, 43*(2), 243–251.

Langner, M. C. (1974). User and user services in science libraries: 1945–1965. *Library Trends, 23*(1), 7–31.
Lillard, J. F. B. (1895). *The medical muse grave and gay: A collection of rhymes up to date, by the doctor, for the doctor, and against the doctor.* New York: I.E. Booth.
Loomis, M. M. (1926). Reference service. *Bulletin of the Medical Library Association Quarterly, 15*(3), 33–34.
Lydenberg, H. M. (1928). Interrelation of medical and public libraries. *Bulletin of the Medical Library Association, 17*(4), 19–21.
Marriage notices. (October 15, 1902). *San Francisco call.*
Marshall, J. G., Sewards, C., & Dilworth, E. L. (1991). Health information services in Ontario public libraries. *Canadian Library Journal, 48*(1), 37–44.
Miles, W. D. (1983). *A history of the National Library of Medicine: The nation's treasury of medical knowledge.* Bethesda, MD: U.S. Dept. of Health and Human Services.
Monahan, H. S. (1955). Service to the lay public: III. Policies set up by medical societies to regulate the use of materials by the lay public. *Bulletin of the Medical Library Association, 43*(2), 252–254.
Morrison, C. (1976). When patients become patrons: patient education in the public library. *Hospital Libraries, 1*(11), 5–7.
Morton, R. S. (1937). *A woman surgeon: The life and work of Rosalie Slaughter Morton.* New York: Frederick A. Stokes.
National Network of Libraries of Medicine. (2015). *Members directory.* Retrieved from http://nnlm.gov/members/.
Paterson, E. R. (1988). Health information services for lay people: a review of the literature with recommendations. *Public Library Quarterly, 8*(3–4), 81–91.
Perryman, C. (July 2006). Medicus deus: a review of factors affecting hospital library services to patients between 1790–1950. *Journal of the Medical Library Association, 94*(3), 263–270.
Poisson, E. H. (1983). *Libraries and the provision of health information to the public* (Unpublished doctoral dissertation). New York, NY: Columbia University.
Powers, A. (Unpublished manuscript). *The ethics and problems of medical reference service in public libraries: Summary and addenda to the September 1979 Bay Area Reference Center Workshop.* Bay Area Reference Center.
Public Library of Cincinnati and Hamilton County, OH. (Unpublished manuscript). *Telephone questions, U.A.R.* (Useful arts room). In collections of the Cincinnati Room, Main Library, Cincinnati, OH.
Radmacher, M. (1963). Questionable medical literature and the library: a symposium: the public library. *Bulletin of the Medical Library Association, 51,* 463–466.
Ranck, S. (1907). Editorial: medical departments in public libraries. *Medical Library and Historical Journal, 5*(3), 210–220.
Richardson, E. C., & McCombs, C. F. (1929). *The reference department.* Chicago: American Library Association.
Rubenstein, E. (2012). From social hygiene to consumer health: libraries, health information, and the American public from the late nineteenth century to the 1980s. *Library & Information History, 28*(3), 202–219.
San Diego County Medical Society. (2013). *Patient quicklinks.* Retrieved from http://www.sdcms.org/PatientQuicklinks.aspx#who-should-i-call-with-a-question.
Schell, C. (1980). Preventive medicine: the library prescription. *Library Journal, 105*(8), 929–931.
Shores, L. (1939). *Basic reference books: An introduction to the evaluation, study, and use of reference materials with special emphasis on some 300 titles* (2nd ed.). Chicago, IL: American Library Association.

Shores, L. (1954). *Basic reference sources*. Chicago, IL: American Library Association.
Sloan, G. W. (1898). [Letter.] *Medical Libraries, 1*(7), 45.
Spivak, C. D. (1899). Medical departments in public libraries. *Philadelphia Medical Journal, 2*, 851.
Spivak, C. D. (1902). Medical departments in public libraries. *Medical Libraries, 5*(4), 17–24.
Stein, E. A., & Lucioli, C. E. (August 1, 1958). Books for public library medical reference work. *Library Journal, 97*, 2110–2113.
Sullivan, W., Schoppman, L. A., & Redman, P. M. (1991). Analysis of the use of reference services in an academic health sciences library. *Medical Reference Services Quarterly, 13*(1), 35–55.
Titley, J. (1970). The library and its public: Identification and communications. In G. L. Annan, & J. W. Felter (Eds.), *Handbook of medical library practice* (3rd ed.). Chicago: Medical Library Association.
The baby book. (1915). Buffalo: Carl J. Ward.
Ulrich, C. (1920). Social hygiene in a large city library. *Journal of Social Hygiene, 22*, 250–251.
United States of America, Bureau of the Census. (1900). *Twelfth census of the United States*. Washington, DC. National Archives and Records Administration, 1900. Year: 1900; Census Place: San Francisco, San Francisco, California; Roll: 102; Page: 7A; Enumeration District: 0111; FHL microfilm: 1240102. Retrieved from Ancestry.com.
Vaillancourt, P. M., & Bobka, M. (1982). The public library's role in providing consumer health information. *Public Library Quarterly, 3*(3), 41–49.
Van Hoesen, F. R. (1948). *An analysis of adult reference work in public libraries as an approach to the content of a reference course* (Unpublished doctoral dissertation). Chicago, IL: University of Chicago.
Walker, W. D., & Hirschfeld, L. G. (1976). Sources of health information for public libraries. *Illinois Libraries, 58*(6), 459–501.
Wallis, C. L. (1949). What the public wants in a medical book. *Bulletin of the Medical Library Association, 37*(3), 251–255.
Wannarka, M. (1968). Medical collections in public libraries of the United States: a brief historical study. *Bulletin of the Medical Library Association, 56*(1), 1–14.
Wickes, F. L. (1934). Library reference service. *Bulletin of the Medical Library Association, 22*(3), 137–143.
Wire, G. E. (1902). Medical departments in rate supported public libraries. *Medical Libraries, 5*(2), 9–12.
Wood, M. S. (1991). Public service ethics in health sciences libraries. *Library Trends, 40*(2), 244–257.
Wyer, J. I. (1930). *Reference work; a textbook for students of library work and librarians*. Chicago: American Library Association.
Yellott, L., & Barrier, R. (1983). Evaluation of a public library's health reference service. *Medical Reference Services Quarterly, 2*(2), 31–51.

Ethical health information: Do it well! Do it right! Do no harm!

Michelynn McKnight
Louisiana State University School of Library and Information Science, Baton Rouge, Louisiana

4.1 Introduction

Sharing health information is a kind and sometimes even life-saving act. To help another—to really help—one must do it well and do no harm. Along with the opportunity for doing good comes the challenge of doing it right. Getting the right accessible health information to the right person at the right time is crucial to the well-being of individuals and society. Richard E. Rubin calls this practical application of ethical principles "information ethics" (2004, p. 327). Unfortunately, many normal conversation patterns may do more harm than the participants may realize. So, one who wishes to provide health information has to develop ethical information habits.

Ethics are not simple rules to follow or to be tested to the breaking point, but are habits that develop into virtues instead of vices. In *Reverence: Renewing a Forgotten Virtue*, Paul Woodruff describes virtue ethics as so ingrained in our behavior that we are motivated by actually feeling better about behaving ethically than not. That capacity is developed not just by reason and learning rules, but from constant practice. "A virtue is a capacity, cultivated by experience and training, to have emotions that make you feel like doing good things…virtue is about cultivating feelings that will lead you in the right way whether you know the rule in a given case or not" (Woodruff, 2001, pp. 61–62). As a vice is a habit of making bad decisions, a virtue is a habit of making good ones. It is easier and more satisfying to make a habitual decision, so ethical habits are important.

> *If the purpose of ethical rules is to give us guidance so that one (or a small set of) overriding rule can give us specific answers to our problems, then it is bound to fail. Virtue ethics advises us to recognize…the complexity of the ethical life and accept that the kind of guidance we can expect will have to be tailored to the nature of the subject matter.*
>
> Athanasoulis (2013, p. 20)

Everyone has habitual ways of reacting to various situations, but the habits of the ethical information provider will make it easier to do it well than badly, to help instead of harm, even if well intentioned. (With this gift, it is not just the thought that counts.) Ethical habits are very important in health information services. There are good and bad ways of relating information to people. Ethics are strengths made stronger by practice or weakened by ignorance or laxity.

The American Library Association (ALA) *Standards for Accreditation of Master's Programs in Library and Information Studies* (the MLIS degree) include program objectives for students to learn "the philosophy, principles, and ethics of the field" (ALA, 2008b, 2014a). Librarians with an ALA-accredited master's degree in library and information science (MLIS) may or may not have taken a semester-long course in information service ethics, but they would have studied ethics as parts of a number of different courses, and they can always profit from a review. Similar responsibilities are common to all in these important conversations.

4.2 Responsibility for the best possible information service

> **Bad example: personal experience may be irrelevant**
>
> *I'm a little nervous about having cataract surgery.*
> *Oh, I had that and it went just fine. It's over quickly and you will see so much better immediately. You shouldn't worry about it.*
> *Well, after my mother had hers, she went blind. Didn't do her a bit of good.*

4.2.1 What's the question?

Casual conversations like that often happen among friends. But they do not provide the best information someone needs. Such answers may be mostly irrelevant to the first person's comment and real information need. It's a temptation to make one's personal experience central to the conversation, but health information providers must do better than that.

Librarians learn not to jump to conclusions too soon, but first to make sure that they understand the person's real question behind the opening statement. Everything starts and ends with listening to the client—not with looking for the information to provide. It starts with a careful, respectful interview and ends with listening to find out if the client understands and is satisfied. Providing good information is terrible service, if it doesn't answer the person's real question. It's all too easy to answer the question one wishes the client had asked instead of the question the client actually has. And that includes being sensitive to the client's existing health information literacy—providing too much, too little, or the wrong information can be worse than not answering at all.

Librarians learn and practice the *reference interview* which looks to most observers to be a casual conversation. In reality, it is a complex process by which the librarian helps the client to explore and refine the question. The librarian never takes the first question as the real question, but understands that it is just an opening statement from which together they can tease out the underlying real problem query. The librarian will ask open-ended questions (i.e., questions that cannot be answered with a "yes" or "no"). He will paraphrase, rather than repeat what the client has said, giving the client the opportunity to correct the librarian's understanding of the real question. A good heuristic for a successful interview is once the real query seems perfectly clear, one should ask one more clarifying question. That added question can sometimes pivot the librarian's perception of the query into a completely different direction (Kern & Woodard, 2011; McKnight, 2010).

Note that understanding the person's question or perceived information need drives this service. This process is different from formal patient education in that the client determines the goal. In patient education the teacher sets the goal and determines whether or not that goal has been met. Teaching one or more people is a different process, with its own purpose, process, standards, and evaluation determined by the teacher. Librarians and other information professionals pay attention to what the client wants to know and the client decides whether or not that question really has been answered (Consumer and Patient Health Information Section of the Medical Library Association (CAPHIS), 2010; McKnight, 1997).

The interview can also reveal how much information the client desires and in what form the client wants it. While working with the client, the librarian can informally assess the client's general literacy and specific health information literacy, so that the resulting search will be for appropriate information in an appropriate format. If a person wants and can understand in-depth information about recent research in the treatment of stage II ductal carcinoma of the breast, a pamphlet on breast self-examination is not going to satisfy that need!

Language and semantic issues are frequent challenges to defining the real question and getting the best possible information. "Librarians and patrons both struggle with the complexities of medical terminology" (Eberle, 2005, p. 32). Medical dictionaries, encyclopedias, and other tertiary sources often are helpful. However, those intended for specialists may not be appropriate for others. Some terms closely resemble terms for completely different things. And there are, few if any, simple explanations for some complex concepts (Ham, 2012). Many health sciences librarians are familiar not only with resources that help translate specialized English terms into more common English terms (National Library of Medicine, 2014a), but also with resources in other languages (National Library of Medicine, 2014b).

Health information seeking—indeed any kind of personal information literacy—has three parts: (1) A person knows that they need to know something that they do not know; they have some sort of question. (2) A person knows how to go about searching for and using information sources that might help answer that question. (3) After a person has found something, they know how to evaluate and use what was found. Librarians and others, who provide information services for people must know not only the most reliable sources, but must be able to match the best sources to the specific person's question. What is the best source? That always depends upon both the original, real question and the person who is asking.

4.2.2 Where should I look for the answer?

> **Bad example—sloppy and unreliable searching:**
>
> *I'll just type what I'm looking for into the search box in Google and take whatever comes up that looks interesting.*

Frequent searchers without the benefit of training in how online information sources are constructed can usually find *an* answer but it may not be the best answer or even a good answer. The naïve occasional or first-time user may not be able to find a relevant answer at all. Without understanding what is actually behind the visible interface, they naturally develop superstitious searching habits. The information professional must ensure the delivery of the answer that is *the best possible* information for the client, not just sometimes, but always—even if it is more difficult to retrieve. It's an everyday responsibility. Responsibility for the best possible service includes constant learning about a wide variety of health information sources and evaluating them for particular purposes.

For example, most health sciences librarians (who typically work in hospitals, health sciences centers, or large clinics) have expertise in using online information systems to search for good evidence to support health care decisions. They frequently use complex, robust bibliographic databases to identify relevant literature. Often they use advanced search techniques to identify research and clinical literature indexed in PubMed/Medline produced by the U.S. National Library of Medicine (NLM) or other such bibliographic databases. To identify excellent Web-based health information from reputable sources for patients and consumers, they frequently use NLM's popular and free portal, MedlinePlus.gov, as well as the Consumer and Patient Health Information Section (CAPHIS) of the Medical Library Association's (MLA) Top 100 list of health Web sites (CAPHIS, 2013a).

NLM also makes it easy to identify libraries and librarians who can help through the free directory of the National Network of Libraries of Medicine (NNLM) on the NNLM Web site at http://nnlm.gov/members/ (US DHHS, NN/LM, 2013). Professional health sciences librarians, especially those with the Academy of Health Information Professionals (AHIP) credential (MLA, 2014) provide an abundant range in information services, including instruction in how to use complex online systems.

Librarians' service ethics (Rubin, 2011) require that even if librarians believe they know the answer to a particular question, they may not offer it without citation to an authoritative and reliable source. Librarians may joke that means that even if asked how to spell "cat," they must give a dictionary reference for the spelling of the word. They work diligently to keep their personal health opinions out of the way so that their clients get the best possible information. They are not professional "know-it-alls" but, more importantly, professional "know-how-to-find-it-alls." Librarians can and do share with their clients their own professional opinions about how best to evaluate and use sources of information. That is an important part of professional service. So

Rainey recommends "Read directly from a source. Do not paraphrase; do not interpret the information; do not give a medical opinion" (1988, p. 65).

Information providers must recognize the big difference between retrieval of published information someone seeks and diagnosing a medical condition or recommending a particular therapy. They can provide clients with a wide variety of health and legal information, but not make any diagnosis or give a personal opinion on what action individual clients should take. They can and do, of course, refer clients to the appropriate professionals who are licensed to give such personal advice (CAPHIS, 2010; McKnight, 1997).

4.2.3 How do I evaluate what I find?

> **Bad example—unfounded assumptions**
>
> This is a good-looking Web page.
>
> They care about me enough to ask to register with my e-mail address.

The most important point of evaluation is, of course, does this information relate directly to the real information need? The people above talking about their own experiences with cataract surgery may not have known what the first person needed. The casual Web searcher may be seduced by an attractive Web page. Registration is often used to create e-mail mailing lists, which may not protect the registrants' confidentiality.

Ethical information service providers evaluate sources more thoroughly. Health information literacy includes careful evaluation of the source of information. It includes expertise in the analysis of information sources for authority, reliability, currency, intended audience, and accessibility.

4.2.3.1 Authority

Who says? Anonymous sources usually are not authoritative. They may contain a mixture of reliable and unreliable information, but in any case, if no one is willing to take responsibility for the information, it may well be irresponsible propagation. It is important to know who wrote or said something and whether or not they have the education, expertise, or experience to be trustworthy on this topic (Smith, 2011; Spatz, 2011; Taylor & Blackwell, 2014). For instance, having a doctor of philosophy (PhD) degree in one discipline does not make a person an expert in all disciplines.

4.2.3.2 Reliability

Is the source based on good scientific evidence or just rumor or personal opinion? Are the authors or publishers biased because they have something to sell? What is their motivation? Why do they want my personal contact information? Have other authoritative scholars reviewed their work? Is this information supported by other authoritative sources (Smith, 2011; Taylor & Blackwell, 2014)? Spatz (2011) divides literature about remedies and therapies into five categories: folklore, quackery, unproven, investigational, and proven. She strongly recommends information from proven therapies based on sound research.

4.2.3.3 Currency

Health information can quickly become out of date when the weight of evidence discredits older beliefs. Every document, even an online one, has a publication date (or an update date) and while the information might be already outdated when it was published it is certain that it will not be any more current than the publication date (Smith, 2011; Spatz, 2011). Be aware that publications for the general public may use knowledge that lags behind more recent research evidence because of how new knowledge is disseminated (see "Intended audience" below).

4.2.3.4 Accessibility

The book or document containing the desired information has to be not only readily available in print or online, but also be in a language and format which the client can use. Can the client understand the terminology? Does the client need a larger typeface (in print or online) or an audio source? Giving a person a reference list is not the same as giving them information about the subject. (Sometimes references may be flawed or even fraud.) Giving a person a Web site address (URL) is useless, if they neither have internet access nor are comfortable with using a browser.

4.2.3.5 Intended audience

For what audience was the document written (Spatz, 2011)? In spite of how the popular media reports announcements of newly reported original research studies, rarely does a single study prove anything. Many studies are usually required to develop a body of research evidence to support new understanding of health issues. It takes time for primary research literature—reports on new research—to go through scholarly peer review (Chung, 2011) and be published in a research journal. Such research journals are primary literature designed to be read by other researchers and scholars (Tu-Keefner, 2014). Secondary health science articles are written later by subject specialists, who in some way summarize the evidence presented by many primary research reports (Glass, 1976). Authors of textbooks and online references base their writings on primary and secondary articles. Others will summarize or translate this relatively new knowledge into everyday lay terms. The publication of new knowledge in a format suitable for the general public or specific nonscholar groups may take years after the original research studies that discovered that knowledge were published. Each of these different kinds of publications are written for different sorts of readers. Publications are also written for a particular audience of readers, be they nurses, children, patients, therapists, moderately educated adults, or other specific groups of people. It is vital to match the source to the person, who needs the information. Therefore a primary research study may be more important to a health science researcher than a health care consumer-oriented publication. And the average adult health care consumer may prefer the latter publication to primary research (Spatz, 2011).

4.3 The right to privacy and responsibility for confidentiality

> **Bad example—betrayal of confidentiality**
>
> *Well, you wouldn't believe what is really wrong with her. She told me that...*

Excellent information services require not only knowledge and expertise but the protection of *both* the client's privacy *and* the confidentiality of any personal information, which the client reveals (Lessick, 2011). The formal codes of ethics for librarians' professional associations include clauses about the importance of protecting both privacy and confidentiality (ALA, 2008a; International Federation of Library Associations and Institutions (IFLA), 2012; MLA, 2010; Special Library Association (SLA), 2010). Privacy and confidentiality, of course, are different concepts; they are not synonyms.

4.3.1 What is privacy?

> privacy. *The condition or state of being free from public attention to intrusion into or interference with one's acts or decisions*
> *Garner (1999, p. 1253)*

> informational privacy. *A private person's right to choose to determine whether, how, and to what extent information about oneself is communicated to others, esp. sensitive and confidential information*
> *Garner (1999, p. 1253)*

Not only is a person's health a private matter, but so is that person's information seeking and decision making about that health. The right to privacy means that no one has to tell an information provider *why* they want information. It means that one does not have to reveal whether the information is for another person or not. Of course, the health information provider often will ask questions to put the information need in context; it makes it easier to get the client exactly what they want. The client may or may not provide background for the question, but the information provider usually can offer the client alternative information sources, which allow the client to choose which is most appropriate (Kern & Woodard, 2011).

In the context of library and information services, anyone can ask questions and use information sources without making any commitment to a particular idea or leaving any traceable trail of that inquiry. The individual client has the right not to reveal to a librarian the substance, purpose, or context of the client's query. The ALA (2014a) has asserted that privacy is an important interpretation of its *Library Bill of Rights* (1996).

Indeed, there are few places, outside one's own mind, where one really has that kind of privacy. In a traditional print library, one can browse and read without revealing anything about oneself. Online information services are notorious for violating that privacy, at many levels, whether or not the user is asked to "register." The so-called "privacy policies" often reveal how little privacy the user of such services can expect (Cleveland & Cleveland, 2009; Rubin, 2004; Spatz, 2011).

4.3.2 What is confidentiality?

> confidentiality, *1. Secrecy; the state of having the dissemination of certain information restricted.*
>
> *Garner (1999, p. 318)*

Confidentiality is the principle that the professional health information provider has a responsibility not to reveal any information confided or revealed by a client in the course of information service. The person's revelation of private, personal information in any way to a practicing information provider does not give the provider the right to pass it on to anyone. The professional also has a responsibility to respect confidential information about donors, colleagues, employees, or others involved in the provision of services (ALA, 1996, 2008a, 2014a; IFLA, 2012; MLA, 2010; SLA, 2010).

Thus privacy is a client's right not to reveal private information. Confidentiality is the information professional's responsibility not to reveal to a third party any information that the client has confided. In the United States, most states and the District of Columbia have statutes protecting the privacy and confidentiality of library clients and requiring that information about what they seek, use, borrow, or acquire (including online sources) be kept strictly confidential. One example is that of Oklahoma (Oklahoma Library Code, 1986). There are, however, other legal exceptions, especially court subpoenas and certain federal security investigations. Librarians often post such laws or privacy statements on library Web sites (Baltimore County Public Library, 2013; Tulsa City-County Library, n.d.).

When it comes to health information, there's a second layer of privacy and confidentiality involved. Health care providers' ethical protection of personally identifiable health care information began long before the widespread use of electronic health records created the need for the privacy and confidentiality law and regulations of the United States Health Information Portability and Accountability Act of 1996 (US DHHS "Health information privacy," "Health Information Portability and Accountability Act"). It is no less important for providers of health information (Cleveland & Cleveland, 2009; Friedman, 2004).

Such ethical guidelines, codes, and laws are not commonly understood by the general public, so it is the information provider's responsibility to explain these principles in appropriate contexts. For instance, if a client seems reticent about asking a question, one may reassure the person that there is such a professional responsibility to keep confidential whatever the client wants to know. Or if a person demands to know what information someone else seeks, one should not punish them for not being aware of professional ethical policies, but instead reassure them that confidentiality applies to everyone (McKnight, 2014).

Different people may seem to ask for the same information when in reality they have different goals. For instance, in a public library, clients supporting opposing political stances on a health issue may ask for material on that subject. Likewise, different teachers in a school considering a new kind of health curriculum may want research about that curriculum for different reasons, and in a hospital different caregivers with different responsibilities may want material about a given patient's condition. Obviously, it is advantageous for politically active people, teachers, and caregivers to be informed. A client may ask about another's request for information relating to the same issue, and even ask about what someone else (with whom they may or may not agree) has requested. The ethical librarian cannot, however, reveal to one client what another has requested. Even though the request itself may be well meaning and should not be offensive to the librarian, it cannot be granted directly. One solution is to answer with "Tell me what you would like to know and I'll make a special search for you," completely avoiding the "What did you give him?" question. Another response could be "Ethically I can't reveal anyone's library information requests, but you could ask that person directly yourself if you wish." If there is an applicable library confidentiality statute in effect in that state or country, it may be necessary to point out that such a revelation might be illegal.

4.4 Providing fair and equitable access

> **Bad example—withholding service unfairly:**
>
> *Well, I'll look up that information for him, but not for her because she's too old.*

Information service should not be denied because of the client's origin, age, background, socioeconomic status, or views (ALA, 1996). It should be obvious that health information providers should be fair. But it may take some extra effort not to behave more like gatekeepers than as access providers. The very people information providers are supposed to help may experience a wide variety of barriers to getting the information they need. These clients may not only have literacy limitations, but also physical impairments, or cognitive challenges. There may be social or cultural norms, which make it difficult for them to get the health information they need to make informed decisions. The skilled information professional reaches out to people experiencing such barriers and make accommodations for them (Harris, 2011).

Equitable access is a right not only of polite, responsible, and highly literate people. Ethical information providers provide the best possible service for people, who procrastinate (the student, who waits until the night before the paper is due), who don't want to use arcane library and information technology systems, or otherwise don't seem to fit the established information delivery systems.

4.4.1 What are some barriers to information access?

The honest and responsible information provider recognizes many kinds of information-seeking barriers. With empathy and creativity, the solutions may be readily at hand or they may require cooperation with other information providers in other organizations.

Some major barriers to information access come from information literacy problems (Rubin, 2004). A person will not seek information that (1) he doesn't think he needs, (2) he doesn't believe exists, or (3) he doesn't think he can find or know how to use. Ethical information providers lead outreach activities to help people in many circumstances. Sometimes they have to take on the very difficult task of translating clinical terminology into more common language while guarding against interpretation based on personal opinion.

No one type of library, agency, mission, or organization is the true prototype for all good information services everywhere. For example, even three commonly accepted values of librarianship—access to information in the public good, promotion of literacy and information literacy, and ensuring preservation of the accumulated wisdom of the past (Groen, 2007)—are not all equally integral to the mission of every library service. Hospital librarians may keep nothing in their local collections more than five years because outdated clinical information can be dangerous. Librarians with a strong archival mission to preserve historical documents may have to severely restrict current access to material in order to protect it for future access. Librarians working for government health agencies may have legal restraints on the release of certain kinds of information. Librarians in schools do not provide service for the general public and do more literacy promotion than some other libraries. Some librarians make literary fiction available for bibliotherapy and some collect none at all. Librarians working for corporations may have to protect information considered part of the company's trade secrets. Librarians in public libraries provide information access to a very wide scope of information and to a much wider demographic mix of people than do school libraries, but they still have a primary mission to serve people to whom the government agency supporting them has the most responsibility. However different the parent organization or agency, librarians and other information providers can and often must know, use, and refer to many different information sources and services—from anywhere in any medium.

Sometimes there really isn't enough time to provide information service for each individual, who wants it immediately. In those times, in order to be fair, information providers must use an ethical form of information need *triage*, an equitable prioritization method respecting the institutional mission or the importance of the request to human health and safety, and not just the order in which the requests were received. The term comes from the practice in battlefield medicine of dividing patients into three groups: those who can safely wait for treatment, those who are most likely to die with or without treatment, and those who should be treated first—the ones who will live with immediate treatment but will die without it. That does not mean that patients in the first two groups are denied treatment, they just may not be the first patients treated (Iserson & Moskop, 2007). The information provider's triage may mean that some requests are handled immediately, some are delayed, and some are delegated or referred to other services. For instance, a hospital librarian would retrieve information needed for an immediate patient care decision before a request based on mild curiosity, and might refer a request for computer help to the information technology help department.

4.5 Intellectual property rights and access to information

> **Bad example—making multiple copies without permission from the copyright owner:**
>
> *I'll just make fourteen copies of this article so that everyone in the group can have one. And I'll keep some extras for me. It's okay because I won't be selling them.*

Another barrier is what happens when a person's information behavior conflicts with intellectual property rights. Plagiarism—claiming someone else's words as one's own or failure to cite sources is disrespect for intellectual property rights; so is copyright violation. For instance, under the Fair Use section of the United States Copyright Act, an individual can make a single copy for her own use—without permission from the copyright holder—as long as certain conditions are met. It does not apply to making multiple copies (Copyright Act, 17 U.S.C. Section 107). Intellectual property includes patents, trademarks, and copyrighted publications in any format or medium, all of which are owned by someone or some entity. Under United States and International laws, authors, publishers, and other copyright owners can very closely restrict access to publication. Interpretation and application of these laws, regulations, and official guidelines are fairly complex. The progression of media from print to online has not changed that; if anything, it is even more complex. Access to many high-value information sources is enforced through licensing agreements which may be much more restrictive than copyright law and guidelines. Thus libraries and others are often purchasing licenses to access electronic material instead of buying copies. Some scientific and scholarly publications are available through one of several models for *open access*—freely available on the Internet (Rubin, 2004; Wiley, 2011).

4.6 Advocacy for information access

> **Bad example—blocking access to information independent of health care:**
>
> *If she wants to know anything about Ebola, she should just ask her doctor.*

Health librarians promote consumer and patient safety and decision making by advocating for consumers' rights to high-quality health information. They promote consumer health information centers and present programs for other librarians and information providers on best practices for such services (CAPHIS, 1996, 2010). Ethical information providers think outside the job not only in relationship to our own profession, but in relationship to information access for all. They are constantly "acting as advocates on the local, national and international levels to promote open access to health information for the public" (CAPHIS, 1996, 2010). They may not only read but also write publications promoting access. They show how they values knowledge services by promoting, preserving, and expanding information services for all (Spatz, 2011). The various information professional associations publicly promote the access rights of their constituencies (ALA, 1996, 2008a, 2014a; IFLA, 2012; MLA, 2010; SLA, 2010). In some situations involving government and commercial interests, they may be the only organized advocates for such constituencies. Librarians have participated for generations in copyright and other intellectual property issues, in the preservation of documentation that might otherwise be lost to future generations, in promoting citizen access to government documents and in resisting efforts to violate readers' privacy (Cleveland & Cleveland, 2009). Public librarians are often involved in literacy initiatives and librarians serving scholars are active in the rapidly developing new models of electronic document access, including various open access models (Kluegel, 2011).

4.7 Providing information versus giving advice

> **Bad example—diagnosing and prescribing without a license:**
>
> *Well, I'm not a doctor but I think you have serious osteoarthritis and if you should start taking glucosamine-chondroitin twice a day.*

4.7.1 Don't play doctor

There is always a temptation to give personal advice when people ask for it. It seems like a helpful thing to do. But the chances of inadvertently harming a person are high. Suggesting to a person a treatment for a disease or injury (which a person may or may not have) can prevent that person from getting the diagnosis and therapy that they need. "While librarians are experts in identifying and providing information, they are not practicing health professionals who interpret information and give advice. It is important that librarians avoid suggesting diagnoses and recommending particular health professionals or procedures" (CAPHIS, 2010).

Providing health information is not the same thing as diagnosing, prescribing therapy, or predicting the course of an individual's illness or injury. States and countries carefully license individuals, who have the education and expertise to diagnose, prescribe, and make a prognosis for individual people. It's not just a legal issue, it is an ethical issue as well. Librarians and other information providers connect clients to factual reliable published information—but do not assess the clients' personal health without a license. Some librarians in the past have been hesitant to provide any health information services at all for fear of overstepping the role of information provider into practicing medicine without a license (Smith, 2014), but that deprives people of the immense value of information access.

So, if a person asks "do you think I might have osteoarthritis?" there are three possible ethical responses for the librarian or information provider. One is to refer (the root of the library word "reference") the person to an unspecified licensed health care provider. Another one is to offer reliable information about osteoarthritis and its treatment which can increase the person's knowledge. The first response may be taken as dismissive. It does no harm, but it doesn't help the person much either. The second is likely to improve knowledge and aid communication with the health care professional, who can diagnose and prescribe. That knowledge may also improve self-care. The best thing to do is a combination of both, making sure that the client understands the difference between published factual information and personal health advice. "Although we are not health professionals, we can provide referral to health professionals and organizations when appropriate" (Liebermann & Ham, 2012).

Representing oneself as qualified to give such personal advice *without* the required education, credential, and license is outright fraud. Some people do distrust formal education and prefer folk or alternative medicine but it can be seriously harmful, if the person does not get a diagnosis or therapy that they really need.

4.7.2 Use disclaimers

Ham notes that "Consumers may be confused about the role of the librarian. They may assume that the librarian can advise them on making health care decisions…" and "It is critical that you do not attempt to provide a diagnosis or recommend a therapy or intervention" (2012). A simple disclaimer can help make that clear (Spatz, 2011). The Consumer and Patient Health Information Section of the Medical Library Association has sample disclaimers on its Web site at http://caphis.mlanet.org/chis/disclaimers.html (CAPHIS, 2013b).

4.7.3 Prescribed patient education and information prescriptions

A licensed physician may prescribe patient education from a licensed patient care provider such as a nurse. The nurse sets the educational goals, carries out the teaching activity, and assesses the patient's learning. Some physicians also practice "information prescription" by assigning reading to a patient (D'Allesandro, Kreiter, Kinzer, & Peterson, 2004; Gavgani & Shiramin, 2013; Ulmer & Robishaw, 2010). Librarians and other information professionals have long been involved in information prescription practices (Frude, 2005; Jones & Shipman, 2004; McKnight, 2014; Timm & Jones, 2011). Government agencies in the United States, the United Kingdom, and other countries have encouraged the practice of information prescription in this century (Gavgani, 2012; Shiner, Hill, King, & Clayton, 2007; NLM, 2004; Neville, 2013).

4.8 Conflicting values, dilemmas, and tough decisions

Ethical judgments are not made in a vacuum. We have serious concerns for social utility, survival of the parent organization, social responsibility, and respect for individuals (Rubin, 2004). Our choices often are not between "good" and "evil" but in determining the greater of two goods or the lesser of two evils. Sometimes we will make decisions that we are later proud of and some we would like to forget.

Dilemmas arise in the context of competing principles and rights. For instance, a person's need for access to information and intellectual property rights, licensing, or cost of access may come into conflict. Some of the most dramatic dilemmas are, of course, clients' access and confidentiality rights in situations that may involve serious harm to the client or to the community.

The information provider may have official or unofficial roles that conflict. Sarah Anne Murphy (2001) writes about conflicts between professional and institutional ethics, especially when institutional policies or politics conflict with individuals' access to information. Tooey and Arnold (2014) surveyed academic health library directors and found that conflict of interest issues with vendors can be troubling when revenue and research funding are at stake. Organizations affiliated with political or religious groups may want some information suppressed. For instance, organizations affiliated with the Jehovah's Witnesses might promote information about substitutes for blood transfusion, but suppress information about why such transfusions might be medically indicated. Roman Catholic institutions may suppress information about elective abortion. Elizabeth Irish (2002) discusses conflicts between personal ethics based on religious beliefs and professional ethics. Conflicts will also arise between personal and institutional ethics. No code or policy is likely to be universally applicable. Some decisions may have serious consequences and some may really be inconsequential.

Mention ethics in a casual conversation among librarians and often the conversation quickly focuses on imagining complex ethical dilemmas. Everyone has a story or a "what if?" scenario about a situation when the professional is faced with a difficult action choice. People like to discuss dramatic stories. Whether they are completely hypothetical or somewhat based on real-life situations, the teller can manipulate the hypothetical details to increase the strength of the conflicting choices. In any case, they usually are about rare, rather than common occurrences.

In an information services textbook, Rubin (2011) presents the hypothetical example of a teenager with tears in her eyes approaching a librarian and asking for book about suicide called *Final Exit* (Humphry, 2002). Does the librarian uphold the access principle or not? With no more information than that students will make quick assumptions and take sides in discussion, even though none of them have read the book. While the book does include descriptions of effective methods of suicide (all of which take a good deal of time, money, and effort to arrange) it constantly reiterates reasons for not committing suicide, including serious consideration of the legal and physical mess left behind. It describes the severely impaired state which an individual may suffer after unsuccessful suicide attempt with some popular methods. The simple fact that suicide is the subject of the book does not, in and of itself, make reading the book dangerous. Most will also assume that the teenager, asking for it, has tears in her eyes because she is depressed and means to commit suicide. The tears could be for other legitimate reasons. She could be distraught as a result of the suicidal death of a friend or relative is struggling to understand it. Her tears could be the symptoms of a cold or respiratory allergy irrelevant to her work on a paper about suicide. It takes a careful reference interview to determine what is really going on. But the point of the scenario in the textbook is to encourage library school students to consider the fact that at times in their practice they will face distressing decisions—and there are no easy solutions for every dilemma.

The more common ethically precarious situation is not the rare dilemma based on obvious conflicts between major ethical principles but rather frequent lapses based on convenience and familiarity rather than professional principles. For instance, toward the end of the day, a tired information provider may take shortcuts, that result is less than the best service. Without paying attention, one can forget to take the time to fully understand the client's question, to do a good search, and to evaluate the information sources found.

Privacy must be respected and confidentiality maintained no matter how gossip seems to bond friendships. Barriers to information access must be reconsidered as well as legal issues concerning copyright. Ethical health information providers constantly advocate for equitable access to information. They know and remember the difference between providing information and giving medical advice.

4.9 Keep learning

Bad example—outdated knowledge:

Oh, I learned all about that years ago.
I don't need to study anything more about it.

Constant continuing education is also an ethical responsibility. Information providers all have much to learn which they do not know, much to review which they have forgotten, and much to catch up on which has changed with revelations from new research. Old reliable information sources are updated or discontinued, communication media change and skills can always be improved. Being complacent and getting too far behind can seriously impair information service (ALA, 1996, 2008a, 2014b; IFLA, 2012; MLA, 2010; SLA, 2010).

Along with the responsibility for one's own constant development comes the responsibility to encourage it in others. That includes teaching colleagues, writing in professional publications, presenting papers at professional meetings, and serving on association committees.

Professional growth thrives in a spirit of cooperation more than in a spirit of isolation or competition. The common model for professional mentoring is that of a more experienced professional guiding someone newer to the profession, but in reality the information give-and-take can run both ways. The more experienced person may learn from insights that the newer person has coming to the practice of information service with fresh eyes, while the newer people can learn from the experienced ones, who have been green and growing for many years.

This chapter began by emphasizing the importance of building the habitual practice of providing the best possible information service. That is a common, every day virtue ethic which makes a difference in peoples' lives.

Do it well! Do it right! Do no harm!

References

American Library Association. (1996). *Library bill of rights*. Retrieved from http://www.ala.org/advocacy/intfreedom/librarybill.
American Library Association. (2008a). *Code of ethics of the American library association*. Retrieved from http://www.ala.org/advocacy/proethics/codeofethics/codeethics.
American Library Association. (2008b). *Standards for accreditation of Master's programs in library & information studies*. Retrieved from www.ala.org/accreditedprograms/sites/ala.org.accreditedprograms/.
American Library Association. (2014a). *Privacy: An interpretation of the library bill of rights*. Last updated July 1, 2014. Retrieved from http://www.ala.org/advocacy/intfreedom/librarybill/interpretations/privacy.
American Library Association. (2014b). *Third DRAFT revised standards for accreditation of Master's programs in library & information studies*. [Released for comment, August 1, 2014]. Retrieved August 30, 2014 from http://www.oa.ala.org/accreditation/?page_id=326.
Athanassoulis, N. (2013). *Virtue ethics*. London: Bloomsbury Academic.
Baltimore County Public Library. (2013). *Privacy statement*. Retrieved from http://www.bcpl.info/services-policies/acceptable-use-statement.
Chung, H. D. (2011). *Peer review in five minutes*. [Video file]. Retrieved from https://www.youtube.com/watch?v=hVN2cudxnug.
Cleveland, A. D., & Cleveland, D. B. (2009). *Health informatics for medical librarians*. New York: Neal-Schuman.

Consumer and Patient Health Information Section, Medical Library Association. (1996). The librarian's role in the provision of consumer health information and patient education. *Bulletin of the Medical Library Association, 84*(2), 238–239. Retrieved from http://www.ncbi.nlm.nih.gov/pmc/articles/PMC299415/.

Consumer and Patient Health Information Section, Medical Library Association. (2010). *The librarian's role in the provision of consumer health information and patient education.* Retrieved from http://caphis.mlanet.org/chis/librarian.html.

Consumer and Patient Health Information Section, Medical Library Association. (2013a). *Top 100 list: Health websites you can trust.* Updated September 20, 2013. Retrieved May 30, 2014 from http://caphis.mlanet.org/consumer/index.html.

Consumer and Patient Health Information Section, Medical Library Association. (2013b). *Disclaimers.* Retrieved from http://caphis.mlanet.org/chis/disclaimers.html. Copyright Act, 17 U.S.C. §107.

D'Allessandro, D. M., Kreiter, C. D., Kinzer, S. L., & Peterson, M. A. (2004). A randomized controlled trial of an information prescription for pediatric patient education on the Internet. *Archives of Pediatric and Adolescent Medicine, 158*(9), 857–862.

Eberle, M. L. (2005). Librarians' perceptions of the reference interview. *Journal of Hospital Librarianship, 5*(3), 29–41.

Friedman, J. (2004). How will HIPAA affect your consumer health information service? *Journal of Hospital Librarianship, 4*(1), 45–51.

Frude, N. (2005). Book prescriptions: a strategy for delivering psychological treatment in the primary care setting. *Mental Health Reference, 10*(4), 30–33.

Garner, B. A. (Ed.). (1999). *Black's law dictionary* (8th ed.). St. Paul, MN: West Group.

Gavgani, V. Z. (2012). Evidence based information prescription (IPs) in developing countries. In N. Sitaras (Ed.), *Closer to patients or scientists?* Rijeka, Croatia: Intech.

Gavgani, V. Z., & Shiramin, A. R. (2013). Physician directed information prescription service (IPs): barriers and drivers. *ASLIB Proceedings, 65*(3), 224–242.

Glass, G. G. (1976). Primary, secondary, and meta-analysis of research. *Educational Researcher, 5*(10), 3–8.

Groen, F. K. (2007). *Access to medical knowledge: Libraries, digitization, and the public good.* Lanham, MD: Scarecrow Press.

Ham, K. (2012). *The consumer health reference interview and ethical issues.* Retrieved from http://nnlm.gov/outreach/consumer/ethics.html.

Harris, F. J. (2011). Reference services for specific populations. In R. E. Bopp, & L. C. Smith (Eds.), *Reference and information services: An introduction* (4th ed.) (pp. 341–383). Santa Barbara, CA: Libraries Unlimited.

Humphry, D. (2002). *Final exit* (3rd ed.). New York: Delta.

International Federation of Library Associations and Institutions. (2012). *IFLA code of ethics for librarians and other information workers.* Retrieved from http://www.mlanet.org/about/ethics.html.

Irish, D. E. (2002). And ne'er the twain shall meet? Personal vs. professional ethics. In G. A. Scott (Ed.), *Christian librarianship: Essays on the integration of faith and profession* (pp. 120–130). Jefferson, NC: McFarland.

Iserson, K. V., & Moskop, J. C. (2007). Triage in medicine, part I: concept, history and types. *Annals of Emergency Medicine, 49*(3), 275–281.

Jones, S., & Shipman, J. (2004). Health information retrieval project: librarians and physicians collaborate to empower patients with quality health information. *Virginia Libraries, 50*(2), 11–16.

Kern, M. K., & Woodard, B. S. (2011). The reference interview. In R. E. Bopp, & L. Smith (Eds.), *Reference and information services: An introduction* (4th ed.) (pp. 57–94). Santa Barbara, CA: Libraries Unlimited.

Kluegel, K. M. (2011). Understanding electronic information systems. In R. E. Bopp, & L. Smith (Eds.), *Reference and information services: An introduction* (4th ed.) (pp. 187–188). Santa Barbara, CA: Libraries Unlimited.

Lessick, S. (2011). On-site and web-based information services. In M. M. Bandy, & R. F. Dudden (Eds.), *The Medical Library Association guide to managing health care libraries* (2nd ed.) (pp. 211–260). New York: Neal-Schuman.

Liebermann, K., & Ham, K. (2012). The consumer health reference interview and ethical issues. U.S. Department of Health and Human Services, National Institutes of Health, National Library of Medicine, National Network of Libraries of Medicine. Retrieved from http://nnlm.gov/outreach/consumer/ethics.html.

McKnight, M. (1997). Field tips: patient education or health information service: what's the difference? *National Network: Newsletter of the Hospital Libraries Section of the Medical Library Association, 21*(3), 10.

McKnight, M. (2010). *The agile librarian's guide to thriving in any institution.* Santa Barbara, CA: Libraries Unlimited.

McKnight, M. (2014). Information prescriptions 1930–2013: an international history and comprehensive review. *Journal of the Medical Library Association, 102*(4), 271–280.

Medical Library Association. (2010). *Code of ethics for health sciences librarianship.* Retrieved from http://www.mlanet.org/about/ethics.html.

Medical Library Association. (2014). *Academy of health information professionals.* Retrieved from https://www.mlanet.org/academy/.

Murphy, S. A. (2001). The conflict between professional ethics and the ethics of the institution. *Journal of Hospital Librarianship, 1*(4), 17–30.

National Library of Medicine. (January 1, 2004). NLM and ACP foundation launch national information program for internists, the information Rx: now the prescription pad will point patients to MedlinePlus. *NLM Newsline, 59*(1–3), 2–3.

National Library of Medicine. (2014a). *Easy-to-read.* Retrieved from http://www.nlm.nih.gov/medlineplus/all_easytoread.html.

National Library of Medicine. (2014b). *Health information in multiple languages.* Retrieved from http://www.nlm.nih.gov/medlineplus/languages/languages.html.

National Network of Libraries of Medicine. *NN/LM members directory.* Retrieved from http://nnlm.gov/members/.

Neville, P. (2013). Prose not Prozac? The role of book prescription schemes in the treatment of mental illness in Ireland. *Health Sociology Review, 22*(1), 19–36.

Oklahoma Library Code, 65. Ok. Stat. §65-1-205. (1986). Retrieved from http://statutes.laws.com/oklahoma/Title65.

Rainey, N. B. (1988). Ethical principles and liability risks in providing drug information. *Medical Reference Services Quarterly, 7*(3), 59–67.

Rubin, R. E. (2004). *Foundations of library and information science* (2nd ed.). New York: Neal-Schuman.

Rubin, R. E. (2011). Ethical aspects of reference service. In R. E. Bopp, & L. C. Smith (Eds.), *Reference and information services: An introduction* (4th ed.) (pp. 29–56). Santa Barbara, CA: Libraries Unlimited.

Shiner, A., Hill, T., King, I., & Clayton, S. (2007). Books on prescription: an example of collaborative working in northern Derbyshire. *CILIP Health Libraries Group Newsletter, 24*(2), 14.

Smith, K. H. (2014). Consumer health information services. In S. Wood (Ed.), *Health sciences librarianship* (pp. 324–343). Lanham, MD: Rowman & Littlefield.

Smith, L. C. (2011). Selection and evaluation of reference sources. In R. E. Bopp, & L. C. Smith (Eds.), *Reference and information services: An introduction* (4th ed.) (pp. 387–410). Santa Barbara, CA: Libraries Unlimited.

Spatz, M. (2011). Health information for patients and consumers. In M. M. Bandy, & R. Dudden (Eds.), *The Medical Library Association guide to managing health care libraries* (2nd ed.) (pp. 321–342). New York: Neal-Schuman.

Special Libraries Association. (2010). *Professional ethics guidelines*. Retrieved from http://www.sla.org/about-sla/competencies/sla-professional-ethics-guidelines/.

Taylor, A., & Blackwell, B. (2014). Organization and management of the reference collection. In J. T. Huber, & S. Swogger (Eds.), *Introduction to reference sources in the health sciences* (6th ed.) (pp. 25–48). Chicago, IL: Neal-Schuman.

Timm, D., & Jones, D. (2011). The information prescription: just what the doctor ordered! *Journal of Hospital Librarianship, 11*, 358–365.

Tooey, M. J., & Arnold, G. N. (2014). The impact of institutional ethics on academic health sciences library leadership: a survey of academic health sciences library directors. *Journal of the Medical Library Association, 102*(4), 241–246.

Tu-Keefner, F. (2014). Bibliographic sources for periodicals. In J. T. Huber, & S. Swogger (Eds.), *Introduction to reference sources in the health sciences* (6th ed.) (pp. 61–64). Chicago, IL: Neal-Schuman.

Tulsa City-County Library. (n.d.). *Confidentiality of customer records*. Tulsa City-County Library Policies. Retrieved from http://www.tulsalibrary.org/sites/default/files/pagefiles/Confidentiality.pdf

U.S. Department of Health and Human Services. (n.d.). *Health information privacy*. Retrieved from http://www.hhs.gov/ocr/privacy/

U.S. Department of Health and Human Services. *Health insurance portability and accountability act of 1996*. Retrieved from http://www.hhs.gov/ocr/privacy/hipaa/administrative/statute/hipaastatutepdf.pdf.

U.S. Department of Health and Human Services. National Institutes of Health. National Library of Medicine. National Network of Libraries of Medicine. (2013). *NN/LM Members Directory*. Retrieved from http://nnlm.gov/members/

Ulmer, P. A., & Robishaw, S. (2010). Information prescriptions: providing health information at the inpatient's point of medical need. *Journal of Consumer Health on the Internet, 14*(2), 138–149.

Wiley, L. (2011). Access-related reference services, legal issues. In R. E. Bopp, & L. C. Smith (Eds.), *Reference and information services: An introduction* (4th ed.) (pp. 209–214). Santa Barbara, CA: Libraries Unlimited.

Woodruff, P. (2001). *Reverence: Renewing a forgotten virtue*. New York: Oxford University Press.

Health information resource provision in the public library setting

Mary Grace Flaherty
School of Information and Library Science, University of North Carolina at Chapel Hill, Chapel Hill, NC, USA

5.1 Background

5.1.1 Public libraries in historical context

The origins of what we have come to know and identify as the public library in the United States began during the Colonial period with circulating and social libraries. The organization has evolved from those somewhat exclusive origins to become a ubiquitous tax-supported institution, serving all members of the community in which it is situated. Our operational definition of the public library comes from the 1876 U.S. Bureau of Education report:

> The 'public library' which we are to consider is established by state laws, is supported by local taxation or voluntary gifts, is managed as a public trust, and every citizen of the city or town which maintains it has an equal share in its privileges of reference and circulation.
>
> as quoted by Shera (1965, p. 157)

Though the first public libraries were established in the mid-1800s, by 1900, there were nearly 1000 public libraries throughout the United States (De la Peña McCook, 2004). While there are commonalities among the institutions across the nation, there are varying degrees of minimum standards and requirements as each state and locality enacted their own laws to establish and govern public libraries within their jurisdictions. Common to the majority of public libraries are the core values of providing access to materials, advancing education, and promoting literacy. Thus, historically, public libraries have been closely linked to promoting democracy by the common perception that they serve to encourage and support an informed electorate and that they help to enable a level playing field, societally speaking.

Currently, the vast majority (88%) of U.S. public libraries are chartered to serve populations of 50,000 or less, and more than half of all public libraries have legal service areas of less than 10,000 (Pearlmutter & Nelson, 2011). Though mostly similar in service philosophy to their urban counterparts in their establishment, the function of the rural public library has differed somewhat. That is, rural public libraries largely originated as traveling libraries, a system where books were sent to areas without library access (De la Peña McCook, 2011); these collections primarily provided

Meeting Health Information Needs Outside of Healthcare.
Copyright © 2015 by M.G. Flaherty. Published by Elsevier Ltd. All rights reserved.

general reading materials that offered moral and cultural support for far-flung communities (Bullock, 1907; deGruyter, 1980). Most materials to which they provided access to were fiction, with very few nonfiction titles to "help [people] to think to some purpose" (Bullock, 1907, p. 9). Thus, the "privileges of reference" referred to in the 1876 Bureau of Education report, such as access to a wide variety of nonfiction and reference materials, may not have been standard throughout library communities as public libraries were being created.

Today, there are close to 17,000 public library locations throughout the United States (Institute for Museum and Library Services, 2013), and in 2012–2013, 86% of Americans 16 years of age and older reported ever having utilized a public library (Pew, 2013). There are very few other or similar publicly funded organizations in the United States, whose primary goal and function is to be accessible and open to all members of society on such an equal basis.

5.1.2 Public libraries' evolving role in the US

The original function of public libraries was to provide access to materials and to act as a community repository for materials, primarily books. But public libraries aren't just about books anymore. While lending "things" certainly very much continues to be a current aspect of their organizational role, more and more public libraries are evolving to respond to community needs for shared space and community centers, along with providing access to information in all types of formats, including print, electronic, and Web-based. Public libraries are now identified as safe havens and organizations that promote and create trust in the communities, where they are located (Public Agenda, 2006; Varheim, Steinmo, & Ide, 2008), and are characterized as information centers that provide access to public computers as well as recreational and educational materials (Johnson, 2010).

Now that resources are available in a variety of formats, libraries are expanding their offerings to include electronic resources, such as downloadable books (Karp, 2011). Other items that are currently being loaned in public libraries across the United States include equipment such as: laptops, kindles, e-book readers, projectors, tools, cake pans, toys, fishing rods, home energy usage meters, and karaoke machines; other nonbook items include art prints, sculptures, and museum passes (Karp, 2011). In Rockport, Maine, a popular item for circulation is a ukulele, which was donated by a patron (Rockport Public Library, n.d.). Many public libraries are becoming popular resources for "seed libraries," where patrons can obtain and exchange varieties of seeds for growing local and native plants (Moffat, 2013). Other public libraries are venturing into health promotion activities in their communities, by organizing walking clubs and loaning pedometers (D. Miller, personal communication, May 1, 2014; Ryder, Faloon, Levesque, & McDonald, 2009).

Public libraries in all types of locales have refurbished spaces with cafes, community rooms, and teen centers (Anonymous, 2008; Peterson, 2005; Toth & Baltic, 2010). Additionally, all types of "new" services are being offered, such as in Cuyahoga County, Ohio, where they offer one-stop passport application and renewal services, including express photo services in many of their branches (Cuyahoga Library, 2010).

In Baltimore, public libraries are working to provide access to fresh vegetables through partnership with local grocers (Owens, 2010). The public library in Chattanooga, Tennessee has created a "maker space" with access to all types of equipment, including sewing machines, printing presses, 3-D printers, and video equipment; they're serving as a model for other public libraries that are undergoing renovations (Maloney, 2014). Financial and digital literacy and other information literacy skills courses are increasingly being offered (New York Public Library, 2014), as well as job training and assistance with resumes and job applications (Sigler et al., 2011). There are public libraries that offer health and fitness opportunities, such as yoga and tai chi classes, health fairs with free screenings, mobile mammography vans, and blood drives (Flaherty, 2013). The Pima County, Arizona, Public Library has a public health nurse on staff (Malachowski, 2014). Public libraries throughout the US are increasingly taking on the varied roles of community resource centers, meeting places, and training providers, in addition to their historically perceived and more traditional roles of materials and information repositories.

5.1.3 Public libraries and health information provision

As public libraries were gaining footholds in communities throughout the United States in the early 1900s, there was animated and somewhat controversial discourse in the medical library literature with regard to medical and health information provision in this setting. At that time, the Director of the New York State Library, Melvil Dewey (1902) argued for providing access to medical information for health-care practitioners in state libraries. Thus, the model of public libraries serving as a point of access for medical and health information is not a new one, whether or not public library staff have universally or enthusiastically embraced the responsibility for this service.

There's a common historical perception among many public library practitioners that there have been three "untouchable" subjects when it comes to reference queries in the public library setting: finance, law, and health (Wood, 1991; see also the chapter by C. Arnott Smith in this volume). In fact, in some public libraries, there may be residual policies that preclude staff from fielding some of these types of reference questions, such as health (Flaherty, 2013). Yet these assumptions, attitudes, and policies evolved *pre-Internet*, before patrons had such extensive access to so much information on all three of these topics. This complicates the interaction between patrons and public library staff, as skill levels for accessing and evaluating the quality of information resources will differ from person to person and library to library.

In contrast, in more specialized library settings, such as legal and medical libraries, authority has been readily ascribed to the professionals who staff them; the law and medical librarian has long been deferred to as the information navigator or expert. These subject specialty librarians almost always possess a postbaccalaureate graduate degree and historically have acted as information interlocutors or gatekeepers.

Public library staff regularly have to answer all types of reference questions. In fact, the U.S. Institute for Museum and Library Services (IMLS) estimates that one in five public library visits involves a reference interaction (IMLS, 2010). While many public libraries generally do not keep statistics on the specific types or subjects of reference questions asked, the public library setting has been shown to be a resource for individuals with health information queries (Flaherty & Luther, 2011). Deering and Harris (1996) found that 60% of survey respondents identified public libraries as one of their preferred resources for health information. Other studies have found that public librarians answered, on average, more than 10 health-related reference questions per week; and between 10% and 20% of all annual reference questions were estimated by library staff to be in the health subject domain (Flaherty & Roberts, 2009; Linnan et al., 2004). Although they may be regularly addressing health reference queries, few public library staff have received training on finding, interpreting, and evaluating medical information (Gillaspy, 2000; Smith, 2011).

The wide variety of organizational structures and function of public libraries throughout the United States also plays a part in how these organizations approach their role of health information provider in their communities. While some public libraries may avoid or decline to take on the role of providing health reference services, others have embraced the role wholeheartedly (Flaherty & Luther, 2011). The range of approaches varies enormously—from deferring questions and referring patrons to other facilities to providing links on library Web sites to subscription health databases or state resources (e.g., nclive.org for North Carolina State residents) or vetted consumer health resources (e.g., MedlinePlus.gov) to full-blown embedded consumer health information centers with dedicated full-time staff offering extensive community outreach (Flaherty, 2013).

5.2 Challenges

The types, sizes, and function of public libraries in the United States vary so dramatically that it is difficult, and perhaps unfair to generalize across the organizations. Because the vast majority of public libraries (88%) have service areas with populations of 50,000 or less, and more than half serve populations of fewer than 10,000 people (Pearlmutter & Nelson, 2011), it may be difficult to compare these smaller, more rural institutions with their urban counterparts. That said, the following exchange likely exemplifies a typical encounter that may take place in a reference interview in any public library, regardless of size or geographic locale. A former consumer health outreach librarian was recounting an experience she had while working in a public library: "One woman called, very upset, with 'cystic fibrosis of the breast.' She had fibroid cysts in her breast. I advised her to call the nurse for the precise name of the ailment…it wasn't within my purview to interpret for her" (J. Benedetti, personal communication, March 30, 2012).

There are many interconnected and sometimes overlapping challenges to consider and contend with, when it comes to providing consumer health information in all types of public library settings. Some of these include the following:

- Some public library staff members' perceptions of the role of public libraries
- Public perceptions of the role of public libraries
- Different types and levels of financial and community support for public libraries

- Different levels of degree requirements for staff across public library settings
- Wide range of quality of health information resources available
- Popular nonfiction's intersection with nonauthoritative health information
- Skill sets of public library staff
- Differences in training opportunities and staff motivation
- Disconnect between public and medical library practitioner communities

The following section includes a discussion of these challenges and some recommendations to consider for addressing or overcoming these challenges in order to turn them into opportunities in public library communities.

5.2.1 Library staff perception of the role of public libraries

Public libraries have been credited with providing most consumer health information services and resources in the US and are often the first place individuals go for fulfilling health information needs (Huber & Gillaspy, 2011). Yet, not much literature is available about the actual resources public library staff are providing in health reference encounters (Flaherty, 2013). To address this dichotomy, and to better understand the current state of health information provision in U.S. public libraries, a series of interviews with 35 randomly selected public library managers were conducted from 2010 to 2014 in two states: New York and North Carolina. While to date this has been a relatively small sample, the process has been elucidating. Some common themes have emerged, especially in smaller library communities, during the discussions that have taken place. Some library managers voiced hesitancy and/or reluctance at providing health reference services with the primary concern and reason for avoiding this service being liability, and the secondary concern of "stepping on toes" of health-care practitioners (Flaherty, 2013).

Additionally, there is the challenge of overcoming, especially in smaller and more rural public libraries, the historical attitude and expectation of: "We just lend fiction" (Flaherty & Luther, 2011). There are a wide variety of other contributing factors, such as the library director's and the staff's attitudes and approach to service provision, library board, and user community expectations, the relationship with local, regional, and statewide library systems, and access to training and funding (Flaherty, 2013). While it appears that some of these factors may be difficult to change, including attitudes and approaches to some service practices, attempts should be made to assure that patrons receive authoritative health information when they seek it out in the public library setting.

5.2.1.1 Opportunity

As public libraries are becoming reenvisioned and reborn in the public's collective psyche as community centers and resource providers in all variety of manners, this is a particularly opportune time for inclusion and marketing of authoritative health information provider as another one of the high-quality services that patrons can expect when they visit the local public library, alongside 3-D printers, fishing rods, and pedometers. The first step is to ensure that all public library staff are comfortable and familiar with providing knowledgeable assistance to patrons with health information; then they can effectively incorporate this service into their information provision models. In the US, national and state-level library and health agencies can help lead the effort to assess public library staffs' current state of knowledge and use of health information resources and to provide necessary support through information sharing and training programs where warranted.

5.2.2 Public perceptions of the role of public libraries

Because they are identified as information providers and community centers that are accessible to all citizens, public libraries are often identified as resources for a variety of needs. After the Patient Protection and Affordable Care Act (2010) was passed, it was announced at the ALA's 2014 Annual Meeting in Chicago that President Obama identified public libraries as one of the organizations where individuals requiring assistance with understanding and interpreting the new health-care legislation could go for help. He called upon librarians to aid in Affordable Care Act (ACA) guidance and health care enrollment; efforts by the author to obtain a copy of the President's statement from either the ALA or the White House were not fruitful (R. Olague, personal communication, April 29, 2014; J. Wright, personal communication, March 20, 2014). Upon hearing of the unavailability of the statement, one public library practitioner who had attended the conference and witnessed Obama's announcement quipped: "What did you expect? That's because it's another unfunded mandate, they're not going to touch it." (M. Jones, personal communication, March 28, 2014). The implication was that public libraries are expected to provide a broad array of ever-expanding services, often without additional funding or resources.

As it turned out, in some cases, patrons did seek out public library staff assistance for aid with understanding the ACA and in enrolling in health-care exchanges (J. McAllister, personal communication, December 12, 2013). In response to expectations, the IMLS stepped up and provided support for training on the ACA for public library staff through programs such as the series of webinars, available through the library support Web site, WebJunction and called: *"Health Happens in Libraries."* Patron reliance on public libraries for help was uneven, however; some public library managers have reported frustration at expending a lot of time and effort to gear up and be ready to answer ACA queries, to have not one patron ask for assistance (S.W. Davis, personal communication, May 12, 2014).

5.2.2.1 Opportunity

Helping patrons navigate the ACA has been a great way for public libraries to showcase the wide variety of services they can provide to their communities. Public library staff can help to coordinate support for the process through: collaboration with local social service agencies, provision of resources and space for signing people up, hosting classes, workshops, etc. Public libraries can also use the ACA as a leap pad for offering programs related to health in their communities.

Today more than ever, there are great opportunities for using public libraries to provide consumers with authoritative health information. The organizational infrastructure exists, with facilities and staff—the time is ripe to investigate ways to exploit it to best effect, to ensure that in this setting, provision of reliable, up-to-date information in all areas related to health is the norm throughout the nation. A discussion of one example of a statewide consumer health initiative appears after this section on "Challenges."

5.2.3 Financial and community support

Public libraries across the United States are governed by a multiplicity of local, county, state, and federal laws. In addition, levels of funding are not the same across public library settings, nor are other forms of external support, such as grants and gifts. For example, some states or localities may offer direct funding opportunities for items such as capital improvement, infrastructure, and staff development or program grants. Public libraries may be members of local and regional library systems with varying levels of involvement with and operational and financial support from state library departments as well. These factors affect not only how public libraries function, but also how they are able to approach service provision. For instance, while libraries in some locations may struggle to find funds to keep their doors open, others may be able to offer a wide variety of services, including dedicated assistance with health information queries and health promotion programming.

One example of an effective model for stable local public library funding, New York State, has been predicated on forming tax districts that support libraries in the same manner as they do schools. Once a library tax district is formed through a majority vote by residents, the residents of that tax district regularly vote on the library's budget and the library is guaranteed a minimum amount of funding from the district. This transparent approach allows for minimum standards for hours, staffing, etc., and ensures a minimum level of service provision in these library tax districts. Additionally, as residents have direct input via their vote on the library's budget, they are likely to be invested in the function and future of the library in their community.

No matter the funding or support structure or size of the library, having a motivated individual in the organization who can advocate, such as the library director/manager or the library board president, on behalf of the library will help to ensure the library can provide a high level of service to their user community.

5.2.3.1 Opportunity

The variety of oversight and organization of state agencies with regard to public library service is a topic that deserves more consideration and discussion than is appropriate here. It is evident, however, that some states could benefit immensely from a thorough examination and subsequent update of existing structures and legislation. For example, in some states (e.g., New York) public library systems are still guided by laws enacted in the 1950s and 1960s when resource sharing was a much bigger challenge than it is today. A revamping and reorganization of statewide library systems with national shared goals for revised, updated minimum standards service provision may be a logical starting point. In this manner, assurance of a minimum level of quality of library service can be enacted and ensured.

5.2.4 Educational requirements for staff

Beyond differing levels of financial and community support, there are also varying levels and a range of professional and educational requirements across communities and states for public library directors/managers and staff. There are no national standards in the US that govern staffing patterns in public libraries (De la Peña McCook, 2011). Some states' standards give recommendations for staffing levels, such as in Illinois and Wisconsin, where recommended staff size is based on population of the legal service area (De la Peña McCook, 2011). Some states have well-defined degree requirements for public library managers based on population served. For example, in New York State, educational requirements are based on population levels of the community or district served by the library. For libraries that serve populations between 2500 and 4999, 2 years of full-time study in an approved academic university or institution is required for the manager; for populations between 5000 and 7499, a bachelor's or 4-year degree is required; for populations above 7500, the manager must hold a public librarian's certificate, which requires the graduate master's degree in library science or equivalent (Carter, 2006).

Because of these varying levels of standards and requirements, there can be a concomitant variation in service levels and expectations of staff with regard to services they provide. There are even variances in terminology for the person in charge of the library. The term "manager" is often used to refer to an individual in charge of a branch location in a regional system; "director" is used to indicate the person in charge of all libraries in the system. This is in no way standard across communities, regions, or states, however. In some cases, the term "director" may mean the only staff member in a small library, and may be used interchangeably with "manager" to describe the individual, who is overseeing the organization. Staff expectations have implications for the services that staff can provide. For example, in some public libraries, staff without a graduate degree have stated that they assume their patrons are better than they are at navigating the Internet for finding any kind of information, including health (Flaherty, 2013). It is unlikely that professionals with the Master of Library Science (MLS) degree agree with this perception, or have this attitude as a starting point for responsibility of providing information and reference assistance to their user community.

5.2.4.1 Opportunity

As there are minimum requirements for teachers, principals, and superintendents, there also should be for public librarians. As a profession, when we reconsider minimum standards for service provision, attention should also be given to minimum certification standards of training and skill levels for individuals, who are qualified to direct/manage a library. The terminology for the tax-supported organization that is a public library should be reconsidered as well, perhaps with nationwide discussions and input from professional organizations such as the American Library Association, federal agencies such as the IMLS, and faculty from schools of information and library science. Through such an effort, an agreed-upon determination of national minimum standards that include staffing and resource requirements could emerge to define and stipulate what constitutes a public library. At present, even the minimum standards for an organization to be defined as a public library vary widely from state to state (Public Library Standards by State, 2015). If the facility doesn't meet the standards, then it should be referred to as a reading room, in order to distinguish between levels of service; in this way, the potential of public libraries throughout the country may be maximized.

5.2.5 Wide range of quality of health information resources

Health and medicine are subject areas that are constantly and often radically changing, and therefore require a commitment of time and effort by public library staff to stay up-to-date in order to ensure patrons' utilization of reliable, high quality, and authoritative information. Before the unprecedented and widely available access to all types of health information afforded by the Internet, health-care providers (e.g., physicians, nurses, pharmacists, specialists, dentists, etc.) were the primary, and in most cases, only, purveyors, interpreters, or providers of health information for most individuals seeking it out. Because we now have such extreme access to exorbitant amounts of a wide range of health information resources, there is a necessity for sifting through all of it to glean what is authoritative, and what is misinformation. Thus, the role of libraries and library staff has evolved. Services run the gamut from resource provision, to creation of a repository for health information materials, ideally, assistance with evaluation of veracity of resources as a primary responsibility to patrons. This is all the more important especially as studies have found that individuals may not be adroit at identifying reliable or accurate health information. For example, the majority of health information seekers (77%) start with a general search engine; they don't check the source and date of the information they find (Pew, 2008, 2013). Also, individuals may not have the adequate skills or background knowledge to evaluate the quality or authority of health information they find online (Eysenbach & Kohler, 2002) and some individuals have been shown to have significant misconceptions with regard to health issues after they've found inaccuracies or misinformation online (Kortum, Edwards, & Richards-Kortum, 2008). Thus, public library staff may have a role to play in providing assistance to patrons in understanding health information resources, even though they may not have confidence or advanced skills in this area themselves.

Commonly used resources, such as Wikipedia, may not be adequate for health information. In a recent study, researchers compared Wikipedia with standard, evidence-based peer-reviewed literature for the 10 most costly medical conditions in the United States. In 9 out of 10 cases (coronary artery disease, lung cancer, major depressive disorder, osteoarthritis, chronic obstructive pulmonary disease, hypertension, diabetes mellitus, back pain, and hyperlipidemia), Wikipedia had wrong assertions and misinformation (Hasty et al., 2014). For the average health-care consumer seeking out information, these findings may have grave implications. Most librarians would not advocate starting with Google and/or Wikipedia for locating health information. Yet research has found that during unobtrusive reference visits to public libraries, when a researcher posed as a patron and asked a health query, she was advised to "just google it" and in another instance she was told that the best place to start was with Google, even though later it was discovered that in both instances the library subscribed to a number of health databases through a library database vendor (Flaherty & Luther, 2011).

Ideally, public library staff would start with a noncommercial, readily available resource such as MedlinePlus, the consumer health Web site produced by the U.S. National Library of Medicine (NLM) as the automatic response to health queries. Research has shown, however, that there are many public library staff who are not aware of or are confused about the NLM and the services it offers (Smith, 2011). It appears that public library staff are more likely to refer patrons to the commercial Web site WebMD as an online resource, and some think this is a resource produced and sponsored by the NLM (Flaherty, 2013). One only need visit any patient waiting room to discover the traction the magazine component of WebMD has made; it is also a common feature in many small public library magazine racks as well.

5.2.5.1 Opportunity

Authoritative, noncommercial consumer health information resources are readily and freely available, but public library staff need to know about these resources before they can advocate for their utilization. Possible approaches for "getting the word out" may include more intensive partnering of university library science programs with public libraries to allow for internships that would include assistance with health information collection development, marketing of resources, etc. Exploration of innovative ways to publicize resources may be warranted. Some ideas might include programs such as contests with patron user groups in public libraries, using some NLM resources (e.g., ToxTown, household products database) and then producing YouTube videos to demonstrate what the patrons have learned. Public Service Announcements through media outlets about online consumer health resources such as MedlinePlus may be another mechanism to exploit for increasing awareness of these valuable information resources.

5.2.6 Popular nonfiction and nonauthoritative health information

Further complicating the issue of health information provision in the public library setting is the overarching mission and mandate of the organization to serve the user community. For many public libraries and their patrons, this means selecting nonfiction materials from popular resources, such as the *New York Times*' best seller lists. Often these popular materials may be in direct contradiction to prevailing medical opinion and practice: for example, some of the books authored by US model, actress, and, until recently, anti-vaccine activist, Jenny McCarthy. In the foreword to her 2008 account of her struggles raising an autistic son, *Mother Warriors*, is the statement by Dr Jay Gordon: "Vaccines can *cause* autism" (McCarthy, 2008, p. xiii, italics original). There is overwhelming scientific and epidemiologic evidence to the contrary, however (Taylor, Swerdfeger, & Eslick, 2014). Nonetheless, a quick search of the global online library catalog WorldCat on May 12, 2014 showed that close to 1800 libraries in the US own this item, and in public libraries, it is classified and often located or shelved in the section for medical sciences. It should be noted that in April 2014, Ms McCarthy recanted her earlier stance and denied having been anti-vaccine (McCarthy, 2014).

Related to this issue, the author conducted research using an unobtrusive reference approach. A researcher posed as a library patron and made visits to 75 randomly selected libraries in three different states in the Eastern US. When public library staff were asked: "Do vaccines cause autism?" a print resource was produced in the majority of cases. Most of the time, the resource did not answer the question and/or provided outdated or outright misinformation on the topic (Flaherty, 2014).

5.2.6.1 Opportunity

Public libraries have an implied responsibility to their user community to provide materials they request, yet it's time to reconsider how public libraries collect health information. At the very least, we must reexamine how popular nonfiction titles that address medical issues are classified or cataloged. Perhaps different ways of grouping or displaying this subset of items should be considered, or a new classification category denoting popular culture's take on health information should be introduced into our classification systems.

Another approach may be to use popular literature as a forum or starting point for talking about and addressing controversial and/or confusing health topics with invited speakers from the health-care community. This approach was used to positive effect in one small library community, when some local residents objected to a book that had been added to the collection that favorably described the creation of the Catskill Watershed, the resource to protect New York City's drinking water quality. When the Watershed was created, there was a great deal of controversy, as many local farms were destroyed (Appleton, 2002). Community members were invited to engage in a discussion panel, moderated by a local professor. The result was a lively, well-attended adult program, with a lengthy and likely contentious social process defused, and many new faces in the library.

5.2.7 Skill sets of public library staff

Providing appropriate health information provision requires a number of skills that are currently and regularly taught in many graduate library science programs in the United States. These skills include the ability to evaluate information resources, identify authoritative sources, and conduct a fruitful and successful reference interview. Yet more than half of all public libraries in the US are run by individuals without a graduate degree (N. Bolt, personal communication, September 9, 2012), so these skills may not be constant across public library settings.

5.2.7.1 Opportunity

Adapting to social changes and community needs is a necessary part of assuring the public library's viability and existence. For instance, many public libraries in the US did not have Internet access or public access computing prior to a concerted effort by and dedicated funding from the Gates Foundation during the last two decades to provide those resources (Price, 2014). Public library and library system staff were required to undergo training for their libraries to be eligible to participate in the initial program, and eligibility for subsequent technology upgrades was predicated on participation in advocacy training workshops to help ensure sustainability once the Gates funding expired (Public Library Association, 2015).

A health information initiative modeled on the Gates approach is worthy of exploration. Given that the overwhelming majority of public libraries are now wired, and public access computing is a common service, the financial investment would be considerably smaller. Grassroots training initiatives on how to find health information and evaluate its credibility, using library science students as trainers and readily available webinars, may be a starting point. Federal agencies that support library initiatives may be leveraged to provide assistance with such an undertaking. These agencies include the regional networks of the NLM—referred to as the National Network of Libraries of Medicine (NN/LM); the U.S. IMLS; and local and regional networks or consortia working with support from state library agencies.

5.2.8 Differences in training opportunities and staff motivation

Another challenge, especially in libraries with already limited resources, is the staff's lack of ability to participate in opportunities for education and training. Often there are limited or no funds for travel, and with limited hours and staff availability, attending off-site seminars or workshops is problematic. Again, different structures of governance and funding can impede large-scale implementation of all types of initiatives.

5.2.8.1 Opportunity

Webinars are one mechanism of providing training that can help address some of the resource limitations. They require no travel or special scheduling, and can be completed on the participant's schedule. Another resource for training and educational opportunities is partnership with community agencies. For example, the local American Red Cross chapter is often willing to host programs, such as cardiopulmonary resuscitation training—for staff and patrons (Conner, personal communication, July 7, 2014). It goes without saying that these initiatives require motivated staff, however, and a need for organizations and organizational oversight that will value and support continuing education.

Another area, where changes can be instituted is in the formal training of our future librarians and information scientists. As not all library and information science curricula offer health information courses and not all master's students will engage in such courses, there is a need to embed the subject elsewhere. For example, in core reference courses, students can learn about conducting reference interviews through using consumer health queries as examples. In knowledge organization and information retrieval courses, indexing with the NLM's Medical Subject Headings can be used, thus introducing and exposing students to medical terminology. In this way, general knowledge of medical and consumer health information can make its way into the core curriculum for all students, engendering exposure by using examples of existing resources as learning tools in the required core courses (Smith, 2006).

5.2.9 Disconnect between public and medical library practitioner communities

At the annual Medical Library Association (MLA) conference in Chicago in 2014, a luncheon round table was held focusing on consumer health issues. Of the 11 attendees, 1 was a public librarian. One might not expect to find any public librarians attending MLA, as there are other conferences that might suit their broad interests better—the annual conferences of the ALA and the biannual conference of the Public Library Association—but these communities don't have a lot of professional opportunities or venues for intersection. The following exchange from the division of the MLA devoted to consumer health, the Consumer and Patient Health Information Section's electronic mailing list exemplifies the divide that can arise between the two communities of practice. In response to the question: "What do you do when patrons donate questionable materials?" one medical librarian answered:

> *I have been donated materials that I would consider "speculative" information and I have said that I need to review the materials first, and if I decide they did not meet our criteria, I could return them to them. Most of the time the patrons tell me I can do whatever I want with them. If I am further questioned, I will tell them about our policy, we only collect information that is timely, relevant, from "quality" sources and that meet the guidelines of evidence-based medicine. So, I would likely not accept these "Moss Reports" since they are not evidence-based or come from a "quality" source. Same reason I refuse to carry anything by Kevin Trudeau or the cancer book written by Suzanne Somers. I leave those sorts of things to our local public library.*
>
> S.M. White, personal communication (2013)

The inference here is that public libraries are more likely and/or willing to carry questionable or low-quality materials when it comes to collecting resources and providing health information. It appears that it would be fruitful for the two communities of practice to have increased opportunities for collaboration and hybridization.

5.2.9.1 Opportunity

As has been demonstrated with efforts in the past (see for instance—Becker et al., 2010; Calvano & Needham, 1996; Guard et al., 2000; Hollander, 1996; Huber & Snyder, 2002; Martin & Lanier, 1996; Ruffin, Cogdill, Kutty, & Hudson-Ochillo, 2005; Spatz, 2000; Wood, Lyon, Schell, & Kitendaugh, 2000), there is an advantage to increasing cooperation and interaction between the medical and public library communities of practice. There are a range of activities and levels of involvement for these opportunities, such as embedding consumer health librarians in public library settings as described in more detail below. As community outreach becomes an ever greater part of the public library manager's organizational mind-set and responsibilities, the definition of "community" can extend to include hospitals and other health-care agencies. In this way, public libraries can create models for collaboration similar to those seen with early literacy efforts through agencies such as Head Start and local preschools, where the children's librarian visits the physical locations of the programs and performs storytime activities as a library outreach program or service. Additionally, regional chapters of professional organizations, such as the MLA, can help to foster collaboration with public libraries, regional library systems, and state library agencies through training opportunities.

5.3 Case study: embedded consumer health librarians in Delaware

In 2003–2004, a statewide initiative was introduced to provide consumer health information to all residents of the State of Delaware. The initiative was a collaborative effort between the Delaware Academy of Medicine, a private, nonprofit health information delivery organization, and a State agency, the Delaware Division of Libraries (A. Norman, personal communication, July 18, 2012). Consumer health librarians were hired and employed by the Academy of Medicine and were physically situated in public libraries in each of the State's three counties (Kent, New Castle, and Sussex) (Flaherty & Grier, 2014). The embedded consumer health librarians were located at the busiest libraries and also performed outreach duties throughout their respective counties (LaValley, 2009). The project was called *Delaware Healthsource*. Funding for the staff positions and expenses was provided via the Health Fund Advisory Committee, the advisory group that provided opportunities for allocation of Delaware's funds from state settlements with tobacco companies (Delaware Division of Libraries, n.d.).

The program was phased in over the course of a 2–3-year period, and the embedded consumer health librarians became very engaged in their public library communities. They provided a variety of services, including confidential reference services; between 300 and 600 individual health reference queries were responded to annually while the program was operational.

Innovative programming included summer workshops centered on healthy nutrition, including using community gardens at the library to grow vegetables; "Wellness Wednesdays" with regular on-site health information consultation opportunities with the embedded librarians; access to free screenings in coordination with local health-care providers, such as cardiovascular health with associated displays and informative literature; children's programs, such as fruit and veggie bingo; online games centered on health and wellness; and physical activity opportunities as a component of winter reading programs. The embedded consumer health librarians participated in regional health events and fairs.

They also assisted the public library staff with assessment and evaluation of the health resources in the libraries. In an initial collection analysis performed in two of the three counties in the state, it was discovered that the average publication date for nonfiction books was 1985 and the consumer health collections were poor in quality. In response to these findings, the Delaware State Library allocated specific funds for updating consumer health collections and enabling floating collections across library settings. The embedded librarians aided in collection development, resource selection and appropriate deselection, and updating.

All types of educational opportunities were provided for patrons and library staff. For example, in one county, the embedded librarian collaborated with a local law school to provide programs on health care and legal issues. Outreach was another important component of the initiative. The team of embedded librarians created a promotional marketing campaign with the tagline: "You ask. We search." They collaborated with local nursing homes, community groups, and hospitals. Developing reference skills of the public library staff was another focus of the program. The embedded librarians trained staff on how to conduct reference interviews and how to use the NLM's health information resources.

The collaborative initiative was active until it was phased out in 2009, due to funding challenges. The initiative succeeded through the hard work, resourcefulness, and enthusiasm of the individuals involved. Speaking about the project, one of the participating consumer health librarians observed: "It was a hard sell to convince the public they could call or come to the public library with their health questions. It isn't part of the collective consciousness that librarians can answer health questions. There isn't an awareness of what a medical librarian does, the public is lacking in understanding in terms of what we can provide" (LaValley, personal communication, November 9, 2012).

She added, because the initiative "was funded creatively, it was a blessing and a curse." Funds were short every year, so she saw the writing on the wall and realized the program wouldn't be sustained. "It was established as viable and necessary within the library population, but state funds trickled and the public libraries couldn't fund it. It was established with private, public and state funding, but no one stepped up when the state funds ran out." She added that in terms of sustainability, the problem was that the model relied on state funding (LaValley, personal communication, November 9, 2012).

In summing up the program and effectiveness of the embedded consumer health librarians, the participant remarked, "They are a viable way to provide consumer health information, but you also need the involvement of other partners, such as senior centers, community centers, hospitals, and the public health community. The efficient and effective model has to include public libraries and community-based organizations, but it's not sustainable in public libraries alone, absolutely not." She went on to say that some libraries have a shortsighted view; while they may talk about outreach, it doesn't happen. "The medical library model doesn't work with the public library alone, if it's only in the library world, it's not going to stick around...There are a lot of people who don't use libraries for health information." One of the valuable outcomes of the initiative was that community organizations were reminded of the value of public libraries. "People don't necessarily know what librarians do, outreach is critical, and it enhances political capital" (LaValley, personal communication, November 9, 2012).

The consumer health initiative did not include a formal evaluation component. In 2012, however, visits were made to half of the public libraries in the state, where a health reference query was posed: Do vaccines cause autism? In two-thirds of cases, authoritative health information was provided to answer the query. It is difficult to tie these results to the initiative, but similar visits in public libraries in Upstate New York and statewide in North Carolina found authoritative information was provided in only one-third of visits when the same health query was posed (Flaherty, 2014).

Though the initiative in Delaware is no longer operational, it can still be used as a model for other states and communities to consider. The integration of the two library communities of practice, the medical and public, appears to have had a positive effect on quality of health information provision in the state (Flaherty, 2013).

5.4 Conclusions

The Delaware initiative exemplifies the act of bridging the medical and public library communities of practice. By embedding consumer health and medical librarians in the public library setting, it is likely that the interaction between the staff had an impact on staff awareness and knowledge of consumer health resources, changing approaches to health information resource provision. Increased and continued collaboration between the public and medical library practitioner communities will help to ensure that public library patrons receive authoritative health information resources when they have health reference inquiries.

As public libraries evolve to become vital community centers, they can be exploited as natural settings for consumer health information provision and health promotion. Full-blown statewide initiatives like Delaware's may not be realistic in all communities, especially given current funding constraints. Other considerations, including state-level support for initiatives, should be explored, perhaps using funds from tobacco taxes or lottery ticket sales. In the same way that the Gates Foundation was successful in wiring and supplying public access computers in public libraries throughout the country, an initiative for assuring reliable consumer health information provision could be undertaken.

In the meantime, there are free and readily available mechanisms that already exist to aid public library staff with all types of health information. In the US, the NLM (www.nlm.gov) and its regional affiliates, the NN/LM (www.nnlm.gov) offer training opportunities at no cost. The IMLS also sponsors a series of webinars on WebJunction: *"Health Happens in Libraries"* (www.webjunction.org). There are examples in the literature of low-cost consumer health projects initiated in public libraries that would take little effort or funding to replicate. For example, consumer health information services were provided in one suburban public library through the efforts of one engaged staff member. With very little financial investment, the staff member turned an under-utilized area in the library into a dedicated area for consumer health information. The public responded very positively to her efforts, and now health promotion is a center piece in the library's offerings (Flaherty, 2014). The tools to enable us to aspire to the 1876 definition of the public library—in which *"every citizen of the city or town* which maintains it has an *equal share in its privileges of reference and circulation"*—are at our disposal. The challenge lies in assuring access and their appropriate use.

References

Anonymous. (May 18, 2008). *Detroit public library opens teen center.* Michigan Citizen p. A9.
Appleton, A. F. (2002). *How New York City used an ecosystem services strategy carried out through an urban-rural partnership to preserve the pristine quality of its drinking water and save billions of dollars.* Retrieved from http://ecosystemmarketplace.com/documents/cms_documents/NYC_H2O_Ecosystem_Services.pdf.
Becker, S., Crandall, M. D., Fisher, K. E., Kinney, B., Landry, C., & Rocha, A. (2010). *Opportunity for all: How the American public benefits from Internet access at U.S. libraries (IMLS-2010-RES-01).* Washington, DC: Institute for Museum and Library Services.
Bullock, E. D. (1907). *Management of traveling libraries (ALA publishing board library handbook No. 3).* Boston, MA: American Library Association.
Calvano, M., & Needham, G. (1996). Public empowerment through accessible health information. *Bulletin of the Medical Library Association, 84*(2), 253–256.
Carter, R. A. (2006). *Public library law in New York state (2006).* Albany, NY: New York Library Association.
Cuyahoga County Public Library. (2010). *Cuyahoga County Public Library offers passport services in seven branches.* Retrieved from http://www.cuyahogalibrary.org/Services/Passport-Centers.aspx.
De la Peña McCook, K. (2004). *Introduction to public librarianship.* New York: Neal-Schuman.
De la Peña McCook, K. (2011). *Introduction to public librarianship* (2nd ed.). New York: Neal-Schuman.
Deering, M. J., & Harris, J. (1996). Consumer health information demand and delivery: implications for libraries. *Bulletin of the Medical Library Association, 84*(2), 209–216.
deGruyter, L. (1980). The history and development of rural public libraries. *Library Trends, 28*(4), 513–523.
Delaware Division of Libraries. (n.d.). *Consumer health information services: Best practices in public libraries report.* Retrieved from http://libraries.delaware.gov/planning/pdfs/ConsumerHealthBestracticesPublicLibraries.pdf.
Dewey, M. (1902). Medical departments in public libraries-A symposium. *Medical Libraries, 5*, 2–3.
Eysenbach, G., & Kohler, C. (2002). How do consumers search for and appraise health information on the World Wide Web? Qualitative study using focus groups, usability tests, and in-depth interviews. *British Medical Journal, 324*, 573–577.
Flaherty, M. G. (2013). *The public library as health information resource?* (Doctoral dissertation). Retrieved from ProQuest dissertations and theses. (Accession number AT3561444).
Flaherty, M. G. (2014). Strategic planning for success. In M. Spatz (Ed.), *The Medical Library Association guide to providing consumer and patient health information* (pp. 27–36). Lanham, MD: Rowman & Littlefield.
Flaherty, M. G. (Unpublished manuscript). *Good, bad and ugly: The wide range of authoritative health information in public libraries.*
Flaherty, M. G., & Grier, P. J. (2014). Statewide initiative to embed consumer health librarians in public libraries: a case study. *Public Library Quarterly, 33*(4), 296–303.
Flaherty, M. G., & Luther, M. E. (2011). A pilot study of health information resource use in rural public libraries in Upstate New York. *Public Library Quarterly, 30*(1), 117–131.
Flaherty, M. G., & Roberts, L. (2009). Rural outreach training efforts to clinicians and public library staff: NLM resource promotion. *Journal of Consumer Health on the Internet, 13*(1), 14–30.

Gillaspy, M. L. (2000). Starting a consumer health information service in a public library. *Public Library Quarterly, 18*(3/4), 5–19.
Guard, R., Fredericka, T. M., Kroll, S., Marine, S., Roddy, C., Steiner, T., et al. (2000). Health care, information needs, and outreach: reaching Ohio's rural citizens. *Bulletin of the Medical Library Association, 88*(4), 374–381.
Hasty, R. T., Garbalosa, R. C., Barbato, V. A., Valdes, P. J., Powers, D. W., Hernandez, E., et al. (2014). Wikipedia vs peer-reviewed medical literature for information about the 10 most costly medical conditions. *Journal of the American Osteopathic Association, 114*(5), 368–373.
Hollander, S. (1996). Consumer health information partnerships: the health science library and multitype library system. *Bulletin of the Medical Library Association, 84*(2), 247–252.
Huber, J. T., & Gillaspy, M. L. (2011). Knowledge/power transforming the social landscape: the case of the consumer health information movement. *Library Quarterly, 81*(4), 405–430.
Huber, J. T., & Snyder, M. (2002). Facilitating access to consumer health information: a collaborative approach employing applied research. *Medical Reference Services Quarterly, 21*(2), 39–46.
Institute for Museum and Library Services. (2010). *Public libraries survey fiscal year 2008.* Retrieved from http://harvester.census.gov/imls/pubs/Publications/pls2008.pdf.
Institute for Museum and Library Services. (2013). *Public libraries in the United States survey.* Retrieved from http://www.imls.gov/research/public_libraries_in_the_united_states_survey.aspx.
Johnson, C. (2010). Do public libraries contribute to social capital? A preliminary investigation into the relationship. *Library & Information Science Research, 32*(2), 147–155.
Karp, G. (August 26, 2011). You can get that at the library? *Chicago Tribune.* Retrieved from http://articles.chicagotribune.com/2011-08-26/features/sc-cons-0825-karpspend-20110826_1_library-card-public-library-lending-period.
Kortum, P., Edwards, C., & Richards-Kortum, R. (2008). Impact of inaccurate Internet health information in a secondary school learning environment. *Journal of Medical Internet Research, 10*(2), e17.
LaValley, S. (2009). Delaware health source: consumer health libraries and health literacy outreach. *Journal of Consumer Health on the Internet, 13*(2), 180–186.
Linnan, L. A., Wildemuth, B. M., Gollop, C., Hull, P., Silbajoris, C., & Monnig, R. (2004). Public librarians as a resource for promoting health: results from the health for everyone in libraries project (HELP) librarian survey. *Health Promotion and Practice, 5*(2), 182–190.
Malachowski, M. (2014). Public libraries participating in community health initiatives. *Journal of Hospital Librarianship, 14*(3), 295–302.
Maloney, J. (June 29, 2014). New York public library looks at innovative models for renovation. *Wall Street Journal.* Retrieved from http://online.wsj.com/articles/new-york-public-library-looks-at-innovative-models-for-renovation-1404090627.
Martin, E. R., & Lanier, D. (1996). Networking consumer health information: bringing the patient into the medical information loop. *Bulletin of the Medical Library Association, 84*(2), 240–246.
McCarthy, J. (2008). *Mother warriors.* New York: Penguin.
McCarthy, J. (April 12, 2014). Jenny McCarthy: the gray area on vaccines. *Chicago Sun-Times.* Retrieved from http://www.suntimes.com/news/otherviews/26784527-452/jenny-mccarthy-the-gray-area-on-vaccines.html.
Moffat, K. (June 23, 2013). *Borrow seeds. Grow plants.* Public Libraries Online, Retrieved from http://publiclibrariesonline.org/2013/06/borrow-seeds-grow-plants/.
New York Public Library. (2014). *Money matters @ financial literacy central.* Retrieved from http://www.nypl.org/help/getting-oriented/money-matters-flc.

Owens, D. M. (April 26, 2010). *Check it out: Get your groceries at the library.* Retrieved from http://www.npr.org/templates/story/story.php?storyId=126282239.

Patient protection and affordable care act. (2010). Pub. L. No. 111–148.

Pearlmutter, J., & Nelson, P. (2011). When small is all. *American Libraries, 42*(1/2), 44–47.

Peterson, C. A. (2005). Space designed for lifelong learning: the Dr. Martin Luther King, Jr. joint-use library. In Council on Library and Information Resources (Ed.), *Library as place: Rethinking roles, rethinking space* (pp. 56–65). Washington, DC: Council on Library and Information Resources.

Pew Internet and American Life Project. (2008). *Health information use.* Retrieved from http://www.pewInternet.org/pdfs/PIP_Health_Aug08.pdf.

Pew Internet and American Life Project. (2013). *Health online 2013.* Retrieved from http://www.pewInternet.org/Press-Releases/2013/Health-Online-2013.aspx.

Price, G. (May 8, 2014). Grants: Gates foundation ending global libraries program over next 3–5 years. *Library Journal.* Retrieved from http://www.infodocket.com/2014/05/08/gates-foundation-ending-global-libraries-program-over-next-3-5-years/.

Public Agenda. (2006). *Long overdue: A fresh look at public and leadership attitudes about libraries in the 21st century* (ERIC Document Reproduction Service No. ED493642).

Public Library Association. (2015). *Turning the page online: Building your library community.* Retrieved from http://www.ala.org/pla/education/turningthepage.

Public library standards by State. (2015). Retrieved from http://plsc.pbworks.com/w/page/7422647/Public%20library%20standards%20by%20state.

Rockport Public Library. (n.d.). *Library cards, loans and renewals.* Retrieved from http://www.rockport.lib.me.us/rockport/libraryservices.asp

Ruffin, A. B., Cogdill, K., Kutty, L., & Hudson-Ochillo, M. (2005). Access to electronic health information for the public: analysis of fifty-three funded projects. *Library Trends, 53*(3), 434–452.

Ryder, H., Faloon, K., Levesque, L., & McDonald, D. (2009). Partnering with libraries to promote walking among community-dwelling adults: a Kingston gets active pilot pedometer-lending project. *Health Promotion Practice, 10*(4), 588–596.

Shera, J. H. (1965). *Foundations of the public library: The origins of the public library movement in New England 1629–1855.* Chicago, IL: Shoestring Press.

Sigler, K., Jaeger, P. T., Bertot, J. C., McDermott, A. J., DeCoster, E. J., & Langa, L. A. (2011). The role of public libraries, the Internet, and economic uncertainty. In A. Woodsworth (Ed.), *Librarianship in times of crisis* (pp. 19–35). Bingley, UK: Emerald Group Publishing Limited.

Smith, C. A. (2006). I am not a specialist: why we all need to be worrying about medical information. *Journal of Education for Library and Information Science, 47*(2), 96–105.

Smith, C. A. (2011). "The easier-to-use version": public librarian awareness of consumer health resources from the National Library of Medicine. *Journal of Consumer Health on the Internet, 15*(2), 149–163.

Spatz, M. A. (2000). Providing consumer health information in the rural setting: Planetree Health Resource Center's approach. *Bulletin of the Medical Library Association, 88*(4), 382–388.

Taylor, L. E., Swerdfeger, A. L., & Eslick, G. D. (2014). Vaccines are not associated with autism: an evidence-based meta-analysis of case-control and cohort studies. *Vaccine, 32*(29), 3623–3629.

Toth, F., & Baltic, S. (2010). Teen center: Crandall public library, Glens Falls, New York. *Voice of Youth Advocates, 33*(1), 38–39.

U.S. Department of Health and Human Services, National Institutes of Health, National Library of Medicine, National Network of Libraries of Medicine, NN/LM Moodle. (2013). *Distance learning opportunities from the National Network of Libraries of Medicine (NN/LM).* Retrieved from http://nnlm.gov/moodle/.

Vårheim, A., Steinmo, S., & Ide, E. (2008). Do libraries matter? Public libraries and the creation of social capital. *Journal of Documentation, 64*(6), 877–892.

White, S. M. (August 23, 2013). *Re: What do you do when patrons donate questionable materials*. [Electronic mailing list message]. Retrieved from http://caphis.mlanet.org/mailman/private/caphis_caphis.mlanet.org/.

Wood, F. B., Lyon, B., Schell, M. B., & Kitendaugh, P. (2000). Public library consumer health information pilot project: results of a National Library of Medicine evaluation. *Bulletin of the Medical Library Association, 88*(4), 314–322.

Wood, M. S. (1991). Public service ethics in health sciences libraries. *Library Trends, 40*(2), 244–257.

Who needs a health librarian? Ethical reference transactions in the consumer health library

Nancy C. Seeger
University Hospitals of Cleveland, Rainbow Babies & Children's Hospital Cleveland, Ohio

6.1 Introduction

Even the healthiest of people find themselves in the role of patient at some point in their lives. And in the increasingly consumer-driven world of health care, individuals will each be called upon to make crucial decisions regarding their own health, or the health of someone they love. Often those decisions will be made somewhat blindly, based on fear, hope, or something that "sounds" like a good idea. And while that will work out for some, others will experience negative consequences from their choices and find themselves in the position of having even more challenging decisions to face as a result—still without the clear understanding needed to make them effectively.

At one time, those facing a health problem or medical crisis relied solely on medical professionals to make a diagnosis and to guide the treatment. Patients rarely questioned their doctors, and doctors were reluctant to give out too much information beyond instructions for daily care required for patients at home. Hospital stays were often longer and more frequent because the level of care needed was not expected of patient families in their homes. This care was something managed only by professional doctors and nurses. Power rested primarily in the hands of the physician, and opinions of medical professionals were considered mysterious and to be respected (Goodyear-Smith & Buetow, 2001; McKinstry, 1992).

During most of the past century, health care was delivered primarily within a paternalistic relationship. Physicians and their patients typically assumed that the doctor knew best what would help the patient most. The mid-1960s brought political and social changes that began to shift the way patients viewed the medical community, and coinciding medical advancements sparked a strong interest in bioethics (Steinhart, 2002). As the science of medicine progressed, medical decisions became more complicated for both doctors and their patients, and issues of rising health care costs played a complex role in the process (Sulmasy, 1992). Disease-related community advocacy organizations began appearing on the scene, offering education to patients, who wanted more direction about medical options (Steinhart, 2002). The new millennium, and the decades leading up to it, brought widening availability of the Internet with its ever-growing sources of health information, and accelerated the shift from that once familiar physician-centered paradigm to one that is more patient-centered (Kaba & Sooriakumaran, 2007). In this evolving model of consumer-directed health care, doctors are expected to include patient input in the

Meeting Health Information Needs Outside of Healthcare.
Copyright © 2015 by N.C. Seeger. Published by Elsevier Ltd. All rights reserved.

treatment decision-making, and patients encountering health and medical concerns are now more likely to look for answers on their own, often with mixed results.

From surveys conducted in 2012, the Pew Internet and American Life Project found that 81% of adults in the United States use the Internet, and of those Internet users, 72% have searched for health information online. When asked which health-related topics had been included in their searches during the last 12 months, slightly more than half of online health seekers (55%) said they had looked for information about a specific diagnosis or medical condition, and less than half (43%) indicated they searched for specific medical treatments (Fox & Duggan, 2013). The survey also indicated that one in three American adults had gone online symptom-searching—specifically to try and determine what medical condition they or someone else might have. Of those who searched for symptoms online, 41% said they had consulted a physician, who confirmed their suspicions (an additional 2% said they received partial confirmation). Eighteen percent said that at a consultation, the physician did not agree or offered a different opinion as to the diagnosis, and 1% of online diagnosers said that a visit to their health care provider was inconclusive. Thirty-five percent of online diagnosers did not follow up with a health care professional at all (Fox & Duggan, 2013).

The Pew reports look at how Americans are using the Internet, but do not indicate the quality of the information found or health outcomes related to that use. The statistics here are concerning, because the reliability and accuracy of online health information varies greatly, and the potential for complicated information to be misinterpreted is likely to be high (Benigeri & Pluye, 2003; Keselman, Browne, & Kaufman, 2008; Schembri & Schober, 2009).

More and more, Americans are given greater control and responsibility for their own health care decisions—along with wider access to an enormous amount of health information—but with very few tools to use it wisely. Health care consumers are often expected to choose between different treatment options; sign consent forms to document their acceptance of these critical decisions; and balance the financial, physical, and emotional impact of this consent. Thus, consumers can find themselves both overwhelmed and ill-equipped to meet these demands. Patients may be confused about their role in the decision-making process; they can fear making the wrong decisions, undermining physicians' authority, or alienating their providers by voicing doubts and concerns (Joseph-Williams, Elwyn, & Edwards, 2014). Even the most confident people can feel confused, lost, and frustrated in the medical maze they enter when becoming a patient or caregiver. This is where a librarian can be of great help. Indeed, everyone needs a health librarian!

Consumer health information is intended for potential or current users of medical services (all of us!). It is designed to be educational, and can help individuals make decisions about health-related behavior and medical treatments. It differs from clinical information—that is, information written by and for medical professionals—in that it is developed with the layperson in mind, involving less technical language and more user-friendly formats. Consumer health information may include resources about prevention, self-care and wellness, diseases and conditions, treatment, health care options, and more.

Consumer health librarians specialize in the selection, evaluation, and provision of consumer health information. The role of the consumer health librarian is multifaceted, and includes assessing individual information needs, searching for relevant, reliable information sources, acquiring those resources, and delivering the information in a format easily understood by library patrons. Consumer health librarians can be found in hospital libraries, where they might also serve in a dual role as clinical medical librarians assisting physicians, nurses, and medical staff with clinical medical research articles and information. But they can also be found in patient resource centers, free-standing health resource agencies, academic and public libraries. They are often medical librarians with a focus on consumer health, but can also be librarians trained for other settings, including academic, public, or special libraries (maintained by businesses, associations, nonprofit entities, or government agencies to collect materials and provide information of special relevance to the work of the organization), with consumer health as a subject specialty. Often academic and public librarians have dual roles serving the students, faculty, and/or the general public with library needs, while also assisting those patrons with specific consumer health information needs. In many cases, consumer health librarians are solo librarians; working alone or with a very small staff, and sometimes volunteers. Their responsibilities may include all facets of the library, including the following:

- **collection development**—evaluating and selecting a balanced variety of printed, digital, and other relevant resources to be included in the library's collection;
- **cataloging**—identifying, describing, and recording what makes a resource or item unique;
- **classification**—assigning subject headings and a call number used to locate a resource or item in the collection;
- **acquisitions**—previewing and purchasing library items;
- **library management**—business planning, budgeting, and overseeing library operations;
- **database management**—maintaining collections of electronic resources;
- **marketing**—devising strategies and creating tools to encourage library use and demonstrate its value;
- **reference**—identifying and meeting the information needs of patrons; and
- **circulation**—facilitating and tracking the loan of library materials.

Primary among the many activities assigned to the consumer health librarian is the provision of reference services. Librarians are in a unique position to build confidence in health care consumers, support shared health care decision-making, facilitate informed consent for medical treatment, and assist patients and their families in becoming active and involved members of the medical team responsible for their care. To meet this challenge effectively, the consumer health librarian must use all the skills and knowledge of the general reference librarian with some significant refinements specific to the health care consumer population.

The Reference and User Services Association within the American Library Association (RUSA) defines reference transactions as "information consultations in which library staff recommend, interpret, evaluate, and/or use information resources to help others to meet particular information needs." Such transactions are the key feature of reference work, which also includes "activities that involve the creation, management, and assessment of information or research resources, tools, and services" (RUSA, 2008).

In his paper, "Ethics and the Reference Librarian," Charles A. Bunge outlines the obligations facing the reference librarian when mastering the art of reference work. These obligations include competence, diligence, confidentiality, independence of judgment, honesty, and candor. Each of these obligations is a crucial factor in the building of trust needed for effective reference transactions (Bunge, 1991). While every reference librarian, regardless of the subject expertise, is obliged to cultivate these characteristics, each takes on a special significance as it relates to the ethical behavior of the consumer health librarian. Evidence of these obligations can be found in the careful study of successful consumer health reference transactions.

Because the goal these obligations strive to meet is the building of trust, a good model to illustrate the rapport required during a reference transaction is the "fiduciary relationship." Rooted in the field of legal ethics, this model describes a relationship of confidence based on trust, and effectively guides the well-executed reference transaction. The fiduciary ethical model for the reference librarian–patron relationship emphasizes the librarian's primary obligation to be worthy of the patron's trust (Bunge, 1991). Nowhere is this more critical than the relationship between the health librarian and a health consumer. Without a foundation of trust, such a reference transaction will be fruitless, and the provision of health information offered without trust risks causing harm as well.

Because they pose inherent practical and ethical challenges, two specific features of the reference transaction will be examined closely in this chapter. These two features, the reference interview and the provision of information, encompass the asking and answering of questions in a consumer health library.

6.2 The reference transaction: asking the right questions, avoiding the wrong answers

"Do you have any questions?" is a question often asked by physicians near the conclusion of a patient's doctor visit. This question aims to open the door for an exchange of information between the patient and health care provider, but it often falls short of its goal. There are many reasons why this can happen:

- **Patient readiness:** Patients just learning of a new diagnosis may have many concerns, but cannot yet formulate the right questions. Often, the real questions do not become conscious until after patients leave the doctor's office, and even when a patient does have questions, it is easier to say a quick "no," than to venture asking the physician a difficult question.
- **Time constraints:** Physicians feel a responsibility to other patients waiting to be seen, and patients may be reluctant to extend the appointment longer than the time allotted.
- **Anxiety/Discomfort:** Physical examinations can put the patient in a vulnerable position, not conducive to asking questions, and make the patient anxious to leave the office. The patient may also feel physically ill, or be experiencing pain, characteristics which can cloud the ability to generate questions.
- **Confusion/Insecurity:** Patients may not trust their own judgment, and may not be comfortable questioning the doctor's authority. Patients may not have understood the doctor's

explanations, feel unsure of what they think about what the doctor has said, and worry that they are giving the wrong impression or asking inappropriate questions.
- **Embarrassment/Shame:** Patients may believe that a medical condition is their own fault, may be embarrassed about choices they have made that could have contributed to their current situation, and may be ashamed to ask questions that could expose such a choice.
- **Limited health literacy:** Patients may not understand new medical terminology, and be unable to pronounce the names of diagnoses, medications, tests, procedures, and treatments. Patients may feel that their questions betray their own feelings of incompetence, particularly if they have forgotten or not understood something that the physician said earlier. Or the doctor might have explained the situation too simply in an effort to be understood, leaving patients without enough information to generate questions.
- **Fear of the unknown:** Some patients may prefer to avoid asking questions because they are afraid to find out that something is wrong, and prefer not to hear bad news.

When patients become library patrons, these concerns do not automatically disappear. The librarian can anticipate that the same obstacles may (or may not) be present when the patient/patron enters the consumer health library, makes an inquiry via telephone, e-mail, or instant messaging. For this reason, the librarian takes steps to recognize and reduce the barriers to communication. This moment marks the onset of a reference interview.

The Dictionary for Library and Information Science defines the reference interview as

> *the interpersonal communication that occurs between a reference librarian and a library user to determine the person's specific information need(s), which may turn out to be different than the reference question as initially posed. Because patrons are often reticent, especially in face-to-face interaction, patience and tact may be required on the part of the librarian. A reference interview may occur in person, by telephone, or electronically (usually via e-mail) at the request of the user, but a well-trained reference librarian will sometimes initiate communication if a hesitant user appears to need assistance.*
>
> Reitz (2004)

RUSA, a division of the American Library Association promoting excellence in library reference services, provides a set of behavioral guidelines for conducting successful reference interviews (RUSA, 2011). These guidelines highlight several attributes that can improve interactions between librarians and patrons. Successful librarians behave in a manner that makes them highly visible and approachable, demonstrate an interest in the patron's topic, and employ effective listening and questioning skills to accurately identify the patron's information needs. Choosing what type of question to ask at what point in the reference interview can make a difference. When expanded information is required for a search, the librarian can encourage the patron to elaborate with open-ended questions. Examples of such questions during a consumer health reference interview might include the following:

- What else can you tell me about your child's illness?
- How much information would you like to have?
- What are your other concerns?
- What kind of resources would make your life a little easier right now?

However, when a patron is throwing a very wide net, wanting to know everything about a broad topic, more closed or clarifying questions can help refine the search. In this situation, examples of questions a consumer health librarian might ask would include the following:

- What do you already know about this diagnosis?
- Are you trying to find out what causes the illness or how it is treated?
- What is concerning you most right now about this situation?
- Do you prefer reading about the treatment options, or would a video be more helpful?

Successful librarians are also skilled at rephrasing and verifying each patron request, explaining the search process, and clarifying needs as the search progresses. Once the search is completed, successful librarians will follow up to ensure the information needs have been met, and refer the patron to additional resources if needed (RUSA, 2011).

The librarian's obligation to competence during the reference interview is marked by the ability to tailor questions, comments, and demeanor to meet the specific needs of each patron from the moment contact is attempted. This is especially true for health-related inquiries, when the information sought may be on personal or worrisome topics. Because of the sensitive nature of consumer health questions, and the potential risk involved if the wrong information is provided, extra care must be taken to ensure that a complete and accurate picture of the patron's information needs is established during the reference interview (Ham & Liebermann, 2012).

The barriers inherent to the health-related reference interview underscore the need for diligence, in order that the librarian can explore all possible directions the interview might take. The patron must be assured that anything disclosed during the interview will remain confidential and free from judgment. The librarian must not only possess the ability to meet the ethical obligations of the fiduciary relationship, but also communicate a commitment to these obligations within the context of the reference interview. To the casual observer, a reference interview appears to be a simple conversation; indeed, the best ones are exactly that. However, the astute consumer health librarian is always aware of the unfolding interaction, and uses many observational skills to gather information throughout the conversation—while remaining steadfastly within clearly established boundaries of respect for privacy, personal space, and the patron's cultural or individual values and preferences (Ham & Liebermann, 2012).

What looks simple can be challenging, because the question that a patron brings to the consumer health librarian is rarely the question causing the most concern. Instead, it is usually the question that allows patrons to dip a toe into the tide to see if they can tolerate the temperature. If the initial response is too cold or too warm, the first question will likely be the only question asked, but if the response is comfortable, the patron may be more willing to venture a bit further into the water, as a foundation of trust begins to develop.

What is comfortable varies from one person to another, so the health librarian must watch for clues every step of the way. Each step guides the next. Providing a welcoming atmosphere, accommodating privacy needs, matching the patron's tone and level of engagement, actively listening to what is being said, and paying attention to nonverbal cues are all steps to establish a foundation for building trust.

Ethical reference transactions in the consumer health library 123

Consider the following interaction that takes place in a small consumer health library serving the general public. Although this library is located within a medical center, a reference transaction like this one could easily occur in a public library or other settings as well.

The patron is a young woman. She sighs frequently, and speaks quietly. She steps up to the reference desk to ask for information about playgroups near her neighborhood. The librarian is immersed in cataloging new materials, but quickly pushes her work aside, looks up, and smiles.

Patron: Do you have any information about playgroups for kids close to me?
Librarian: Sure, I can find some information about that. Where do you live?
Patron: I live in _____, and I don't have a car so it can't be far away, and needs to be on a bus line.
Librarian: Ok. Do you have young children?
Patron: Yes, two of them. I need to get out of the house. (*sighs*)
Librarian: You really do have your hands full! But they grow up so fast. They'll be grown up before you know it. Let's find some really fun things for you to do with your children. What kinds of groups are you interested in joining?
Patron: (*shrugs and looks away*) I don't know, just something to get me out of the house. It's depressing to be at home, but I don't feel like going out.
Librarian: Well. How about some story hours at your local library? I can look up the times for those. And there is a pool in your neighborhood, as well as a local park. They have activities too. And the YMCA has many programs...
Patron: Oh you don't need to go to all that trouble...
Librarian: I don't mind at all! That's what I'm here for. I'll be happy to look that up for you. It will just take a few minutes. Would you like to wait? Or I can also mail it out to you, if you give me your e-mail or mailing address.
Patron: Thanks. I can come back for it a little later.

The librarian starts generating a list of local activities, and looks up to ask for a phone number she can call when the information is ready, but the patron has already left the library. She finishes her research, and prints out the list with contact information. But the young mother never comes back for it.

Providing a welcoming atmosphere with a smile and focused attention is a good start, but in this case, the librarian failed to match the patron's tone and level of engagement. The patron was not ready to get excited about the information she was requesting, and was subsequently overwhelmed by the librarian's enthusiastic response. The many options provided were more confusing than helpful, and the librarian's immediate request for contact information may have felt like a breach of privacy. The librarian missed some nonverbal cues that might have told her that playgroups were only a small piece of the information that was really being sought. So although the librarian was anxious to be of help, the patron gained nothing.

Here is a scenario that begins exactly the same way, but then takes a different turn.

The librarian is once again immersed in cataloging new materials, but looks up and smiles when the young mother enters the library:

Patron: Do you have any information about playgroups for kids close to me?
Librarian: Sure, I can find some information about that. Where do you live?
Patron: I live in _____, and I don't have a car so it can't be far away, and needs to be on a bus line.
Librarian: Ok. Do you have young children?
Patron: Yes, two of them. I need to get out of the house. (*sighs*)
Librarian: Ok. I'll be happy to look that up for you. It will take a few minutes. Do you want to look at the books while you wait? Or come back later? I can also send the information to you if you prefer.
Patron: I'll wait. (*She wanders over and begins perusing the book collection*)

The librarian does some searching, and generates a list of potential group activities for mothers and young children. When she is finished, she prints out the list with contact information and brings it over to the young mother.

Librarian: Here you go! I found several story time activities and playgroups right in your neighborhood.
Patron: Thank you. I'll try these.
Librarian: Is there anything else I can help you find?
Patron: No. That's okay. Thanks. Bye.

It seems as though this interaction was successful, and it was. The librarian followed the patron's lead, and was helpful without being overwhelming. And the patron left with exactly what she requested. However, the information she requested was not the information she really wanted. The librarian missed some of the cues that offered opportunities for active listening. This becomes clear in a third scenario.

The librarian is still immersed in cataloging new materials, but looks up and smiles when the patron arrives:

Patron: Do you have any information about playgroups for kids close to me?
Librarian: Sure, I can find some information about that. Where do you live?
Patron: I live in _____, and I don't have a car so it can't be far away, and needs to be on a bus line.
Librarian: Ok. Do you have young children?
Patron: Yes, two of them. I need to get out of the house. (*sighs*)
Librarian: How old are they?
Patron: I have a toddler—she's almost 2, and a newborn boy. (*sighs again*)
Librarian: It can be difficult to have small children at home. A new baby is lots of work.
Patron: Yeah. I get really depressed—especially since I had my baby. My doctor suggested joining some playgroups, but I told her I don't have the energy.

Librarian:	Getting out of the house is not as easy as it sounds when you have a new baby.
Patron:	(*nods and looks away*) The doctor says maybe I have postpartum depression. She gave me a prescription for it, but I'm afraid to take it. So I thought I'd just try to find some playgroups or something first to get me out of the house.
Librarian:	Ok. I'll find you some information on a few playgroups and activities in your neighborhood. It'll just take me a minute. While I'm looking, I can show you our books about postpartum depression if you like.
Patron:	(*nods*) Ok.

They walk over to the bookshelves, and the librarian pulls out one book.

Librarian:	This book explains a little about how postpartum depression affects women and how the medications work. And there are others here too. I'll be finished with the search in a few minutes.

While the patron sits down to look through the book, the librarian goes back to the computer and generates a short list of playgroups with contact information, and also finds a local support group for new mothers with postpartum depression. She takes the information she found to the patron.

Librarian:	Here you go. I also found a local support group for moms with postpartum depression. Would you be interested in something like that?
Patron:	Thank you. I'm not sure about a support group though.
Librarian:	They aren't for everyone, but some people find them very helpful.
Patron:	(*looks over the list*) I'll think about it. Can you give me the number just in case?
Librarian:	Sure. Stick it on your refrigerator, and it'll be there if you decide to try it. Is there anything else that I can find for you right now?
Patron:	No, but can I check out this book?
Librarian:	Absolutely. Did you find any others that might be useful?
Patron:	Oh. I didn't even look. Can I look around a little bit?
Librarian:	Please do! Take your time. Bring me whatever you want to borrow when you're ready. And if you want some help choosing a book, let me know.

Later the patron brings two books to be checked out. The librarian takes her information, checks out the books, and explains the due date and return process.

Librarian:	If there is anything else I can help you find, please give me a call.
Patron:	I will, and I'll bring the book back after my next appointment. Maybe I'll bring the kids with me next time (*sighs again*), if I can get myself together enough to get them both dressed and out of the house.
Librarian:	I'd love to meet them.
Patron:	(*smiles*) Thanks again. Bye.

Each of these scenarios began in exactly the same way, but resulted in very different outcomes. In the third scenario, the librarian probed a bit further by asking a more specific question about the age of the patron's children. This helped to provide a clearer picture of her information need, while still allowing for privacy. The librarian then responded to several nonverbal cues with active listening, thus demonstrating a willingness to assist without making judgments or creating expectations. As a result, the patron was able to express her more urgent need for information on postpartum depression, and receive more relevant resources than in the previous scenarios. Because the librarian was matter-of-fact about the patron's disclosure, the mother felt comfortable increasing the level of engagement, and is more likely to return with more requests for information as her needs evolve.

A successful reference transaction lays the groundwork for a fiduciary relationship between librarians and patrons. The ultimate goal is to create an ongoing relationship, where the patron continues to use the services of the consumer health librarian with repeat visits and new questions. But as the relationship develops, many ethical challenges can arise with the potential to thwart trust and confidence.

6.3 Looking for the answers: symptom-checkers and self-diagnosing

One of the most common conundrums for the consumer health librarian is meeting the information needs of patrons who are looking for a diagnosis. Patrons seeking the services of a consumer health librarian may not know what information they need. Often they come with a list of symptoms, wanting to find out what's wrong. It may be that they have not seen a doctor at all; or they have visited the doctor, but no diagnosis has been made yet. Sometimes, a patron might be dissatisfied with the current diagnosis, and suspects that the doctor has missed something. Patients using online health resources for self-diagnosis may delay care, self-treat, or go against medical advice based on poor or incomplete information, perhaps without telling their doctors.

The use of online consumer health resources to facilitate self-diagnosis with patrons poses a significant ethical dilemma for the consumer health librarian. Librarians generally have no clinical medical training, but they are experts in searching for medical information, know the terminology, have access to details about most diagnoses, and are aware of—even if they do not typically endorse—the many symptom-checking tools available online (Table 6.1).

Choosing from lists of applicable symptoms found using these online tools will generate a list of potential causes for the symptoms chosen. It is possible for a librarian to run this search for—or in partnership with—a library patron to generate a plausible list of possible diagnoses in a matter of minutes. While this method of searching is simple, it typically yields less than adequate results. There are many variables involved in diagnosing an illness, but online symptom-checkers force users to choose the closest adjectives offered in a series of checklists.

Ethical reference transactions in the consumer health library 127

Table 6.1 Selected symptom-checking tools available online

Name of tool and source	Basic features
Family doctor's search by symptom tool www.familydoctor.org/familydoctor/en/health-tools/search-by-symptom.html American Academy of Family Physicians (AAFP)	• Begins with a single general symptom • Uses flowcharts of Yes/No questions • Results in a single general diagnosis • Suggests appropriate actions to take • Indicates when to see a doctor • Requires reading, and following a flowchart
KidsDoc symptom checker www.healthychildren.org/English/tips-tools/Symptom-Checker/ American Academy of Pediatrics (AAP)	• Uses graphics to indicate location of pediatric symptoms • Pop-up menu lists topics related to physical location of symptom • Menu selection links to topic description and other resources. • Requires reading and selecting symptom-related topics
Mouth healthy dental symptom checker www.mouthhealthy.org/en/Symptom-Checker American Dental Association (ADA)	• Uses graphics to indicate location of dental symptoms • Uses a checklist of symptoms • Results in a list of possible dental conditions with descriptions and actions to take • Requires reading, but includes some photos of dental conditions
Mayo Clinic symptom checker www.mayoclinic.org/symptom-checker/select-symptom/itt-20009075 Mayo Foundation for Medical Education and Research	• Begins with a single presenting symptom • Provides separate symptom lists for adults and children • Uses checklists of related factors/symptoms • Results in a short ranked list of common possible causes • Offers drop-down lists of related factors/symptoms for each potential cause • Links to condition descriptions, and actions to take. • Includes lifestyle and home remedies, and questions for the doctor. • Requires reading and selecting symptom descriptors
Video symptom checker www.everydayhealth.com/symptom-checker Everyday Health Media, LLC	• Commercially-produced tool • Uses video of a doctor speaking to the viewer • Asks a series of detailed symptom questions • Includes sample photos of symptoms • Results in a short list of possible causes • Indicates when to see a doctor. • Requires some reading, but primarily audio-visual tools

Continued

Table 6.1 **Selected symptom-checking tools available online—cont'd**

Name of tool and source	Basic features
WebMD symptom checker www.symptoms.webmed.com/ WebMD, LLC	• Commercially-produced tool • Uses graphics to indicate location of symptoms • Pop-up menu lists related symptoms to select • Results in a ranked list of possible diagnoses with descriptions, and actions to take. • Includes questions to ask the doctor. • Requires reading and selecting symptom descriptors

Consider the following situation:

A woman's son calls her from his college dorm to say he has been sick with a very sore throat, a skin rash, and a recurring very high fever. His temperature nears 104° every afternoon, but goes away during the night when he breaks out in a sweat. Her son hasn't seen a doctor yet, and she encourages him to go to the campus health center. But in the meantime, she wants to do some investigating on her own. From what her friends have said, and her own past experience, she thinks he probably has mononucleosis, but wants to see what else might possibly be happening.

When searching using the WebMD tool with these symptoms, everything goes smoothly until it is time to describe the fever. There is no option to choose for a daily afternoon fever. Then the skin rash is difficult to pinpoint. Is it lattice-like with red spots? Or blotchy with rosy-pink raised bumps? And although there is no way to know if her son has been exposed to an infectious disease, it's highly possible given that he lives in a college dorm setting.

Once the best guess options have been chosen from the checklists, the resulting list does include mononucleosis—which seems to confirm her suspicions. However, it is twelfth in a list of more than 70 potential diagnoses. Viral pharyngitis, influenza, contact dermatitis, strep throat, and the common cold are among the possibilities that rate higher than mononucleosis. Thirteenth on the list is rheumatic fever, and the remainder of the list includes medical conditions ranging from hepatitis and lyme disease, to thyroid cancer and the plague.

When using a symptom-checker tool, the end result is often a useless list that is no more reliable than a mother's gut instinct. A list of diagnoses like this one may cause the delay of treatment for a serious disease, or needless panic over a self-limited illness. When facilitating a search such as this, a librarian may lend credibility to the resource, and risks implicitly suggesting that the results are meaningful—even if the patron is warned of the symptom-checker's limitations through the librarian's own written and verbal disclaimers.

A pair of consumer marketing researchers (Yan & Sengupta, 2013) found that symptom-matching strategies like those used in a symptom-checker tool may lead consumers to overestimate the likelihood of getting a serious disease because they are focused on their own symptoms and overlook the fact that it is highly unlikely (although possible) that those symptoms are caused by a serious medical condition. But if consumers fear the worst when it comes to their own health, they tend to be much more objective when it comes to others. When it is someone else complaining of heartburn, friends or family members are more likely to be the voice of reason, and assume it indicates indigestion or a little acid reflux, but their own heartburn feels like a heart attack. The authors of the study conclude that the best way to keep this from happening is to consult a doctor, because doctors not only have the clinical skills to accurately diagnose an illness, but also see their patients' symptoms more objectively.

In some cases, a patient's self-diagnosis can influence the doctor's clinical examination. When patients come to the doctor's office having already drawn their own conclusions concerning a possible diagnosis, the doctor may bypass the usual steps taken to rule out differential diagnoses and treat based on patient reporting (Avery, Ghandi, & Keating, 2012).

Conventional wisdom holds that self-diagnosis is a risky venture that can lead to poor outcomes. Research has historically supported this idea, and health care professionals have discouraged the practice as unreliable (Ahmad, Hudak, Bercovitz, Hollenberg, & Levinson, 2006; Jutel, 2010). Consumer health librarians are cautioned against attempting to provide diagnoses or propose treatment in any way (Ham & Liebermann, 2012). However, patients and health consumers continue to search the Internet for their own diagnosis or a diagnosis for someone else with surprising regularity (Fox & Duggan, 2013), and if those diagnoses are sometimes wrong, at other times they are right, or at least helpful in finding the correct diagnosis.

What makes this issue of self-diagnosis especially complex is that online searching for a diagnosis can work well, or certainly seems to, for many online health seekers. A recent poll conducted by Wolters Kluwer (2012) indicated that two-thirds of Americans who seek medical information on the Internet tend to trust the information they find, and 63% say that they have never misdiagnosed themselves.

Given the rise of new technology, the phenomenon of Internet self-diagnosis is unlikely to trend down. As more and more consumers get online, and become more Internet-savvy, the health information online continues to grow exponentially, and become more organized. New tools used to seek that information, including potential diagnoses, are continuously being developed as more efficient, user-friendly applications. Some of these tools are designed with medical professionals in mind, and provide highly clinical health information, while others are geared toward laypersons and consumers, presenting information in a more plain language format. Most of these tools are available to anyone with access to an Internet connection. Physicians search online, and so do their patients, because often enough, it seems to work.

Health consumers engage in symptom searching and seek information about various diagnoses online more often than other Internet health information-seeking activities (Fox & Duggan, 2013)—evidence that these are significant health information needs. It begins to seem counterproductive, then, for consumer health librarians, professionals who want to meet these information needs and develop an open, confident, and trusting fiduciary relationship with their patrons, to continue discouraging the practice of symptom searching and self-diagnosis. That seems a little like preaching abstinence to teenagers without also teaching safe sex!

A more sensible approach for consumer health librarians than outright prohibition of symptom-checking might be that they use their skills to help patrons conduct more effective symptom searches. By taking the opportunity to demonstrate how symptom-checkers actually work, librarians can honestly address the strengths and vulnerabilities of these tools, and assist patrons in evaluating the search results they find. Reference transactions handled in this way establish trust by respecting and validating the patron's need to search for answers, without attempting to diagnose the *cause* of their symptoms. Reaching that level of trustworthiness, the librarian can comfortably suggest that the patron follow up with a physician to discuss what they have found, and decide with their health care providers just what action to take. That is exactly what eventually happened with the mother in the previous scenario, who suspected her son in college may have mononucleosis.

After conducting her search on the WebMD symptom-checker, the mother was overwhelmed by the number and the wide range of potential diagnoses she found, and asked the librarian for assistance. She wanted to allow her son to have some independence in managing his own health, but was concerned that this illness might be serious enough to warrant a trip home to see his primary care physician.

After explaining the weaknesses inherent in symptom-checking tools, and the risk involved in self-diagnosis, the librarian suggested that she conduct her search in a couple of different tools to get a feel for how they work.

Librarian:	Some symptom-checkers, like WebMD, offer very detailed descriptors in an attempt to narrow the search, but if you aren't sure about a symptom, or can't find the exact word in the list that describes the symptom, the results will be less accurate. Other tools, like the Mayo Clinic symptom-checker, use broader symptom terms and match them with related common symptoms. It gives you a more general list of the most common possibilities.
Patron:	That makes sense. I had to guess about some of his symptoms, like what his rash looks like, because I can't see it for myself. He said, "It's just a rash, Mom." (*laughs; rolls her eyes*)
Librarian:	(*laughs with her*) Sounds just like a young man in college!
Patron:	It's frustrating because every time I talk to my son, he says something different. But he doesn't usually complain like this—even when he's sick. That's why I'm thinking I might just want him to come home, but I hate to overreact. I really don't know what to do about this.
Librarian:	Using symptom-checkers really is just a guessing game. The best answer you can hope for from using these tools is some good questions to ask the doctor. But it can help you figure out what you need to ask your son specifically about his symptoms. It will help him learn how to describe them too. And being able to describe symptoms is important when he does get to the doctor—it can help find the right diagnosis. Let's try your search in the Mayo Clinic symptom-checker too.

Because it was the first symptom her son described, the patron chose **sore throat** as the primary symptom, and went on to select the following accompanying symptoms from a list:

- fever
- chills or sweats
- enlarged, tender lymph nodes in neck
- difficult or painful swallowing
- skin rash
- muscle aches (she wanted to include joint pain, but that option was not available)

The following list of possible causes resulted:

1. Mononucleosis
2. Strep throat
3. Tonsillitis
4. Influenza (flu)
5. Epiglottitis
6. Common cold
7. GERD

Mononucleosis was the closest match, but did not include difficult or painful swallowing as a symptom. Strep throat and tonsillitis did include it, but did not include chills or sweats. Since both of these symptoms were among those her son complained about consistently, his mother thought that one of the first three causes was most likely. None of the results contained all of the selected symptoms.

Patron:	I think it is more than just a cold or the flu, but all of these make more sense than the plague! (*laughs*)
Librarian:	These are what might be the most common causes of the symptoms you chose. But remember that you didn't have anywhere to include the joint pain, and there wasn't any way to be specific about the kind of skin rash, or how high the fever gets. That could make a difference in the results.
Patron:	What's GERD?
Librarian:	We can look that up, but what are the matching symptoms?
Patron:	(*clicking on the link to show a list of associated symptoms*) Oh, this one has difficulty swallowing, but that's the only one that matches.
Librarian:	If you click on the link for GERD, it will take you to a description of it.
Patron:	(*clicks on GERD and reads a bit of the article*) This doesn't sound like what he has at all. I'm going to read some of the other ones.

After reading overviews about the other diagnoses, the patron said that the first 3 were all possibilities, but none of them mentioned painful joints, and her son had complained about that quite a bit.

Librarian:	I wonder if it would help to do it again and start your search with joint pain so that you are sure it's included.
Patron:	(*starting a new search*) Was that one of the choices in the beginning? Yes! Here it is.
Librarian:	Go for it. And feel free to print any information you'd like to take with you. It will print out up by the desk. Let me know if you get stuck.

The librarian returns to the reference desk. A little while later, the patron approaches the desk.

Patron:	Well. That made a big difference. A lot more came up when I started with the sore joints. None of them really match exactly when I look carefully. But some of them seem close, and look more serious—like lupus, and some kind of arthritis. I'm not sure my son will tell the doctor about the joint pain because it started later.
Librarian:	It really is a guessing game, and that can be frustrating. So many symptoms are common to lots of different illnesses, and every person is unique. Not everyone will have every symptom associated with an illness, and some people will have symptoms that are unexpected. It's also possible to have more than one diagnosis at the same time, and they can mask each other. That's why it is really so important to talk to the doctor about anything you find.
Patron:	I see that. Now I have a lot of questions—not just about mono! I'm going to call my son and see how he is feeling now, and what he found out at the health center—if he went. If the campus doctors think it's mono, I won't rush up there like an overreactive mom! (*laughs*) But I do think I want him to see his doctor at home too. It seems like he has symptoms that I didn't see in this stuff I read.
Librarian:	That sounds like a reasonable plan. His doctor might be able to quickly rule out many of those possibilities you came across in your search.
Patron:	I did print out a few pages, just to help me remember everything.
Librarian:	(*gets the print-outs, puts them in a folder, and hands it to the patron*) Here you go. Let me know what else you need. Call or stop in anytime. Or you can e-mail me too. My contact information is on the folder.
Patron:	Thanks so much for all of this.
Librarian:	You're welcome. I hope you and your son, and his doctor figure this out soon so he can feel better and get back to class.
Patron:	(*smiles and nods*) Me too. Thanks.

In this scenario, the librarian kept the patron from being misled using a symptom-checker by simply pointing out what this tool can and can't do. While searching for a potential diagnosis for her son's illness, the patron discovered that while her hypothesis of mononucleosis was a reasonable one, it wasn't the only one, and did not really match her son's symptoms completely. At the same time, the process gave the patron a clearer picture of her own concerns, and provided her with several new more focused questions for her son, and his doctor. Most important, the patron left feeling more confident with a plan to help manage her son's medical problems. And that plan included consulting his doctor.

6.4 What did the doctor say? Health literacy and deciphering a whole new language

Risky self-diagnosis is only one of the pitfalls encountered in the consumer health library. Finding out what ails them or a loved one is only the first step for most consumers. Once a diagnosis is established, there is a name for the ailment, but patients and their families may know little or nothing about what it means. Many times, patrons entering the library are not able to spell or pronounce the medical diagnosis or understand the terminology used to describe it. This is true regardless of socioeconomic status, educational background, support systems, or communication skills. Health literacy is an obstacle for every new patient and caregiver.

Health literacy is defined as "the degree to which individuals have the capacity to obtain, process, and understand basic health information and services needed to make appropriate health decisions" (U.S. Department of Health & Human Services, 2014a). This definition may be purposefully broad out of necessity, because when it comes to meeting the information needs of health care consumers—one size doesn't fit all.

There are now many health literacy programs on the market teaching health educators and health care providers how to assess and interact with patients suspected of having low health literacy (U.S. Department of Health & Human Services, 2014b). There are also a variety of systems and applications designed to help those in the health care field to convert complex clinical information into "plain language" that an "average" health consumer can comprehend (Office of Disease Prevention & Health Promotion, 2005) (Table 6.2).

While these tools can be helpful, they are not an answer for the many concerns surrounding health literacy. When someone enters the health care system, there is a whole new language to learn, and process to be navigated—often under great stress. Consumer health librarians can help mitigate the stressors with reliable information about the diagnosis, its causes, symptoms and prognosis, as well as direct patrons to resources to help support them through the process.

Knowing more about the diagnosis with individualized information, and having some resources for developing a support system, prepares patients and caregivers to enter what is often a maze of treatment options. The expertise of a librarian can help here too, with careful searching to develop a range of accepted treatment options for a specific diagnosis, and provide consumers with detailed information about the pros and cons of each option. This kind of information can be very valuable in helping patients make decisions, and feel confident with whatever treatment plan is eventually followed. But to do this well, the librarian must choose materials to match individual patron needs, and deliver the information contained in those materials in a way the patron can effectively absorb. That process may—at times—work similarly from one patron to the next; more often, librarians encounter patrons with widely differing needs, which are not always easy to determine.

Health care consumers—even those who are very competent in school, or on the job, in their respective areas of knowledge—are often uncomfortable when discussing personal, complex, and unfamiliar medical topics, particularly with medical professionals.

Table 6.2 **Selected health literacy and plain language tools and resources**

American Medical Association (AMA)	Health literacy and patient safety: help patients understand www.ama-assn.org/ama/pub/about-ama/ama-foundation/our-programs/public-health/health-literacy-program/health-literacy-kit.page
America's Health Insurance Plans (AHIP) Trade Association	Health literacy: a toolkit for communicators http://www.ahip.org/healthliteracy/toolkit/
Agency for Healthcare Research and Quality (AHRQ)	Health literacy universal precautions toolkit www.ahrq.gov/professionals/quality-patient-safety/quality-resources/tools/literacy-toolkit/healthliteracytoolkit.pdf)
National Cancer Institute (NCI)	Clear and simple: developing effective print materials for low-literate readers www.cancer.gov/cancertopics/canccrlibrary/clear-and-simple
Centers for Disease Control and Prevention (CDC)	Simply put: a guide for creating easy-to-understand materials www.cdc.gov/healthliteracy/pdf/Simply_Put.pdf
Health Literacy Innovations, LLC (HLI)	Health literacy advisor online demo http://www.healthliteracyinnovations.com/products/demo/
Group Health Research Institute (GHRI)	PRISM–program of readability in science and medicine: readability kit www.grouphealthresearch.org/capabilities/readability/ghchs_readability_toolkil.pdf

After an encounter in a clinical setting receiving clinical information from their physicians, patients may leave the doctor's office in a daze of confusion, and need to go elsewhere to gather information they can use effectively. Some medical centers and health-related social service agencies house small consumer health libraries (sometimes called resource centers) that provide such information in a nonclinical setting. These spaces offer a safe place for patrons to express frustrations, tell their stories, explain special needs, and receive help understanding written information, medical terminology, and complex health-related forms. Consumer health librarians can also provide the same services in local public libraries, medical libraries in hospitals and academic settings, often in medical schools.

Deciding how much information to provide, how much detail is desired, and what form of delivery is most appropriate are all important pieces of the reference transaction. Librarians can candidly offer information in a variety of formats, for example, print, audiovisual, computer-based applications. These materials can be created with varying degrees of language complexity, including plain language, bilingual/multilingual, easy-to-read, charts and tables, and complex clinical formats. Librarians working in an accepting manner, independent of judgment, will be able to see quickly what works best for each patron.

Not only does each patron come with different skills, abilities, and backgrounds, but where they are in navigating the process of managing their medical condition differs too (Longo & Woolf, 2014). Some patrons start off requesting very general information, and do not require more advanced information until later in the process; others prefer to have everything that can be found on a topic from the beginning. With some topics, that can be quite a lot! Yet, with other topics, for example, new, controversial or nonstandard treatments, or rare diagnoses, it may be difficult to find reliable information that is accessible to the patron. For example, the parents of a child diagnosed with a more common form of childhood cancer, such as leukemia, will find a plethora of resources on various reputable Web sites (e.g., American Cancer Society, National Cancer Institute, and links found through Medline Plus Health Topics). This information can be helpful, but sometimes overwhelming. However, when a child is diagnosed with a form of cancer that is typically seen only in adults, finding information that is really useful for that child's parent can be quite a challenge. Because some cancers are so rare in children, the diagnosis will not have been well-studied as a pediatric disease, and various treatment attempts may only be described in isolated reports and case studies of a few children. For example, while there may be many resources for adults with gastric cancer, the course of the pediatric disease can be very different from an adult illness, and require modified treatment. Adult-based information will not be useful for the parents of a child with gastric cancer, and meeting their information needs is more of a challenge for the librarian. In fact, sometimes the most information the librarian can provide is the knowledge that few—if any—reliable resources are currently available.

Some patrons find the general information written for a patient population to be unsatisfactory. This is especially true for patrons dealing with a chronic illness, or a complex, rare medical condition. These patrons may have already devoured all of the standard information, and are now seeking professional medical journal reviews and clinical trials. The goal of the consumer health librarian is to match the right information with the right patron at the right time.

Even when provided with information that is readily understood, related to established diagnoses, medical conditions, treatment options, and prevention strategies, patrons can become very discouraged or overwhelmed by what they find—or don't find—since nearly every search will turn up biased, contradictory, and sometimes questionable information. When that happens, instead of withholding information, this provides an opening for the health librarian to help consumers develop strategies for evaluating and balancing the information gathered. Learning to evaluate the quality of health information is a crucial skill in developing health literacy, and one that is often ignored by health care providers when it comes to advising patients. A quick route to ensuring patients receive reliable health information is for the doctor to provide it directly, and warn patients not to trust what they read on the Internet. However, for those patients wanting more information than health care providers are able to offer, the Internet is a powerful draw. Giving patients the skills to evaluate the quality of what they discover online may have a greater, more lasting impact on patient access to information that is useful, practical, and reliable.

There is a wide variation in the quality of health information in any format. Information that has not been updated recently, or does not show a current date, may or may not continue to be relevant. Sources of information may be questionable due to lack of authority or conflicts of interest. Web sites that do not provide contact information or details about the information sources can look professional, but may have content of poor quality. Health-related opinions have their uses, but must not be confused with health-related facts, and health information that is summarized, synthesized, or paraphrased may fall far from its source. This is especially true of journalistic pieces such as news articles, commentaries, and press releases, where much consumer health information seeking begins.

Consumer health librarians can be instrumental in helping health consumers acquire the skills needed to evaluate the resources they have found. Tools for evaluating health information are available online, and a good place to start is with the resource list, *Evaluating Health Information*, provided by Medline Plus (http://www.nlm.nih.gov/medlineplus/evaluatinghealthinformation.html). This list contains a variety of tools for different needs, and librarians can choose the best tool to be used for a specific patron. But handing the patron an evaluation tool is rarely enough. A more thorough approach involves mediated practice, where the librarian takes Web sites, articles, and other information retrieved online, and navigates the evaluation process cooperatively with the patron, modeling the strategies librarians use to evaluate health information. Developing such skills can help consumers avoid the wrong answers and choose current information from high-quality sources.

Health information based on little or no research is problematic for both librarians and patrons. A treatment option found online that offers an appealing outcome can seem like a hopeful cure, but without sound supporting evidence, there is no way to evaluate the accuracy of such information. And many consumer health library patrons, wanting to be hopeful, are vulnerable to wishful thinking.

6.5 When the answers have questions: experimental treatments and integrative medicine

Often patrons come into the library seeking a cure for what ails them, and those answers can be as complex and elusive as the diagnosis. This is particularly true when standard treatments for established diagnoses have been exhausted and unsuccessful, or an illness is recurring or no longer responding to earlier treatments. These consumers may have heard about, and come in looking for a procedure, a medication, a therapy, a diet, or a dietary supplement that falls outside the boundaries of standard medical practice, or one that has been suggested by their physician, but is currently considered an experimental treatment. They want to explore alternatives, and this requires learning complicated clinical terminology, and gaining familiarity with medical language and its associated jargon. Consumer health librarians are helpful here too, providing medical dictionaries, encyclopedias, and introductory information about experimental trials and alternative treatments.

Patients, who have exhausted standard medical treatment without success, or wish to try something outside current standard treatment guidelines, may consider trying an experimental or investigational treatment. This can mean joining a clinical trial program. Clinical trials are research studies that explore new ways to prevent, diagnose, or treat disease. Treatments being tested in a clinical trial might be new medications or a different combination of drugs, new surgical techniques or medical devices, or updated methods of applying existing treatments. Clinical trials typically compare a new treatment (i.e., medication, device, procedure, or therapy) with an existing treatment. The goal of these investigational studies is to determine if a new test or therapy is effective and safe to use (National Institutes of Health, 2014). Clinical trials are closely monitored, and can provide an opportunity for patients to receive cutting-edge treatments before they are widely available, but because of their experimental nature, carry inherent risks. Clinical trials may require participants to consent to the possible use of medications or procedures that have not been fully approved by the Food and Drug Administration, and the outcomes of treatment are unpredictable.

In some studies, one group of participants are assigned to receive a placebo (an inactive version resembling the medication being tested), and the experiences and outcomes of that control group are measured against those in the experimental group, who do receive the new drug. In "blind" studies, the participants don't know which group they have joined, and must be aware that they may or may not receive the experimental medication (National Institutes of Health, 2014). Those that do receive the new treatment risk experiencing unintended side effects or results. Also, insurance companies may or may not cover the use of experimental or investigational medications or treatments (National Heart, Lung, & Blood Institute, 2012).

Patients considering joining a clinical trial will quite naturally have questions about the safety of experimental treatments. To help patrons weigh the risks and benefits of these investigational treatments, consumer health librarians can gather information about specific clinical trials for patrons, and provide general information about how scientific medical research works, but they cannot offer reassurance that these treatments will be safe or effective. The very nature of the experimental studies precludes knowing the outcomes in advance. Although researchers do everything possible to reduce the chances of possible harm during investigational studies, there is no guarantee that any experimental intervention will work or is completely safe. However, candid information about clinical trials presented in an objective manner allows patrons to make informed decisions about consenting to experimental treatment, clearly knowing the risks involved.

Some patients may be concerned that the doctor might be heavily invested in the research, and encourage patient participation for the benefit of the study. The librarian is a neutral party in the research process, and can provide an unbiased viewpoint. Medline Plus offers various resources useful in explaining the risks and benefits involved in experimental treatments on its *Clinical Trials* (http://www.nlm.nih.gov/medlineplus/clinicaltrials.html) health topic page, additional information and a searchable database of clinical trials available around the world can be found at https://clinicaltrials.gov/.

Health consumers are also regularly seeking natural, holistic health alternatives to standard treatment options. Growing interest in complementary and alternative medicine has led the health care industry to take a closer look at these often controversial approaches. The National Center for Complementary and Alternative Medicine (NCCAM) defines complementary and alternative medicine (CAM) as "a group of diverse medical and health care systems, practices, and products that are not presently considered to be part of conventional medicine" (NCCAM, 2011). NCCAM further divides these complementary health approaches into two subgroups:

- Natural products (herbs/botanicals, vitamins and minerals, probiotics, and dietary supplements)
- Mind and body practices (acupuncture, chiropractic treatment, healing touch, hypnosis, massage therapy, meditation, tai chi/qi gong, yoga, etc.)

In recent years, the use of CAM has become widespread, despite controversy among the science-based medical community over the use of such practices. Trends supporting strong, scientific evidence-based medical care have typically bypassed the use of more ancient, natural remedies, and CAM has not received the rigorous study, testing and oversight that conventional medicine has been expected to withstand (Kantor, 2009). However, the 2007 National Health Interview Survey indicated that about 4 in 10 American adults (38%) and 1 in 9 children use CAM approaches (NCCAM, 2013). Because of its growing popularity among patients, the medical community has begun to study the uses and effects of CAM, and examine ways to integrate natural products and mind–body practices with more conventional techniques (Ventola, 2010). Although this integrative medicine approach is being implemented in a growing number of health care systems (Ruggie, 2005), it continues to receive less attention than conventional practices (Kantor, 2009).

Consumers who have heard about, or discovered CAM information on their own, may not receive support or further information from physicians, who usually have not been trained in the use of complementary health practices. The consumer health library is a logical destination for health consumers seeking access to information on CAM. Finding such information can pose a difficult ethical dilemma for librarians. Many concerns exist related to these health approaches, and consumer health librarians must be careful to balance differing views when assisting patrons seeking a wide range of treatment options. Once consumers begin seeking complementary and alternative health solutions, the search becomes even more murky and contradictory than it can with conventional health information searches, due to the relative scarcity of scientific research and a high degree of inconsistent uses and unsupported claims of efficacy. In some cases, the use of CAM may delay needed conventional treatment. Some CAM practices may also work in direct opposition to standard accepted medical practices, and have the potential to interfere or interact with conventional medical treatments leading to potentially dangerous results (Kantor, 2009; Lim, Cranswick, & South, 2011; Ventola, 2010; Vohra, Brulotte, Le, Charrois, & Laeeque, 2009). In addition, many CAM treatment options have not been well regulated (Cohen, 2003), and are typically not covered by medical insurance. Factors like these sometimes go unnoticed by patients considering treatment options (Tarn et al., 2014).

While librarians are experts in searching for available information, they cannot find research that doesn't exist. Stringent reliable research on CAM therapies is difficult to find. Although there is a growing interest in evidence-based study of CAM, such research is relatively new to the field of scientific medicine. Many claims made by CAM practitioners are based on anecdotal reports (Cowan, 2014), or nonconventional assumptions related to holistic health and healing that have not been supported in medical literature (Jonas, Eisenberg, Hufford, & Crawford, 2013). This doesn't mean that such practices are necessarily dangerous or ineffective; it only means that safety and success have not been adequately measured. While it seems beneficial to wait until such proof exists before embarking on an alternative treatment plan, patients and caregivers facing urgent decisions often do not want to wait. In these cases, consumer health librarians can still be helpful—just as they can for every other health information request—by providing information on the current state of CAM research, assisting patrons in evaluating the CAM information they do find, and encouraging them to inform their doctors about any alternative treatments they may be using. The NCCAM (http://nccam.nih.gov/about) is a reliable resource for those seeking information about complementary health approaches. NCCAM looks at the known safety issues of various therapies, identifies CAM treatments that have shown inconclusive outcomes, and exposes many unsupported claims. NCCAM supports scientific research about complementary and alternative treatments, and maintains a database of evidence-based CAM information.

6.6 Conclusions

Today's health consumers are expected, and often choose, to be active participants in their own medical care. This move away from a paternalistic view of medicine toward a more patient-centered approach creates challenges for both patients and their health care providers. Health literacy is a key feature of patient autonomy, shared decision-making, and informed consent, and requires that health care consumers understand the medical issues they face, and can communicate concerns and needs effectively.

When consumer health librarians engage in ethical reference transactions to meet patron information needs and improve health literacy, they are valuable assets to patients, who come to the library seeking useful health information. Such reference services are based on a trustworthy fiduciary relationship, focused on the individual needs of patrons, and designed to encourage wise health consumerism. To that end, the consumer health librarian strives not only provide useful and reliable information, but also, whenever possible, to help patrons understand the tools available to access information, effectively use those tools, and carefully evaluate the information they find. Whether patrons are seeking basic or very detailed information about the prevention, diagnosis, prognosis, or treatment in any health-related or medical topic, the consumer health librarian can facilitate a productive and useful search by walking through the process with them, explaining and clarifying each step along the way, and ultimately helping each patron to discover what works, and what doesn't work, for them. This kind of mediated searching and evaluation transaction places consumers in a better position to accept the limitations of health information provision, and begin to realize that as new information is gathered, it typically leads to more questions rather than definitive answers. Librarians can help consumers take those questions, and put them into a format effective for further searching. An incomplete question (or no question at all) typically leads to incomplete, useless, or off-topic information. This is true when seeking information online, conducting a reference interview, consulting with a medical professional, and even engaging in general conversation!

Consumer health librarians can also use their knowledge of health literature and expertise in questioning to facilitate effective doctor–patient communication when they help patrons discover their own purposeful questions. The consumer librarian can enable patrons to become their own advocates by helping them generate a list of specific questions based on what they really want to know from their health care providers. Patients who have identified their own information needs and can express their true concerns can then begin to effectively participate in their own medical care, and the care of their loved ones. Armed with usable information and new confidence, these patients will be ready the next time a doctor asks, "Do you have any questions?"

References

Ahmad, F., Hudak, P. L., Bercovitz, K., Hollenberg, E., & Levinson, W. (2006). Are physicians ready for patients with internet-based health information? *Journal of Medical Internet Research, 8*(3), e22.

Avery, N., Ghandi, J., & Keating, J. (2012). The 'Dr Google' phenomenon–missed appendicitis. *The New Zealand Medical Journal, 125*(1367), 135–137.

Benigeri, M., & Pluye, P. (2003). Shortcomings of health information on the internet. *Health Promotion International, 18*(4), 381–386.

Bunge, C. A. (1991). Ethics and the reference librarian. In F. W. Lancaster (Ed.), *Ethics and the librarian* (pp. 45–62). Urbana-Champaign, IL: University of Illinois, Graduate School of Library and Information Science.

Cohen, M. H. (2003). Complementary and integrative medical therapies, the FDA, and the NIH: definitions and regulation. *Dermatologic Therapy, 16*(2), 77–84.

Cowan, R. P. (2014). CAM in the real world: you may practice evidence-based medicine, but your patients don't. *Headache, 54*(6), 1097–1102.

Fox, S., & Duggan, M. (2013). *Health online.* Washington, DC: Pew Research Center, Retrieved from http://www.pewinternet.org/2013/01/15/health-online-2013/.

Goodyear-Smith, F., & Buetow, S. (2001). Power issues in the doctor-patient relationship. *Health Care Analysis, 9*(4), 449–462.

Ham, K., & Liebermann, J. (2012). *Consumer health reference interview and ethical issues.* Retrieved from http://nnlm.gov/outreach/consumer/ethics.html.

Jonas, W. B., Eisenberg, D., Hufford, D., & Crawford, C. (2013). The evolution of complementary and alternative medicine (CAM) in the USA over the last 20 years. *Forschende Komplementarmedizin, 20*(1), 65–72.

Joseph-Williams, N., Elwyn, G., & Edwards, A. (2014). Knowledge is not power for patients: a systematic review and thematic synthesis of patient-reported barriers and facilitators to shared decision making. *Patient Education and Counseling, 94*(3), 291–309.

Jutel, A. (2010). Self diagnosis: a discursive systematic review of the medical literature. *Journal of Participatory Medicine, 2*, e8. Retrieved from http://www.jopm.org/evidence/research/2010/09/15/self-diagnosis-a-discursive-systematic-review-of-the-medical-literature/.

Kaba, R., & Sooriakumaran, P. (2007). The evolution of the doctor-patient relationship. *International Journal of Surgery*, *5*(1), 57–65.
Kantor, M. (2009). The role of rigorous scientific evaluation in the use and practice of complementary and alternative medicine. *Journal of the American College of Radiology*, *6*(4), 254–262.
Keselman, A., Browne, A. C., & Kaufman, D. R. (2008). Consumer health information seeking as hypothesis testing. *Journal of the American Medical Informatics Association: JAMIA*, *15*(4), 484–495.
Lim, A., Cranswick, N., & South, M. (2011). Adverse events associated with the use of complementary and alternative medicine in children. *Archives of Disease in Childhood*, *96*(3), 297–300.
Longo, D. R., & Woolf, S. H. (2014). Rethinking the information priorities of patients. *JAMA*, *311*(18), 1857–1858.
McKinstry, B. (1992). Paternalism and the doctor-patient relationship in general practice. *The British Journal of General Practice*, *42*(361), 340–342.
National Center for Complementary and Alternative Medicine (NCCAM). (2011). *Introduction*. Retrieved from http://nccam.nih.gov/about/plans/2011/introduction.htm.
National Center for Complementary and Alternative Medicine (NCCAM). (2013). *Use of complementary and alternative medicine in the United States*. Retrieved from http://nccam.nih.gov/news/camstats/2007/camsurvey_fs1.htm.
National Heart, Lung, & Blood Institute. (2012). *Clinical trials*. Retrieved from http://www.nhlbi.nih.gov/health/health-topics/topics/clinicaltrials/.
National Institutes of Health. (2014). *NIH clinical research trials and you: The basics*. Retrieved from http://www.nih.gov/health/clinicaltrials/basics.htm.
Office of Disease Prevention & Health Promotion. (2005). *Plain language: A promising strategy for clearly communicating health information and improving health literacy* (issue brief). Washington, DC: U.S. Department of Health & Human Services.
Reference and User Services Association (RUSA). (2008). *Definitions of reference*. Retrieved from http://www.ala.org/rusa/resources/guidelines/definitionsreference.
Reference and User Services Association (RUSA). (2011). *Guidelines for behavioral performance of reference and information service providers*. Retrieved from http://www.ala.org/rusa/resources/guidelines/guidelinesbehavioral.
Reitz, J. M. (2004). *Dictionary for library and information science*. Westport, Conn: Libraries Unlimited. Retrieved from http://www.abc-clio.com/ODLIS/odlis_r.aspx.
Ruggie, M. (2005). Mainstreaming complementary therapies: new directions in health care. *Health Affairs*, *24*(4), 980–990.
Schembri, G., & Schober, P. (2009). The Internet as a diagnostic aid: the patients' perspective. *International Journal of STD & AIDS*, *20*(4), 231–233.
Steinhart, B. (2002). Patient autonomy: evolution of the doctor-patient relationship. *Haemophilia*, *8*(3), 441–446.
Sulmasy, D. P. (1992). Physicians, cost control, and ethics. *Annals of Internal Medicine*, *116*(11), 920–926.
Tarn, D. M., Guzman, J. R., Good, J. S., Wenger, N. S., Coulter, I. D., & Paterniti, D. A. (2014). Provider and patient expectations for dietary supplement discussions. *Journal of General Internal Medicine*, *29*(9), 1242–1249.
U.S. Department of Health & Human Services. (2014a). *Health literacy: Overview*. Retrieved May 21, 2014, from http://www.health.gov/communication/literacy/Default.asp#overview.
U.S. Department of Health & Human Services. (2014b). *Health literacy: Tools*. Retrieved May 21, 2014, from http://www.health.gov/communication/literacy/Default.asp#tools.

Ventola, C. L. (2010). Current issues regarding complementary and alternative medicine (CAM) in the United States: part 1: the widespread use of CAM and the need for better-informed health care professionals to provide patient counseling. *P&T: A Peer-Reviewed Journal for Formulary Management, 35*(8), 461–468.

Vohra, S., Brulotte, J., Le, C., Charrois, T., & Laeeque, H. (2009). Adverse events associated with paediatric use of complementary and alternative medicine: results of a Canadian paediatric surveillance program survey. *Paediatrics & Child Health, 14*(6), 385–387.

Wolters Kluwer. (2012). *Survey: Consumers show high degree of trust in online health information, report success in self-diagnosis*. Retrieved from http://www.wolterskluwerhealth.com/News/Pages/Survey-Consumers-Show-High-Degree-of-Trust-in-Online-Health-Information,-Report-Success-in-Self-Diagnosis–.aspx.

Yan, D., & Sengupta, J. (2013). Influence of base rate and case information on health-risk perceptions: a unified model of self-positivity and self-negativity. *Journal of Consumer Research, 39*(5), 931–946.

Consumer health information: the community college conundrum

Anne Chernaik
MSLIS, College of Lake County, Grayslake, IL, USA

Community colleges have a unique position in academic librarianship due to their dual obligations to their communities and academic programs. Little has been published in the professional literature of librarianship about community college libraries and health information. This chapter is informed by a combination of a literature review, the author's personal experience, and results from a small survey of community college librarians; the goal is to present community college perspectives on consumer health information.

Throughout the growth in health information-seeking behavior by consumers, libraries have increasingly served as centers for exploration and sites of information expertise (Cooper & Crum, 2013; Huang, 2006). Most notably, public libraries have served as centers for health collections directed toward general consumers (Snyder, Huber, & Wegmann, 2002) with libraries at colleges and universities (academic libraries) lagging somewhat behind due to the emphasis on faculty and students (Hollander, 2000). In 2000, Hollander outlined the questions often raised about academic libraries and the general public: are they included in the mission, allowed access, given assistance, or offered other services?

While often aggregated in discussions in the general category of "academic libraries," 2-year colleges, or community colleges, are more directly tied than their 4-year counterparts to the geographic area and population surrounding them. These colleges are comfortable with flexibility of service to a wider range of populations (Osika & Kaufman, 2012). Typical educational opportunities found at community colleges are highlighted in the box below.

Community Colleges: Educational Options

Transfer programs—Designed for students who plan to pursue advanced degrees at 4-year colleges and universities. Often, students complete the equivalent of the first 2 years of a baccalaureate degree at the community or junior college and then transfer to a college or university to complete a 4-year degree.

Career degrees—Designed to prepare students to work in a specific field immediately after graduation. The focus is on providing occupational skills for a technical or paraprofessional career. Students may still transfer to a college or university to complete a 4-year degree, but that is not an emphasis in the curriculum.

Developmental education programs—Designed for students who have been identified as needing additional coursework and skills to be successful in college-level courses, particularly in the areas of mathematics, reading, and writing. Courses do not count toward a degree.

Librarians at community colleges are constantly challenged to meet the health information needs of patrons seeking different kinds of education, possessing different levels of understanding and purpose, and often without any specialized training in the subject area. Consider for a moment the variety of patrons to be found in the community college library: residents who live or work in the county; students in vocational, adult education, or general liberal arts programs; and faculty and staff. Although a community college is an academic institution, it has a closer tie to the local community than even public universities. A community college is vital to the local community, both in urban and rural settings, serving as a center of culture as well as an educational institution reactive to the specific educational needs of its community. Community college libraries reflect this close relationship by offering resources and services to community members as well as enrolled students, and maintaining collections that reflect the unique diversity of the local population as a whole.

Differences based on this diversity can be seen even between branches of libraries which are located in different areas of the community college district. For example, one branch may serve a large Spanish-speaking urban population and have hospitals as the primary employer for the area. The courses offered at this location, and therefore the resources offered at the branch library, will differ greatly from another branch located in a rural area focused on farming.

Most professional literature available relating to health information provision, for consumers or otherwise, is either for a general audience across library types, or directed toward librarians at health institutions, public libraries, or 4-year colleges or universities. Materials written specifically for the community college library tend to focus on subject matters such as student learning and success, marketing, teaching, and other organizational challenges. Community college libraries are relatively underexplored in this domain. Who do we serve with health information, and what are their primary concerns? How are libraries dealing with staffing issues and maintaining quality and consistency in training? What access is provided to the public, and how are access issues addressed? How are collection development decisions made? What are different methods for organizing information to better serve the needs of each audience? How do librarians use our role as teachers to minimize confusion, given the variability of health literacy in our user communities?

This chapter specifically addresses these questions in relation to community college libraries. It examines the different populations that might approach librarians with health information questions and for what reasons. The chapter also discusses levels of access within a community college library to materials, electronic sources, and services by different patron types. It surveys staffing models as well as knowledge and training concerns as barriers to health information service. It points out key issues with collections including health literacy, acquisitions, and organization. Lastly, suggestions and strategies are offered to cope with identified barriers.

7.1 The community college setting

Community colleges are typically 2-year institutions dedicated to open access to higher education for their communities and committed to serving the unique needs and purposes of the areas and people they serve (Mullin & Phillippe, 2013). Such institutions can also be known as junior colleges, technical colleges, city colleges, or county colleges. They offer publicly funded educational opportunities designed specifically to meet the needs of their own communities. Typically, these schools offer vocational/occupational or career certificates, associate's degrees in various subjects, developmental coursework for students needing additional skills to become ready to succeed in college-level courses, and various personal enrichment and lifelong learning opportunities. Students may also enter the college on a transfer track with the goal of transferring credit to a 4-year college or university and earning a bachelor's degree. In 2014, the American Association of Community Colleges (AACC) reported a total number of 1132 community colleges across the United States. In 2011–2012, over 1 million associate's degrees were awarded (an increase of 71% from 10 years prior), with the largest percentages in the area of liberal arts and sciences, general studies, and humanities (33%), and health professions and related programs (21%) (U.S. Department of Education, National Center for Education Statistics, 2014).

Generally, community colleges have an open-door admissions policy, accepting all applicants who have earned a high school diploma, graduate equivalency diploma, can demonstrate the skills to succeed in the college, or are willing to take developmental courses to improve their skills in order to take college-level courses. Entrance exams such as the Scholastic Aptitude Test or American College Testing Exam are not required for admission but students may still be required to take school placement exams or provide evidence of college readiness in the areas of math, reading, and writing in order to take certain classes (Parsad & Lewis, 2003).

Cost for a college education at 2-year institutions is usually lower than at 4-year institutions. Maintaining a lower cost for tuition helps to keep educational opportunities accessible to greater numbers without leading to heavy educational debt. According to the U.S. Department of Education' National Center for Education Statistics (2013), average tuition, fees, and room and board rates for full-time undergraduate students in 2-year public institutions was 51% lower in the 2011–2012 academic year than their 4-year counterparts. Also, most community colleges in the United States are commuter schools, as only about 25% of institutions offer on-campus student housing (AACC, 2013).

However, the funding model for these community institutions is changing. The reduction in state, local, and federal funding combined has led to increased reliance on tuition. Unfortunately, the National Student Clearinghouse Research Center (2014) estimates enrollment decrease of 7.4% at 2-year public institutions between Spring 2011 and Spring 2014.

Declining enrollments lead to budget cuts at community colleges as a whole as well as in the library. The library must try to support the same number of programs and services with fewer fiscal resources. Similar to all public academic institutions, budgets are often highly dependent on the fiscal health of the area (Crookston & Hooks, 2012). For example, the AACC reported (2003) that the top three funding sources for public community colleges were state appropriations at 38%, tuition and fees at 20%, and local appropriations at 17%. The same analysis by the AACC in 2014 revealed the top three sources of revenue to be tuition at 29.5%, state appropriations at 28.1%, and local appropriations at 17.3%. Past dependency on public money to keep tuition low is changing for community colleges in the new economic environment, no less than for their 4-year university counterparts. The increasing reliance on tuition strains community college's ability to maintain their open-door mission. Loss of public funding could have lasting effects on the affordability and accessibility of community colleges.

7.1.1 Community college students

Student populations in a community college are more fluid than in other academic institutions as students more frequently change courses of study, miss terms, and may change back and forth from full time to part-time enrollment status (Crosta, 2014). These students also have variable reasons for enrolling, whether it be to transfer to a 4-year institution, train in a specific career certificate or associate's degree program, gain key skills to enter the workforce, participate in lifetime learning activities, or go through remedial courses to ready themselves for transfer (Miller, Pope, & Steinmann, 2005). In fact, Mullin (2012) argues that the very mission of accessibility to all in the community means that community colleges do not create a "student body" as seen in 4-year institutions.

In the fall semester of 2012, 12.8 million students were enrolled at community colleges across the United States, approximately 45% of all US undergraduates (AACC, 2014). Of those enrolled, 51% were White, 19% Hispanic, 14% were African-American, and 6% Asian/Pacific Islander. The mix of students includes those straight out of high school, career changers, lifelong learners, and first generation college students. Some may already have higher education degrees and are returning to take another career path, and others are attending college for the first time. Students also tend to be older than the "traditional" undergraduate ages of 18–22; their average age is 28, and 57% of enrolled students are between 22 and 39 (AACC, 2014).

It is difficult to judge the patron base of any community college based solely on general medical statistics given how closely the college is tied to individual communities. Understanding the population served by a community college means looking beyond the simple numbers to the actual educational needs of the area. The enrollment numbers also can never capture the numbers of the public who use the college either for library services, theater programs, workforce development, community gardens, art shows, or any number of other programs offered.

7.1.2 The community college library

The community college library, like all academic libraries, is very closely tied to the mission of its parent institution. The purpose of these libraries, as compared to those at most 4-year institutions, is more focused on the information needs of the students and supporting teaching needs of faculty. Teaching is emphasized over research, and faculty often carry more teaching hours as part of their loads than those at other institutions (Mullin & Phillippe, 2013). Thus, community college collections typically do not have research faculty and graduate students to support. In fact, while the nature of community college libraries can be described as academic, it is closely aligned with the public library in its open-door policy and focus on local needs (Evans & Saponaro, 2005).

When you walk into a community college library you can expect to find a mixture of physical materials: books, magazines and journals, audio-visual items, and even microfilm. Increasingly, with the emphasis on electronic sources such as databases and e-books, you'll also find computers, media rooms, and charging stations. Materials have broad coverage and are at the appropriate level for first- and second-year college students. Collections, whether physical or digital, are always curriculum-focused and reflect the programs offered at the college. Transfer programs require a more general selection of materials; career degree programs in which students expect to be stepping directly into the workforce require information that is highly specific and up-to-date; and developmental students working on college readiness need items that support their reading interests and levels. A brief tour of the library should give the visitor an idea of what types of courses and degrees are offered. For example, you might find an entire section of legal books for a paralegal program, a robust collection of children's books for early childhood education courses, automotive pamphlets and materials for the automotive technician certificate, a "bone room" with anatomical models for anatomy and physiology courses, and even collections of rock specimens for geology.

Material collections also serve an interesting mix of personal needs and interests. You might find best-selling fiction and nonfiction, graphic novels, self-help books, and cookbooks. While community college libraries collect items that fit the needs of the curriculum, there is also a strong desire to serve the personal needs of both the primary patron base, students, faculty, and staff, and the secondary patron base, residents of surrounding communities.

While there is too much variety in organizational structure to permit a standardized depiction of the community college library, all libraries of this type share some similar characteristics of purpose and service. Staffs are mainly small, and may include a combination of student workers, clerical staff, paraprofessionals, librarians, and management. Librarians are required to hold a master's degree in library science, are usually hired as generalists rather than for the needs of a specific academic subject specialty (i.e., history or biology); in some institutions, the librarians are also faculty members. Community colleges may also have multiple campuses serving different areas of their districts. Physical libraries may or may not exist at each of these branches; and various library functions such as research help may be limited or unavailable to students at those sites.

7.2 Health information needs at the community college

There are many variations in patrons to be found in the community college library, any one of which might have a health information need: nursing and allied health students and faculty, students in transfer or career programs unrelated to medical or life sciences, or the general public, including residents who live or work in the district serving a community college.

7.2.1 Medical academic

Many community college libraries cater to specialized curricula, such as nursing and allied health, in addition to general transfer programs (Ennis & Mitchell, 2010). These programs include associate's degrees and/or certificates in dental hygiene, emergency medical technology, health information technology, massage therapy, medical assisting, medical imaging, nursing, phlebotomy, surgical technology, radiation oncology, and pharmacy. In 2012, 21% of all associate's degrees awarded by postsecondary schools in the United States were in the health professions and related programs (U.S. Department of Education, National Center for Education Statistics, 2014).

The medical academic student and instructor requires materials that contain information far beyond what the typical layperson would need to understand. This content often relies on large amounts of technical jargon, has a complicated writing style, and assumes the audience already has knowledge of the subjects. Curricula for medical and allied health programs are driven by accrediting bodies which require library resources of sufficient academic quality to meet the standards of their organization. For example, the Accreditation Commission for Education in Nursing requires a full 60-min conference with a librarian during accreditation site visits; accreditors take a tour of the library and evaluate the adequacy and relevance of learning resources. For academic purposes, nursing students regularly need to have access to and use books such as *Chronic Disease Epidemiology and Control*, rather than *High Blood Pressure for Dummies*. While both books are worthwhile additions to a collection, each has a distinctly different audience.

Community college librarians face a wide range of health information questions from both students and faculty. These run the gamut from an overview of a disease, to creation of care plans, to an evidence-based medicine article on standards in tracheostomy care. Patrons working toward specific medical academic purposes require a level of service that necessitates greater understanding by the information professional, not only of the medical resources available, but a general understanding of medical terminology. Librarians at community colleges also need to have a solid understanding of curricular program requirements and the roles of allied health professionals in the workplace. Nursing students do not require the same resources as does a medical student; the tasks and knowledge needed to succeed in those careers are different.

7.2.2 Nonmedical academic

A community college is host to myriad numbers of vocational, career, and transfer courses requiring library services. Even students taking courses outside the clinical medical fields will need help navigating health information for class projects. For example, students enrolled in English or Speech courses may approach the librarian for health information on side effects of drugs for projects on legalization of marijuana, medical consequences to environmental disasters, or even ownership rights for stem cells. Often, these courses allow students to select their own topic of interest. Approaching the assignment from a medical perspective may serve two purposes—completing a required component of a course, and exploring a topic of personal interest to the student.

In many ways, the nonmedical academic population needs many of the same resources and services as does the general public. Even though the student's questions tend to arise from an assigned paper or project, they nevertheless often want resources that might be designated in professional librarian practitioner literature as "consumer health information" or "popular works." These materials provide broad overviews of a health topic, and are written in plain language that a layperson can understand without highly technical jargon often found in scholarly medical resources. Many times, interactions require some of the same questions and skills that one might use with a community member coming in for a personal health reason.

It might be easy to assume that a student asking for medical information for a class assignment needs to use the same resources as those enrolled in nursing and allied health programs. However, it is essential to determine first what class the assignment is for and the student's current understanding of medical terminology. While some students may indeed be comfortable utilizing highly technical resources, others may need a more general introduction to a health topic. In fact, the choice of assignment may even have arisen due to personal circumstances or family illness, requiring librarians to proceed with compassion and delicacy while balancing the need to analyze and evaluate the information for an academic purpose.

7.2.3 Consumer health

Health-care consumers in the general public use libraries as centers for gathering information, whether out of general interest or because of personal concerns. While the primary purpose of the community college library may be the academic needs of students and faculty, it contains a wealth of information that also addresses the questions of health consumers.

7.3 Issues in health information provision

A survey was sent in May 2014 to the listserv for the Community and Junior College Libraries Section (CJCLS) of the Association of College and Research Libraries, a division of the American Libraries Association, the predominant professional library association. The CJC-L listserv focuses on discussions of issues, concerns, and services of community and junior college libraries (CJCLS, 2010). There are approximately 1640 subscribers to this open, unmoderated listserv™. Members were asked several open-ended questions regarding health information services, collections, and access at their libraries (see Appendix A). Sixteen librarians responded to the survey, which was active for 2 weeks. Respondents' comments are cited as CJCLS Survey in the discussion below.

7.3.1 Staffing: organization and quality

Personnel organization within the community college library, as it relates to health information, is as variable as the institutions themselves. Neither survey responses nor the published literature on community colleges and health information reveals a typical pattern. Cooper and Crum (2013) indicate that some librarians begin as generalists, or informal contacts, for various departments, and then slowly develop a formal liaison role defined as a primary contact between the library and health programs, which is very like my own personal experience. Respondents (CJCLS Survey) did mention that various staff led efforts in the health information area; some were named as liaisons to health science programs, and some employees were specifically designated as health science librarians or even academic program librarians or library directors. One very lucky library actually has a health sciences campus where all of the librarians on staff are designated as health sciences librarians. A cohort of health sciences librarians are likely to have the knowledge and training to provide service to both the academic students and general populace.

Why is this important? According to the Library Statistics Program (2012), there are on average four librarians at 2-year public colleges across the US, with one librarian and other professional staff per 1000 FTE (full-time equivalent) students. Crumpton and Bird (2013) note that there are very few or even only one librarian at some community colleges, and these librarians have a broader spectrum of responsibilities than peers at 4-year institutions who can more easily specialize in any subject area, such as health. There is a trend toward generalists, and even primary contacts can lack training in the subject area. This can lead to many potential problems when called on to serve patrons looking for health information.

> Consider this scenario: a patron approaches the desk and specifically asks for a nursing drug reference book that lists the street names and side effects of drugs. The librarian then supplies the student with exactly what was asked for—a nursing drug reference book. Extremely dissatisfied, the patron insists the item was not what she asked for. With both librarian and student frustrated and not sure which direction to go in, the librarian contacts the health sciences librarian. When the student is questioned about her need for drug street names and side effects, she reveals she is not an allied health student, but writing a paper for her human services and addictions course on uppers and downers. From there the health sciences librarian is able to find exactly what the patron really wanted—a resource on the physical and medical effects of psychoactive drugs.

There was nothing "incorrect" about the initial reference interaction described in this scenario. The key was discovering not the particular *type* of information, but the particular *purpose* of the information. It is sometimes our job in consumer health to turn a search for a highly technical, professional term into a general search, or in this case replace "drug names" with "uppers and downers" (Ennis & Mitchell, 2010). Often, nonclinical students need to start with a more basic overview to build a proper paper on a disease, condition, etc. and then move to the more academic materials found in our nursing resources.

Beyond the physical reference and informational services offered by staff, there are virtual services available through chat, e-mail, or in the online classroom. "Embedded" librarians in academic settings, including community college settings, are staff dedicated to work with specific courses, often within the virtual classroom. These librarians can also encounter health questions that may be for academic assignments, but still need to be treated in a more personal way.

> For example, one student contacted a librarian via the discussion board with a question about heart attacks. Given the same interaction face-to-face, the librarian might have discovered through gentle questioning and interviewing that the student was interested in writing her English paper on the subject because she lost her father to heart disease. The same information would be very difficult to determine in an online environment in which conversations are readable by other students. This student might actually need scholarly materials for the assignment, complemented by consumer health resources for the personal aspect.

Generalist librarians confront the stresses of maintaining familiarity with several collections and understanding the needs of patrons. With limited staff, many tasks, and multiple subject areas to familiarize themselves with, it is easy to focus attention on the scholarly materials in health information. Yet Smith (2008) notes the need for providers of health information to have public librarians' customer service skills, coupled with academic librarians' knowledge of health resources. This is doubly important for the community college setting where collections and the patron base can mimic patrons of both academic and public libraries simultaneously. The average community college library has at least one branch campus (Library Statistics Program, 2012). How do we ensure that all community college librarians have the training, knowledge, ability, and confidence to serve consumer health patrons? Can we create a model of service in which all librarians on staff feel comfortable working with health information patrons, or does the very nature of the information require a subject specialist?

7.3.2 Access: users and resources

Providing access to knowledge plays a predominant part in the mission of any library. However, some libraries restrict usage of library facilities, resources, electronic information, and services to certain people. Libraries in the K-12 public school system often restrict access to students, teachers, and administration. Public libraries may be open for anyone to enter, but borrowing of materials is limited to patrons in the district.

The general public is often unclear about whether they can even use the library at a college or university, and to which specific resources they have access. Hollander (2000, 2003) found that most health sciences libraries actually do offer at least some level of service to the general public including reference, circulation of materials, and use of library computers, but lack the staff, money, and time to offer full service to those outside of the college or university who might want to use the library. The public does seem more likely to visit publicly funded academic libraries than private institutions. The particular challenge for community colleges is the very close ties to the local areas and the feeling of ownership people have about "their" college.

The community college library is commonly open to the general public as well as the students, faculty, and staff at the college. Materials are, of course, available to current employees and students. Circulation of materials to the public, however, is highly variable depending on the institution. Seventy-five percent (12) of respondents to the survey (CJCLS Survey) indicated the public is able to check out materials at their community college libraries, either through use of a special community card, or through an agreement with local public libraries. The remaining 25% (4) did not allow nonstudents to check out items, but three of these four libraries were open to the public and made library materials available for unlimited use within the library itself.

If community colleges are spread across multiple campuses, books and other physical items may only be available at certain locations, since many of the branch libraries do not contain the breadth of materials found at the main campus library. Patrons using these branch libraries have two choices: they can limit themselves to the materials physically located at that library or they can go to a different campus, something that is not always an option for students who have limited ability to commute.

Advanced library users may know that they can have items sent from another location to the campus for pickup, but that process takes time and knowledge of the library's online catalog system. Many community college students are not prepared to effectively use the library and are not yet comfortable with the environment. Students may not yet have the skills necessary to attend college or, as older students whose education may have been interrupted, may be overwhelmed by the changes in how libraries work since they last attended school. The potentially sensitive nature of health information items is another issue. Students may be reticent to request stigmatizing or embarrassing items from another location, especially knowing the item will be handled by several library staff members and will often be tagged with the patron's name.

> For example, an older gentleman enters the small library at a branch campus of the local community college. A little reserved, he refuses help from the librarian and goes to the computer used for looking up items in the online catalog. He types slowly and is obviously having difficulty using the online system. He is even more nervous as additional people enter the library. After seeing his increasing frustration, the librarian again approaches him. After a few moments he looks around and whispers that he's looking for a book on male incontinence. The librarian shows him how to use the online catalog using keywords to find materials the library has access to or owns. After a quick look the librarian finds a few books that would work but they are all available at another campus. She then offers to help him look on consumer health-oriented Web sites for general works giving the basics, treatment options, etc. but he insists that he only wants a book. When she offers to have items sent from the main library to the branch for him to pickup he refuses. He does not want all those library workers to know about his problem. He asks for directions to the main library and leaves.

Consider also that many consumer health sources purchased by community colleges, or in fact many libraries, are materials such as sourcebooks, encyclopedias, and other reference materials. Depending on library policy and how the books are cataloged, these items may or may not be available for circulation. Community colleges are in a unique position when it comes to selecting their resources for readability and scholarly level of the items collected. Reference materials can often fall into the midrange, providing a solid scholarly background to medical information, yet are not too advanced for the knowledgeable consumer or a beginning health programs student to understand. While the purpose of the circulation restriction may be to ensure availability of the reference resources for everyone, such a policy may actually prove to be a barrier to its use.

Finally, access to physical materials in the library does not take into account the outside commitments that much of the community college population face. According to the AACC (2014), 22% of students enrolled as full-time students and 41% enrolled part-time at community colleges were also employed full time during the 2011–2012 school year. An additional 40% of full-time students and 32% of part-time students work at least part-time. Miller (2010) also reports that 27% of enrolled students are parents and 16% are single parents—double the rate of their counterparts at 4-year institutions. Most community college students are also commuting to class rather than living on campus. With such limits on time, a visit to the library for any reason may not be feasible. How then, with limited student time on campus, and limited familiarity with library resources, do we ensure that our patrons are able to find and use materials for their health information needs?

7.3.3 Electronic resources

Electronic resources, whether freely available on the Web or within licensed database products, play an important role in providing consumer health information. Library terminals are available at most community college libraries for access to electronic resources. All but two library survey respondents (CJCLS Survey) report that they provide computer terminals and/or on-campus database access to the general public. Five responding libraries require guest passwords for log-in access; others are completely open and available. A patron affiliated with the college has a significant advantage in *off-campus* accessibility of electronic resources.

As is typical in academic libraries, however, licensing agreements limit database access from off-campus to currently enrolled students and college faculty and staff. Access may also be limited by personal resources. According to the Pew Research Center, only about 67% of community college students own a desktop computer, 70% own a laptop, and 78% have broadband Internet access at home (Smith, Rainie, & Zickuhr, 2011). Community college librarians can make any number of excellent sources available to students, but if the students are unaware that resources exist, or are unable to navigate the search interface, the information remains out of reach. How do we publicize and educate our patrons about what the library has available?

7.3.4 Library services

Services offered to those not affiliated with community colleges are highly variable. All 16 libraries surveyed (CJCLS Survey) allow the public to use the library, but place different limits on the services offered. Eleven of the sixteen give community patrons the ability to check out materials for home use, usually through reciprocal agreements with local public libraries or through use of a community card. Reference services in person are likely to be open to any and all who approach, sometimes simply because librarians do not ask patrons for identification before we help them. Access to in-depth research consultations, one-on-one appointments, document delivery and interlibrary loan, and virtual services tend to be determined differently by each library. Unfortunately, while we may desire to expand services and access to provide everything to everyone, budgets and other considerations have placed this outside of our capability (Borman & McKenzie, 2005).

7.3.5 Acquisitions and collection development

The primary guidelines for selection at community college libraries are the specific needs of the curriculum and programs at the college, combined with an understanding of the literacy and scholastic levels of the students. For the medical and related subject areas, selection must also consider any specific standards of accreditation committees, because many allied health programs require accreditation by a national organization.

For example, a medical imaging program is designed to prepare students for a career as a radiologic technologist. This program may be nationally accredited by the Joint Review Committee on Education in Radiologic Technology (JRCERT) which requires a review of learning resources in support of student learning during reviews of accreditation (JRCERT, 2014). Nursing programs may be accredited by the Accreditation Commission for Education in Nursing (ACEN). ACEN (2013) emphasizes quality, currency, accessibility and requires that resources ensure students meet the objectives of educational units. There is a close connection between program faculty and librarians in ensuring that library materials fit the college's educational objectives.

When clinical programmatic faculty are closely involved in the maintenance of a medical collection, there is potential for restrictions on purchase of general health materials, and texts on alternative medicine, as being too basic or inappropriate for students (Baker & Manbeck, 2002). Yet a full and vibrant collection must not only teach mammography to radiologists, but also describe what happens during a mammography for readers who are worried patients or family members. Given the dual nature of a community college library, and the wide range of purposes for visiting the library, some degree of balance must be reached between the highly academic and more popularly accessible items. When a collection is broad in scope, it provides patrons with basic health information as well as more technical and detailed information in one place—something public library patrons would normally need a hospital or health sciences library to obtain (Baker & Wilson, 1996).

Further problems arise in deciding which sources and materials are appropriate. Wessel, Wozar, and Epstein (2003) found that even public librarians often worried about exactly what constitutes a core general health collection. Collection librarians are the staff in charge of deciding which materials should be added to the library. They may be more comfortable using selection tools such as book reviews, recommended lists, and vendor catalogs geared toward scholars rather than a general-interest readership. Nursing materials, a staple for many community college libraries, often provide consumers an excellent option for comprehensible scholarly sources (Baker & Manbeck, 2002). The Medical Library Association's (MLA) Consumer and Patient Health Information Section (CAPHIS) provides a wonderful series of resources including classic titles and recommendations for materials in foreign languages: http://caphis.mlanet.org/chis/collection.html.

This is a very real problem at any institution looking to provide a basic health collection to their patrons. In the building of any popular and/or consumer health-oriented material collection which also serves an academic audience, it is essential to ensure and promote the quality of information. This is not to say that the materials must endorse only current medical establishment theories or ideas, but instead that collections avoid outdated or potentially harmful material. Quality and well-defined selection criteria are essential. In the community college setting, librarians must be ready to provide accrediting bodies on site visits with the logical reasoning for any popular works included in the medical collection.

> For example, during a typical accreditation visit, the librarian is required to give the site visit team a packet of information outlining all health and medical resources; describe the method of selection and cooperation with the subject faculty; and physically show the team the book stacks where the collection resides. During the tour, one member pulls a consumer health item out and asks why it was included on the shelf, given it was not appropriate for students in their program.

Survey respondents (CJCLS Survey) indicated a wide mix of policies regarding the purchase of medical and health resources. In general, a majority of the community college budget is spent on books supporting allied health curricula. For those librarians who do add popular works to the collection—only three indicated they did not—most purchase consumer health resources specifically targeting other courses within the college, such as English, Speech, and health and wellness, or consider specific assignments by various departments. One respondent did mention the use of grant funds for collection of popular health materials, but collection budgets for noncurricular materials is often an issue that cannot be fixed by librarians themselves (Borman & McKenzie, 2005). It can be of little surprise that community college libraries, like many other library types, can simply not afford to expand consumer health collections (Benedetti, 2002), especially since there is often less than a third of the general funding to spend on students as private research universities (Mullin & Phillippe, 2013).

7.3.6 Organization of materials

All respondents to the survey (CJCLS Survey) reported that their consumer health and medical books are fully integrated into one collection, rather than housed separately. Most indicated the materials were cataloged and shelved by Library of Congress Classification "R," the class for medicine. Baker and Manbeck (2002) suggest that integrated collections limit the capability to browse, but it appears community colleges prefer to have their health materials in one place to allow all patrons to find both consumer-oriented and professional materials on similar subjects (Hollander, 2003). With the exception of reference materials designated for in-library use only, and course reserves materials attached to certain classes, many community college libraries try to keep most materials located together. This ensures that students looking for materials are able to locate one item via the online catalog and browse materials on a similar subject on the shelf nearby.

Unfortunately, fully integrating popular health information and academic collections can create problems finding books and other materials at the appropriate level for a patron's purpose. Readability scales are not built into any description of these items; it is very easy for nonmedical academic searchers and the typical member of the general public to select a title that is beyond their capability to fully understand. Conversely, nursing and allied health students may easily select materials that are not rigorous enough for their assignments.

7.4 Health literacy in the community college setting

7.4.1 Patrons

Basic college readiness and literacy is a serious concern for community colleges. Approximately 60% of students enrolling in community colleges needed to take at least one developmental course in math or reading/writing, over a period of 3 years, in order to develop the skills to complete entry-level courses (Bailey, 2009). Studies have indicated that these students have less success than peers who do not have to complete developmental coursework (Russell, 2008). Librarians at community colleges need to understand the different needs and abilities of their patrons in order to provide the best service. A reference encounter presuming that person asking a question is necessarily a college-ready student in search of scholarly articles for a research assignment could lead to a very dissatisfied patron—and would completely disregard the characteristics of the community college.

It is not enough, however, to determine the basic literacy or course level of a student when looking at health information. The US Institute of Medicine's report *Health Literacy: A Prescription to End Confusion* revealed that even the highly educated may not be able to fully understand and process the technical jargon of the medical field (Nielsen-Bohlman, Panzer, & Kindig, 2004). For our purposes, it is best to use the MLA's definition of health literacy as the ability to recognize a need, find sources, evaluate the data, and use the information in a productive manner (2015). Many students and consumers alike may feel comfortable enough to search the Web on their own, but McInnes and Haglund's (2011) study on the readability of health information online indicates that much of the supposed consumer health information on the Web is not understandable to people with low levels of basic literacy. Baker and Wilson

(1996) mention that those who find or are provided information above their own reading level will not only be unable to understand what they are given but are likely to become even more confused. In general, our world does not lack for information but instead has too much of it to process, health or otherwise.

7.4.2 Librarians

Because community college libraries tend to have smaller staffs than other academic libraries, librarians have a wider range of responsibilities. The generalist nature of community college libraries means that most time and energy is spent in maintaining broad, not subject-specific, knowledge in curricular areas of the college. Full-time and part-time librarians may thus feel inexperienced and unqualified to respond to health questions (Kouame, Harris, & Murray, 2005; Wessel et al., 2003).

> For example, a student asks for help finding information on "parental nutrition." Assuming she is looking for information on either what pregnant mothers should eat or how parents' food choices affect their children, the librarian asks further questions to help narrow down appropriate resources. While the patron grows increasingly confused over the questions she is now being asked, the librarian struggles to determine how to help. He takes a moment to think and asks the student what she is learning about that led her to ask for more information. She responds that she is interested in learning more about feeding tubes and heard the phrase "parental nutrition." Finally understanding the problem in communication is one of terminology, he asks if she might mean "parenteral nutrition," that is, feeding delivered intravenously. Looking quite relieved, the patron says "Yes" and the librarian is then able to direct her to appropriate materials.

The patron eventually received the information she wanted. The reason for the success of the interaction was the librarian's recognition that the word "parental" was being used instead of "parenteral." Misspellings and bad pronunciation are not uncommon with health questions (Borman & McKenzie, 2005). A solid knowledge of medical terminology significantly raises the chances of a positive outcome for the question. Luckily, search tools are constantly improving in their ability to help with predictable mistakes. A Google search for the phrase "parental nutrition" will actually yield results for "parenteral nutrition." Unfortunately, this is still not a solution to the root problem, and a lack of training continues to be of concern for all library types.

7.5 The future for community colleges and health information

Kirkwood and Riegelman (2011) argue that the community college will play an essential role in health education of the public in the next 10 years through the integration of health objectives into general education offerings. How can the community college library better help in this potential growth area? Can we serve consumer health information-seekers, whether academic or not?

The first positive step in this direction occurs when there is full agreement by the library staff and administration to actively provide that level of health information service to the community as a whole. The practical limitations imposed by budgets, staff size, time, comfort level with health information, and facilities could easily lead to a passive "solution": Provision of the most basic level of resources and services, while referring all needs beyond the basic level to another library. A decision to whole-heartedly pursue the objective of service in consumer health is not to be made lightly. It should be determined with the full awareness and potential problems—as well as an understanding of the steps that need to be taken in order to ensure the quality in service.

Active outreach in health and wellness to the community may be an appropriate goal for the library. But we must consider the current climate and vision of the community college as a whole as well as the very real budget, staff, and time limitations that may stop such projects (Duesing & Near, 2004). Does the college administration support such efforts? Does the community need a center for consumer health information, or are other institutions already effectively meeting this need? Does a consumer health information project align with the strategic goals of the college? It will be much easier to argue for support of consumer health services if the project objectives link to the goals of the college. Plan for budget expenditures, and gather statistical data from circulation of popular works in health, reference interactions, and attendance at related library workshops. Visit departmental and divisional meetings of any nursing and allied health programs that could provide insight, volunteers, or letters of support. Meet with student government groups for their input, ideas, and support. A proposal with solid support from constituents of the college and clear objectives will make the project more feasible for the administration to consider.

Once a decision has been made to move forward, it is time to consider how to answer some of the questions raised within this chapter:

- Who are the primary populations your library serves, and what are their major health concerns?
- How could you deal with staffing issues and maintaining quality and consistency in training?
- What type of access should you provide to the public and how are any issues to be addressed?
- How does the library ensure that patrons are able to find and use materials for their consumer health information needs?
- How are collection development decisions for popular works, versus academic medical sources, going to be made?
- Is there a different method for organizing information that can better serve the needs of each audience?
- What training is needed to ensure that all librarians have the knowledge, ability, and confidence to serve consumer health patrons? Could or should a generalist be expected to be able to keep up with consumer health information provision training?
- Should there be a marketing campaign to the local community to publicize new or additional services in the area of consumer health?

7.5.1 Staff training

Maintaining quality of information services depends on the knowledge and confidence of reference librarians and the ability to conduct in-depth and sensitive reference interviews that use neutral questions to determine literacy (Kefalides, 1999) and prior knowledge of the health subject matter (Baker & Wilson, 1996). Generalist librarians may feel unprepared to handle consumer health questions without training in resources and techniques for establishing a respectful interaction with patrons who may be reticent to share their information needs. One solution may be to refer all consumer patrons to designated librarians, but such a model is not feasible in libraries with very small staffs or who rely on part-time librarians during late evening and weekend hours.

Developing training and resources for all librarians, however, brings with it some other considerations. One option is to designate a day, either on a weekend or a time when the library is closed, where all librarians can come together for a training session. One librarian, ideally the health sciences liaison or someone with experience in the health sciences, would be in charge of organizing the training session. Research indicates that consumer health seminars that incorporate training with discussion, time to use new skills, and questions and answer sessions were considered successful by the attendees (Snyder et al., 2002; Younghee, 2013). Snyder et al. (2002) describe a particular training method that combines exposure to reference tools, databases, and books with a number of suggested Web tools.

For community college libraries without access to health science librarians, the staff of the National Network of Libraries of Medicine (2013) are sometimes available for members of the organization (a free membership) to give a presentation free of charge at training events. Members also have access through their Web site (http://nnlm.gov) to a rich resource of handouts, educational resources, and training opportunities from various institutions such as the National Library of Medicine (NLM), the National Center for Biotechnology Information, the MLA, and the National Institutes of Health. In fact, the Greater Midwest Region of the National Network of Libraries of Medicine (2014) recently offered two Community College Library Awards to provide training and professional development workshops, promote the use of health resources in curriculum, or develop health information instruction for students.

Of course, many libraries do not have the ability to set aside a full day for everyone to participate in this type of training. Libraries with access to tools such as LibGuides by Springshare, a content management system that allows users to create online subject and topic guides to information, can create a guide to resources with links to tutorials, databases, and more that can be used to guide individualized training which staff can work through on their own time. Examples of these resources include:

- *Understanding Medical Words* http://www.nlm.nih.gov/medlineplus/medicalwords.html
- *Health Information on the Web* http://nnlm.gov/hip/
- *Interactive Health Tutorials* http://www.nlm.nih.gov/medlineplus/tutorial.html

Another excellent resource is the CAPHIS, a section of the MLA focused on librarians looking to serve the information needs of patients and consumers. Anyone, MLA/CAPHIS members or not, can find quality information on funding, collection development, literacy, patient education, reference, recommended Web sites, and more on the section Web site (http://caphis.mlanet.org/).

7.5.2 Access

Even without major changes in library policies, there are small ways in which community college libraries can encourage the use of the consumer health materials within the library. Use marketing pieces to publicize materials and services and make the college and surrounding community aware of the community college library as a source for quality health information. Libraries with limited budgets for marketing can again look to the National Network of Libraries of Medicine (NN/LM), a US organization devoted to improving access to health information. Members have access to materials and a collaborative network of libraries and information centers. Other small steps can include marketing through social media and personal visits with faculty around the college to spread the word among their students.

7.5.3 Materials

The community college library can offer health-care consumers a single place to find popular works, general overviews of health topics, as well as more scholarly resources meant for nurses and other clinicians. The first step is to review material usage rules for the public. It can be very difficult to change circulation policies for books, even if you wish to move from limited to open circulation for all patrons. Material checkout limits are often based on problematic situations from the past that led to loss of materials. Arguing for change when materials are generally open for in-library use may be difficult. Have a frank discussion with the different departments in the library including circulation, administration, and reference to determine the reasons behind current policies. Stress the benefits to the community and the problems with placing barriers to material usage. Six-Means (2011) specifically notes the positive effects of circulation of consumer health materials to allow a patron more time to come to a decision on important health concerns.

Sharing among campuses is limited by the method of delivery between libraries. Many rely on van services either run by the college or a local consortia of libraries. Van delivery and pickup is limited to specific days and times so if a transfer request comes into the library after the van pickup occurs, the patron must wait for the next pickup. Branch libraries can choose to maintain a small core collection of general information if there is room in the budget.

Duesing and Near (2004) suggest developing partnerships in order to help ease some of the problems with limited access and time constraints. Community college districts often have several different public libraries and health science centers within their area. Meeting with these constituents and developing resource-sharing and visiting agreements could prove highly beneficial in increasing the amount of material and expertise available to patrons.

7.5.4 Electronic resources

Vendors for licensed databases, containing journal articles published in journals and trade publications not freely available on the Web, limit usage (especially off-campus access) of products by those not affiliated with the college. Computer terminals may require log-ins and have restrictions on public use. If log-ins are required, libraries need to ensure that guest passes are available to those who want to use the library and are not currently registered students. Often even students who are taking a semester off will not have a working log-in. Such passes could be daily or time limited. Printing and copying services also need to be addressed. For institutions with print-management systems, patrons need to be encouraged to purchase print cards or simply send the information to themselves electronically. Occasionally administration is concerned about noncollege patrons using scarce technology resources. It may be possible to designate one or a few public-access terminals to serve general patrons.

Another solution to improving access to materials in this age of digital information is to purchase resources in e-Book format so the information is available to library users across campuses or even at home. Yet, low availability of electronic books in recommended health sciences titles limits options in this format (Husted & Czechowski, 2012).

7.5.5 Services

The key to quality consumer health information service is agreement between the librarians and library administration on the parameters of services to be offered. Each service needs to be highly individualized to the needs, community, and policies of each college. Is the library staff comfortable with agreeing to be available for all health information questions, regardless of patron purpose or level of need? Do the librarians have the necessary training to serve in that capacity? Is there a dedicated health sciences or consumer health librarian? Could that person have set hours where they are available to answer consumer health questions, either as drop-in or via appointment? Partnering with the health clinic and allied health nurses is one way to develop peer contacts.

There needs to be some consideration of the needs of this special population. Community college students may be unable to come into the library due to school and family obligations, transportation issues, or because the on-campus offerings take place at times they cannot attend. Librarians could easily create tutorials or short recorded presentations on Web and database resources for the library Web site. Online pathfinders and topic guides are also excellent tools to help both librarians and patrons navigate through the medical collection for consumer health information. For example, one online consumer health information content guide can host several tutorials on medical terminology, how-to media clips for using electronic resources, links to sample catalog searches and subjects for consumer health, examples of physical and virtual materials that might be useful, and contact information for further help.

7.5.6 Collections

The creation of a collection development policy for consumer health information and popular works is an essential step in moving forward with a consumer health information service. Policies provide a solid foundation for determining what and when consumer health information will be purchased, as well as how these materials fit in with the health sciences collection as a whole. When a consumer health liaison or health sciences librarian does not do collection development, the policy allows for input into what the collection should look like while removing the need to be involved in each individual decision. Finally, a written policy clearly communicates to patrons, faculty, and accrediting bodies the purpose of collecting works written for the layperson, in addition to the more scholarly materials for allied health programs.

Intellectual access to the collection is also an important consideration. Work with appropriate technical services staff to brainstorm methods of identifying consumer materials within the online catalog. Some libraries use the "popular works" identifier in the online catalog (Hollander, 2000). Librarians may also consider tagging physical items with a "consumer health" sticker to differentiate from the more scholarly works.

7.5.7 Health literacy instruction

Health literacy is a challenge that community college libraries cannot solve alone. Nevertheless, as teaching institutions, we must be concerned about increasing the ability of all consumers researching health information not only to search generally, but to assess the quality of what they find. Consumers need to be educated on appropriate search terminology and shown how to effectively find and evaluate reliable resources (McInnes & Haglund, 2011; Younghee, 2013). Health information instruction can have a real impact on the surrounding communities (Wessel et al., 2003), especially when offered in a central location that is a trusted educational institution open to all. Reaching out to college health professionals such as clinic staff and health sciences faculty when developing instructional sessions is a natural expression of partnership and can increase the effectiveness of such workshops (Hallyburton, Kolenbrander, & Robertson, 2008). Core health information offerings can include topics and tools such as:

- **Evaluating health information on the Web**—Patrons should know how to use basic questions to evaluate whether a health Web site is accurate and helpful. Workshop attendees can learn to look at things such as Web site authorship, intended audience, frequency of Web site updates, and whether the site promotes fact or opinion.
- **MedlinePlus** (http://www.nlm.nih.gov/medlineplus/)—Developed by the US NLM, this Web site is specifically designed for health consumers, patients, families, and caregivers. Overviews of health topics, interactive tutorials, and drug information are just a few of the invaluable tools available at MedlinePlus. Much of the site is also available in Spanish.
- **Top 100 List: Health Web sites You Can Trust** (http://caphis.mlanet.org/consumer/)—This list, curated and maintained by the CAPHIS of the MLA, provides consumers and librarians with general information on a variety of health topics, ranging from senior health to complementary and alternative medicine.

Health information workshops should be made available not only to the student body and the general public but also for library staff continuous improvement (CAPHIS, 2010).

Short instructional sessions, focusing on one or two core objectives, can impact the health literacy of those who attend. These can also be incorporated into any existing library seminar series, run in conjunction with programs run by Student Activities or Student Development, or hosted in partnership with the health clinic or wellness coordinators. Short workshop sessions can focus on training in specific tools; longer programs can include active learning activities and assessments.

7.6 Conclusions

United States community colleges are in a unique position in the academic world. These institutions provide access to education through an open-door policy that is flexible and responsive to the needs of the surrounding communities. Budget constraints and a primary mission to serve those affiliated with the college can curtail the library's ability to serve as a resource to the community in consumer health. Still, in an era where consumers are increasingly in search of quality health information that they can understand, it is important for community college libraries to respond to the need. This chapter attempted to move community college libraries forward in the discussion of consumer health, outline some basic functionalities and concerns, and provide potential ideas for moving forward with implementation of consumer health information services.

Appendix A: Community and Junior College Libraries Section (CJCLS) of the Association of College and Research Libraries

Listserv survey questions

1. Does your library have a designated health sciences librarian or a departmental liaison for medical and allied health programs? If not, who is in charge of collection development in the medical area?
2. Is your library open to the public? Can they (the public) check out materials or only use them within the library? Is the public able to use your medical databases on-campus? And if so, do they have to log-in to the computers or are they open?
3. Does your library collect medical and health resources for the layperson (i.e., consumer health) as well as at the academic level for your medical and allied health students? If so, how do you balance purchasing for these two very different levels of information?
4. If you have a dedicated consumer health collection, is it integrated into the collection or in its own area?

References

Accreditation Commission for Education in Nursing (ACEN). (2013). *ACEN 2013 standards and criteria: Associate*. Retrieved from http://www.acenursing.net/manuals/SC2013_ASSOCIATE.pdf.
American Association of Community Colleges (AACC). (2003). *Public community college revenue by source*. Retrieved from http://www.aacc.nche.edu/AboutCC/Trends/Pages/publiccommunitycollegerevenuebysource.aspx.

American Association of Community Colleges (AACC). (2013). *On-campus housing*. Data Points. Retrieved from http://www.aacc.nche.edu/Publications/datapoints/Documents/CampusHouse_8.28.13_final.pdf.
American Association of Community Colleges (AACC). (2014). *2014 fact sheet*. Retrieved from http://www.aacc.nche.edu/AboutCC/Documents/Facts14_Data_R3.pdf.
Bailey, T. (2009). Challenge and opportunity: rethinking the role and function of developmental education in community college. *New Directions for Community Colleges, 2009*(145), 11–30.
Baker, L., & Manbeck, V. (2002). *Consumer health information for public librarians*. Lanham, MD: Scarecrow Press.
Baker, L., & Wilson, F. (1996). Consumer health materials recommended for public libraries: too tough to read? *Public Libraries, 35*(2), 124–130.
Benedetti, J. M. (2002). Strategies for consumer health reference training. *Health Care on the Internet, 6*(4), 63–71.
Borman, C., & McKenzie, P. J. (2005). Trying to help without getting in their faces. *Reference & User Services Quarterly, 45*(2), 133–146.
Community and Junior College Libraries Section, American Library Association (CJCLS). (2010). *CJC-L listserv*. Retrieved from http://www.ala.org/acrl/aboutacrl/directoryofleadership/sections/cjcls/cjclswebsite/listserv.
Consumer and Patient Health Information Section, Medical Library Association (CAPHIS). (2010). *The librarian's role in the provision of consumer health information and patient education*. Retrieved from http://caphis.mlanet.org/chis/librarian.html.
Cooper, D., & Crum, J. A. (2013). New activities and changing roles of health sciences librarians: a systematic review, 1990–2012. *Journal of the Medical Library Association, 101*(4), 268–277.
Crookston, A., & Hooks, G. (2012). Community colleges, budget cuts, and jobs: the impact of community colleges on employment growth in rural U.S. counties, 1976–2004. *Sociology of Education, 85*(4), 350–372.
Crosta, P. M. (2014). Intensity and attachment: how the chaotic enrollment patterns of community college students relate to educational outcomes. *Community College Review, 42*(2), 118.
Crumpton, M. A., & Bird, N. J. (2013). *Handbook for community college librarians*. Santa Barbara, CA: Libraries Unlimited.
Duesing, A., & Near, K. (2004). Helping consumers find reliable health information on the internet: an overview of one library's outreach projects in Virginia. *Journal of Consumer Health on the Internet, 8*(3), 53–67.
Ennis, L. A., & Mitchell, N. (2010). *The accidental health sciences librarian*. Medford, NJ: Information Today.
Evans, G. E., & Saponaro, M. Z. (2005). *Developing library and information center collections*. Westport, CT: Libraries Unlimited.
Hallyburton, A., Kolenbrander, N., & Robertson, C. (2008). College health professionals and academic librarians: collaboration for student health. *Journal of American College Health, 56*(4), 395–400.
Hollander, S. (2000). Providing health information to the general public: a survey of current practices in academic sciences libraries. *Bulletin of the Medical Library Association, 88*(1), 62–69.
Hollander, S. (2003). Academic health sciences libraries: an underutilized resource for patients and consumers. *Journal of Consumer Health on the Internet, 7*(4), 1–6.
Huang, L. (2006). A new type of consumer health library. *MLA News, 384*, 15.

Husted, J., & Czechowski, L. (2012). Rethinking the reference collection: exploring benchmarks and e-book availability. *Medical Reference Services Quarterly*, *31*(3), 267–279.

Joint Review Committee on Education in Radiologic Technology. (October 2014). *2014 JRCERT standards for an accredited educational program*. Retrieved from http://www.jrcert.org/programs-faculty/jrcert-standards/.

Kefalides, P. T. (1999). Illiteracy: the silent barrier to health care. *Annals of Internal Medicine*, *130*(4), 333–336.

Kirkwood, B., & Riegelman, R. (2011). Community colleges and public health: making the connections. *American Journal of Preventative Medicine*, *40*(2), 220–225.

Kouame, G., Harris, M., & Murray, S. (2005). Consumer health information from both sides of the reference desk. *Library Trends*, *53*(3), 464–479.

Library Statistics Program, National Center for Education Statistics. (2012). *Library comparison report: Public, 2 year*. [Data file]. Available from http://nces.ed.gov/surveys/libraries/compare/default.aspx.

McInnes, N., & Haglund, B. A. (2011). Readability of online health information: implications for health literacy. *Informatics for Health & Social Care*, *36*(4), 173–189.

Medical Library Association. (2015). *What is health information literacy?* Retrieved from https://www.mlanet.org/resources/healthlit/define.html.

Miller, K. (2010). *Child care support for student parents in community colleges is crucial for success, but supply and funding are inadequate*. Institute for Women's Policy Research Fact Sheet. Retrieved from http://www.iwpr.org/publications/pubs/child-care-support-for-student-parents-in-community-college-is-crucial-for-success-but-supply-and-funding-are-inadequate.

Miller, M. T., Pope, M. L., & Steinmann, T. D. (2005). A profile of contemporary community college student involvement, technology use, and reliance on selected college life skills. *College Student Journal*, *39*(3), 596–603.

Mullin, C. M. (2012). *Why access matters: The community college study body*. AACC Policy Brief 2012-01PBL. Retrieved from http://www.aacc.nche.edu/Publications/Briefs/Documents/PB_AccessMatters.pdf.

Mullin, C. M., & Phillippe, K. (January 2013). *Community college contributions*. American Association of Community Colleges Policy Brief 2013-01PB. Retrieved from http://www.aacc.nche.edu/Publications/Briefs/Documents/2013PB_01_gray.pdf.

National Network of Libraries of Medicine: Greater Midwest Region. (2013). *Provide outreach – exhibits and presentations*. Retrieved from http://nnlm.gov/gmr/outreach/.

National Network of Libraries of Medicine: Greater Midwest Region. (2014). *Community college library award*. Retrieved from http://nnlm.gov/gmr/funding/communitycollege/communitycollegecfa.html.

National Student Clearinghouse Research Center. (2014). *Current term enrollment report – Spring 2014*. Retrieved from http://nscresearchcenter.org/wp-content/uploads/Current-TermEnrollment-Spring2014.pdf.

Nielsen-Bohlman, L., Panzer, A. M., & Kindig, D. A. (2004). *Health literacy: A prescription to end confusion*. Washington, DC: National Academies Press.

Osika, B., & Kaufman, C. (2012). 'Mobilizing' community college libraries. *Searcher*, *20*(9), 36–46.

Parsad, B., & Lewis, L. (2003). *Remedial education at degree-granting postsecondary institutions in fall 2000*. NCES 2004-010, Retrieved from http://nces.ed.gov/pubs2004/2004010.pdf.

Russell, A. (2008). *Enhancing college student success through developmental education*. Washington, DC: American Association of State Colleges and Universities. Retrieved from http://www.aascu.org/policy/publications/policymatters/2008/developmentaleducation.pdf.

Six-Means, A. (2011). *To lend or not to lend, an important consideration for all consumer health libraries*. Retrieved from http://caphis.mlanet.org/chis/lending.html.

Smith, C. A. (2008). Consumer health information. In M. S. Wood (Ed.), *Introduction to health sciences librarianship* (pp. 429–458). Binghamton, NY: Haworth Information Press.

Smith, A., Rainie, L., & Zickuhr, K. (2011). *College students and technology*. Retrieved from http://www.pewinternet.org/2011/07/19/college-students-and-technology/.

Snyder, M., Huber, J. T., & Wegmann, D. (2002). Education for consumer health: a train the trainer collaboration. *Health Care on the Internet, 6*(4), 49–62. http://dx.doi.org/10.1300/J138v06n04_05.

U.S. Department of Education, National Center for Education Statistics (NCES). (2013). *Average undergraduate tuition and fees and room and board rates charged for full-time students in degree-granting institutions, by level and control of institution: 1969–70 through 2011–12*. [Table 381] Digest of Education Statistics, 2012 (NCES 2014-015). Retrieved from https://nces.ed.gov/programs/digest/d12/tables/dt12_381.asp?referrer=report.

U.S. Department of Education, National Center for Education Statistics (NCES). (2014). *Associate's degrees conferred by postsecondary institutions, by sex of student and discipline division: 2001–02 through 2011–12*. [Table 321.10] Digest of Education Statistics, 2013. Retrieved from https://nces.ed.gov/programs/digest/d13/tables/dt13_321.10.asp.

Wessel, C., Wozar, J., & Epstein, B. (2003). The role of the academic medical center library in training public librarians. *Journal of the Medical Library Association, 91*(3), 352–360.

Younghee, N. (2013). The development and performance measurements of educational programs to improve consumer health information (CHI) literacy. *Reference & User Services Quarterly, 53*(2), 140–154.

Contexts

Health information delivery outside the clinic in a developing nation: The Qatar Cancer Society in the State of Qatar

Ellen N. Sayed[1], Alan S. Weber[2]
[1]Director, Distributed eLibrary, Weill Cornell Medical College in Qatar, Doha, Qatar;
[2]Premedical Department, Weill Cornell Medical School in Qatar, Doha, Qatar

8.1 Introduction and background

This chapter investigates the provision of nonclinical health information resources, provided to consumers, in the developing nation of the State of Qatar by examining the information-oriented activities of the Qatar Cancer Society (QCS) in the context of neighboring countries with similar culture and belief systems. The Society provides Internet and paper-based information on cancer, support services (both emotional and financial), community services, and referrals to other services for all residents of Qatar. The Society arose from a need to provide cancer information and counseling to supplement the overburdened public hospital medical system, Hamad Medical Corporation (HMC), after Qatar's population grew from 800,000 to 2.1 million within one decade. Counseling and psychiatric services are in short supply, with only 25 attending psychiatrists and 63 beds in Qatar (Kronfol, Ghuloum, & Weber, 2013). Very little scholarship exists on the role of these kinds of institutions within the health care systems of the Arabian (Persian) Gulf or the Middle East and North Africa (MENA) region. The chapter also outlines the important role that specialized libraries can potentially play in information provision on health matters outside of the clinical setting and in collaboration with health organizations such as QCS.

Specialized medical libraries and medical schools are a relatively new phenomenon in the Arabian Gulf, and their full potential as a health care partner with hospitals and private practices has not yet been realized. Before Qatar's only medical school—the Weill Cornell Medical School in Qatar—opened in 2001–2002, most physicians were expatriates trained in Egypt, Lebanon, India, and Syria. In 2008, only 14% of medical professionals in Qatar were Qatari nationals (HMC, 2008, p. 2008). In addition, health associations such as QCS have recently gained increased importance due to the evolution of modern health care internationally, which is now emphasizing a multidisciplinary approach and patient-centered care. In this new model, nonclinical health partners contribute to patient well-being, patient education, and counseling to help cancer sufferers and their families understand their disease and choose the right treatment and follow-up care options.

The Interprofessional Education model—in which students are trained in mixed professional groups to recognize and appreciate the different competencies on the health care team—is growing in popularity in medical education and has been studied in depth. A World Health Organization (WHO) report pointed out that "[C]ollaborative practice happens when multiple health workers from different professional backgrounds work together with patients, families, caregivers and communities to deliver the highest quality of care. It allows health workers to engage any individual whose skills can help achieve local health goals" (2010, p. 7). This approach opens up further resources for the patient, the patient's family, and the community and allows further scope for contributions from nontraditional health information providers, such as nongovernmental organizations (NGOs), nonprofit organizations, public and academic libraries, Internet-based companies, publishers, alternative healers, and nonclinical organizations such as the QCS.

8.2 Qatar

Qatar is one of the smallest countries in the world by land area (11,571 km^2), similar in size to Gambia and Vanuatu, and located on the Arabian side of the Persian (Arabian) Gulf. The nation is situated on the Arabian Peninsula along with Saudi Arabia, Oman, the United Arab Emirates (UAE), and Yemen. Qatar is part of a political alliance called the Gulf Cooperation Council (GCC) along with the countries of Bahrain, Kuwait, Oman, Saudi Arabia, and the United Arab Emirates, who all share a similar Muslim cultural heritage. Qatar, which joined the Organization of Petroleum Exporting Countries in 1961, has been a substantial producer of petroleum since the 1960s, but in addition was also the third largest producer of natural gas in the world in 2012 at 160 billion cubic meters. Qatar is also the second largest exporter of gas after the Russian Federation (IEA, 2013). The high price of hydrocarbons in the last decade coupled with Qatar's enormous reserves has resulted in extraordinary GDP growth and government financial surpluses. While much of these surpluses have been invested abroad, large internal investments have also been made in infrastructure, health care, social services, education, and basic government salaries and benefits. Thus Qatar is in the unique situation of possessing ample public health funds coupled with leadership who are intent on improving health indicators and health outcomes for both citizens and expatriates through upgrading medical infrastructure, training personnel in its newly launched medical, allied health, and nursing schools, and launching informational health campaigns.

8.2.1 Qatar Vision 2030

To move the economy away from natural resources production toward the creation of knowledge, and to build what has been called a "knowledge economy," the State of Qatar has embarked on a strategic development plan entitled *Qatar Vision 2030*. A knowledge economy is based on developing salable intellectual properties—patents, rights, and proprietary processes—and obviously rests on the pillars of education, research and human capacity development to foster innovation and new ideas (General Secretariat for Development Planning (GSDP), 2008, p. 6). The State of Qatar has developed a coherent national strategy to carry out a variety of knowledge economy projects, many of which are underway and producing tangible results. In the past decade, the State of Qatar has overhauled its K-12 educational system; established its first medical school—a branch campus of Weill Medical College of Cornell University located in New York, NY (WCMC-Q) as well as a nursing school (University of Calgary, Qatar); inaugurated a national research funding agency called Qatar National Research Fund; and established the technology and science business incubator Qatar Science and Technology Park (QSTP).

Thus Qatar's newly launched *National Cancer Strategy* and *Qatar National Cancer Research Strategy* issued by the Supreme Council of Health (SCH) form part of the larger developmental vision of Qatar 2030: not only to improve health outcomes in Qatar, but also to provide a badly needed scientific evidence base for understanding the genetic, behavioral, and environmental factors responsible for cancer in Arab and Qatari patient populations. Part of this evidence base includes research into socioeconomic factors and cancer patient behaviors, including their knowledge of risk factors; information-seeking behaviors; information provision; and the role of allied health sciences and nonclinical health care partners in providing more satisfactory outcomes, not only in the decrease of morbidity and mortality, but also in the improvement of the patient experience. Qatar is one of the few nations in the world with a clearly defined comprehensive national cancer strategy with targeted numerical outcomes indicators, assigned roles for various ministries, agencies, and individual institutions, and a final target date of 2016 for implementation of all new cancer policies. The *Cancer Strategy* even contains suggested penalties for non-compliant stakeholders and for missing targeted outcomes. As a preexisting cancer organization in Qatar, the QCS was integrated into the National Cancer Strategy primarily in the roles of volunteer coordination, patient education, and counseling.

8.2.2 Demography

The citizens of Qatar—numbering between 250,000 and 300,000 official passport holders—are primarily Sunni Muslims with origins in the Nejd and Rub al Khali (Empty Quarter Desert) regions of Saudi Arabia, the Persian Empire, or from freed East African and Sudanese slaves. Qatar abolished its slave trade in 1952 under pressure from the British government who controlled Qatar's external affairs until 1971. Recent genetic testing has confirmed the existence of three main subtypes of Arab, African, and Indo-Iranian origin in Qatar. The demographic structure of Qatar is one of its unique features: Qatari nationals make up about 14% of the total population of 2.1 million, and only 6% of the total workforce (Weber, 2013, p. 50). The primary countries of origin of Qatar's expatriates are India, Pakistan, Nepal, Philippines, and the nearby Arab countries, with less than 5% consisting of highly paid Western professionals. Most expatriates are concentrated in the oil and gas and service industries, and a considerable but unknown percentage of non-Qatari workers can be found in the health care and knowledge industries, such as information and communications technology (ICT) and education. Many of the lower paid expatriates are non-Arabic speakers and are culturally distinct from *Khaleeji* (Gulf) Arabs, which can sometimes create cultural and language barriers to health care provision.

8.2.3 Health care in Qatar

Except for periodic visits from American Mission Hospital doctors from nearby Bahrain, Qatar did not have a formal medical care system until the opening of a hospital in Al Jasra in 1947 (Gotting, 2006). Previously, Qataris relied on traditional healers called "hakims," who prescribed traditional remedies and on "Quranic medicine" such as prayers, honey, cautery, and *hijama* or cupping (bloodletting). Today, all Qataris can access any public or private hospital in Qatar for free under the new *Seha* insurance scheme, and employers are required to pay the premiums of all non-national employees, which effectively provides universal health coverage for everyone in Qatar. The main public hospital, Hamad General Hospital, opened in 1982 and was accredited by the American organization Joint Commission International in 2006 and 2011. The official dedicated cancer hospital in Qatar is Al Amal hospital, which has been renamed as the National Center for Cancer Care and Research. This new organization will centralize the following services: cancer research, patient care, and medical training and research in oncology. Common diseases in Qatar include cancer, cardiovascular disease, obesity, diabetes, and respiratory illness due to the desert environment, lack of rain, and oil industry well flaring. The country also has an extremely high motor vehicle accident rate (Bener, Abdul Rahman, Abdel Aleem, & Khalid, 2012) due to poor road infrastructure and loosely regulated driving behavior.

8.3 Methods

Structured expert interviews with three QCS officials were carried out by the authors in March 2014 with further follow-up in April and May. The objective was to determine the role that a representative nonclinical health organization in Qatar (QCS) plays with respect to health information provision to the general public outside of the hospital setting. Respondents included Dr Mahaseen Oukasha, PhD, head of the Health Education Department, Heba Ali Nassar, health educator, and Manal Jihad Abu Zayed, health educator. Officials reviewed the information collected by researchers and corrected any factual or interpretative errors. All QCS officials had professional qualifications and work experience in the field of nursing in the Arabian Gulf, Jordan, and Egypt. This qualitative assessment of a nonclinical health support organization in Qatar also drew on the published materials of the QCS; a literature review on cancer in Qatar using PubMed MEDLINE; references from the second author's review of Arabian Gulf health information literacy (2015), and national policy planning documents of the State of Qatar, such as the *National Cancer Strategy: The Path to Excellence* (2011) and *Qatar National Cancer Research Strategy* (2012). The researchers also drew on their professional experience working in a medical college in Qatar as the director of the Distributed *e*Library (DeLib) and as a humanities teacher in the Premedical Department.

8.4 Sources of consumer health information in the GCC

Several field-specific studies have focusing on physicians' relation to health information in various medical domains, for example, primary care physicians (Al-Harbi, 2011; Al-Sughayr, Al-Abdulwahhab, & Al-Yemeni, 2010; Altuwaijri, 2011). No comprehensive research has been conducted on consumer health information resources or nonclinical sources of health information in the region with the exception of Weber et al. (2015). No clear picture emerges from the published literature. Even a simple qualitative meta-analysis of the literature retrieved by the authors is difficult due to the paucity of material. A systematic review of health information literature in the Arabian Gulf carried out by the second author (Weber et al., 2015) identified 283 articles, primarily in the area of online health (ehealth). However, due to the very similar political and economic history of the GCC nations of Bahrain, Kuwait, Oman, Qatar, Saudi Arabia, and the UAE, as well as their demographic structure, useful information about health information and information literacy in Qatar can be obtained from studies in other GCC countries. This information must be used with caution, however, since there are also significant differences among the GCC states—for example, the population of Saudi Arabia is over 12 times that of Qatar, Bahrain and Oman have dwindling oil supplies and more diversified economies, and Sunni Islam is not the dominant branch of religion in Oman (main sect Ibadhism) or Bahrain (a majority Shia nation).

The sources of health information that are used and trusted may vary considerably from one Arabian Gulf nation to the next. For example, in Qatar, with a high mobile phone and computer penetration rate and easily accessible government health and services portals (*hukoomi*), Internet health information may possibly be used more frequently than in Oman or Yemen, two countries with lower technology usage. In those countries, which have substantial rural populations, religious and folk healers along with family members may be frequent sources of health information along with licensed practitioners (Weber, 2011).

Several health informatics researchers in the Gulf have noted the lack of reliable Arabic language health information in any media; thus, despite high technological literacy in the UAE and Qatar, online sources are possibly not being utilized by Arabic speakers. The King Abdullah Health Encyclopedia (www.kaahe.org), a project launched in Saudi Arabia, similar to the *Web*MD and Mayo clinic Web sites, now provides health information for the general public in Arabic (Alsughayr, 2013).

QCS educators offer individual counseling to diagnosed cancer patients and also plan public information campaigns, visit schools to deliver lectures and workshops, and develop Web resources. Thus these educators have first-hand knowledge of the information needs of a broad cross section of cancer patients and their families in Qatar. QCS educators interviewed for this study indicated that the most common information sources about cancer in Qatar originate in advice from friends and family members in addition to the patient's physician. Much more research into the question of health information-seeking behaviors, information provision, and the specific kinds and formats of information available needs to be done in Qatar, since these questions are critical for policy planning. For example, for information provision to be successful in the State of Qatar, government agencies and clinical and nonclinical organizations need to determine more clearly the preferred information sources used by Qataris and expatriates, and specifically how different socioeconomic and ethnic groups are obtaining and using health information.

Anecdotal reports gathered by the authors suggest that GCC residents generally first seek information from conversations with physicians and hospital personnel, followed by family, friends, and knowledgeable elders and religious leaders in the community: Imams, Muftis, Sheikhs, Sheikhas, and Hakims. *Sheikh* (feminine *Sheikha*) is an Arabic general honorific title for an older, knowledgeable person. *Imams* and *Muftis* are religious leaders who have gained some kind of formal learning in religious or legal studies. Since Islam forms a comprehensive and integrated system of life, religious scholars are frequently consulted on medical matters, especially mental illness, which in Islamic belief may originate from *djinn*, or devils, who are attacking or testing the sufferers' religious faith. *Hakim* is a general term for a wise man or medical healer. Finally, printed sources and the Internet are consulted. These preferences may be related to the oral nature of indigenous Bedouin culture. Also, an unknown number of patients seek out folk healers for medical advice, particularly in Oman and Saudi Arabia for such cures as traditional herbal remedies, cautery (*wasm, kowie,* or *kaii*) and cupping (*hijama*) which form part of Prophetic Medicine (*tibb al-nabawi*). A cross-sectional study of traditional healers in Riyadh reported that 42% of respondents had consulted a traditional healer using the following therapies: "the Holy Quran (62.5%), herb practitioners (43.2%), cautery (12.4%), and cupping (4.4%)" (Al-Rowais, Al-Faris, Mohammad, Al-Rukban, & Abdulghani, 2010, p. 199). Weber also documented active folk medical practices in modern Oman (2011).

A 2009 study of 450 Saudi women, however, indicated that medical magazines and visual media were the preferred sources of information on health awareness. The order of source preference was: "first, medical magazines, second, visual media, third, health centers, fourth, other non-medical publications, fifth, the internet, and sixth, relatives and friends" (Al-Ghareeb, 2009). In AlGhamdi and Almohedib's study of online health information-seeking behaviors in Saudi Arabia, 93% of respondents indicated that they first sought health information from their physicians (2011, p. 292) before searching the Internet. Also according to their study, although the majority of respondents found Internet health information beneficial, only 8% said that they would always trust the information found on the Internet (AlGhamdi & Almohedib, 2011, p. 292).

A study in Kuwait focused on information about tooth avulsion—the complete displacement of the tooth out of the socket—found that for the general public, "The Internet, health care professionals, and TV were the three most preferred sources of information...across all groups, regardless of the sociodemographic characteristics" (Al-Sane, Bourisly, Almulla, & Andersson, 2011, p. 432).

Teachers are often viewed as role models in the community, and their teachings are taken seriously. In Bahrain, however, a study indicated that school teachers may not be the best source of health knowledge for the general public since only 50% could correctly demonstrate common knowledge about five prevalent health problems: bronchial asthma, sickle-cell anemia, hypertension, diabetes, and smoking. The authors indicated that lack of appropriate training was the primary cause for this, and recommended educating school teachers about prevailing health issues in society (Alnasir & Skerman, 2004, p. 537).

According to the authors of a Saudi Arabian study, mothers in Saudi Arabia received most of their knowledge about childhood diarrhea from nonhealth professionals and both their knowledge of the condition and practices in treating it were deficient (Moawed & Saeed, 2000). The authors, a nurse holding a PhD degree, and a physician with an MD degree, recommended that health professionals should be the preferred means for transmitting knowledge about this health issue to mothers.

A similar study on mother's knowledge of pediatric disease by Al-Ayed in Saudi Arabia demonstrated that mothers obtained most of their information from family members and that mothers' knowledge of their newborn's problems was not sufficient (Al-Ayed, 2010, p. 26). In addition, Al-Ayed cited previous research demonstrating the unavailability and lack of effectiveness of health education in Saudi schools, and argued that health care institutions have not been proactive in providing proper educational services: "At present, knowledge on child health matters taught in schools in the Kingdom is inadequate. Health care institutions play a limited role in health education. There should be proper effective practical means of disseminating information on child health matters among mothers in our community" (2010, p. 22).

Thus the published literature indicates that GCC residents find health information from different sources, including immediate family, their physicians, health magazines, books, and the Internet. There is no indication that they would consider asking a librarian for assistance due in part to the scarcity of public libraries in the region. This picture then, is what one may expect to find in Qatar as well.

8.5 Barriers to health care in Qatar

The existing literature and anecdotal evidence indicate that patients in the GCC region generally do not turn to the Internet or libraries as their primary sources for health information; many turn to their family and friends first along with their health care providers. A handful of studies from Qatar shed further light on this issue. Saleh, Bener, Khenyab, Al-Mansori, and Al Muraikhi (2005) found that Qatari women had misconceptions about the causes and treatment of urinary incontinence. Despite their condition, they did not seek information or assistance from clinicians due to guilt or shame. In addition, their health care providers did not inquire about this condition (2005). A study of health information-seeking in the context of a cervical cancer diagnosis found that most women relied on family and friends for health information, rather than a physician. Based on similar studies of Arab populations in the region which found low rates of Pap smears despite adequate knowledge of their benefits, the researchers hypothesized that the fear and embarrassment associated with cervical cancer screening may discourage both patients and physicians from discussing the issue (Al-Meer, Aseel, Al-Khalaf, Al-Kuwari, & Ismail, 2011, p. 858). A similar situation was found for breast cancer examinations and screening. Younger, educated Qatari women were more likely to perform breast self-exams, be examined by a health care provider or have a mammogram. While Qatari women are generally well informed about breast cancer, screening levels remain low. Physicians can reduce barriers to early detection by recommending screening (Bener et al., 2009).

A study by Donnelly et al. (2013) of Arabic-speaking women in Qatar found that women, who believed that cancer is God's punishment or bad luck, were significantly less likely to practice breast cancer screening of any kind. Nonsupport from husbands or male relatives had no bearing on the decision not to have a mammogram. On the other hand, women, who wanted to know if cancer was present, trusted their physician, or had no gender preference for their health care provider were more likely to undergo breast cancer screening. The most common reason for women not to practice breast cancer screening was that it had not been recommended by their physicians. These authors recommended that care options be provided in alignment with Arabic cultural norms, that physicians be trained to discuss screening with their patients, that the community, such as Imams, be involved, and that cancer survivors be involved in awareness campaigns to lessen fears. According to El Hajj and Hamid, pharmacists in Qatar have expressed interest in participating in breast cancer awareness, but found that they lacked education on the subject, time, and the patient education materials to share with their customers. They also found that their customers were not expecting the pharmacists to be involved in breast cancer awareness (El Hajj & Hamid, 2011, p. 76). It is with this cultural setting in mind that the QCS staff must approach the public as they address cancer prevention and education.

8.5.1 Truth-telling

A pattern of physicians not inquiring about women's health issues or recommending appropriate screening in Qatar is emerging in the peer-reviewed medical literature. These are barriers for women's seeking of health information and cancer screening. Another barrier is posed by the truth-telling practices by physicians in Qatar. In a cross-sectional study of 131 physicians practicing in Qatar, the majority (88.5%) reported that they would disclose a cancer diagnosis, but 66.4% also reported that they would make exceptions to disclosure in rare cases. Factors taken into consideration when deviating from truth-telling were: the emotional stability of the patient (74%), age (68.8%), and perceived intelligence (67.7%). Other factors considered, but to a lesser degree in importance, were gender (26%) and religion (25%) (Rodríguez del Pozo et al., 2012, p. 1471).

Although physician behavior in the Arabian Gulf has not been widely studied, Rodríguez del Pozo et al. (2012) observed that "only Arab or Muslim physicians, or doctors who had studied in the region, reported that their usual policy was not to tell" (p. 1472). Those who did disclose a cancer diagnosis were on the average 5 years younger than those who chose not to. However, 70.2% of participating physicians said their practice might change with the results of empirical research (Rodríguez del Pozo, 2012, p. 1472). The obvious implication for health information-seeking in this population is that patients from whom diagnosis is withheld will not seek any form of health information either from the physician or outside the clinic, and are therefore deprived of potential valuable information sources on coping with the sequelae of the disease.

8.6 The Qatar Cancer Society

In light of the attitude toward cancer in Qatar, the QCS has an important role to fill. The importance assigned to the Society by the SCH in promoting awareness of cancer in Qatar highlights its central role in educating the Qatari population about cancer.

QCS is a nonprofit, NGO founded in 1997 in the State of Qatar. Although the society operates under the umbrella of the Ministry of Labor and Social Affairs of the State of Qatar, its funding derives primarily from private donations, grants, companies and institutions in Qatar, and from fund-raising events. The society does receive some government support as well. QCS began entirely with volunteer staff, while permanent staff were hired beginning in 2011.

The officially stated objectives of the QCS are as follows:

1. Global awareness of cancer and how to prevent it
2. Financial support to cancer patients, who are unable to afford treatment[1]
3. Make recommendations and provide the necessary plans to fight all cancer types[2]
4. Submission of projects on health policy and a comprehensive national program to fight cancer
5. Coordination between the various stakeholder's cancer treatment and follow-up emerging from other countries[3]
6. Determine the size of the problem, including the incidence of the disease and the extent of its spread and the number of deaths resulting from it[4]
7. Prepare and support research and special studies of cancer and access to the latest therapeutic means to cope with this disease (QCS, n.d., p. 2)

8.6.1 Services of the QCS

The central QCS office is located in downtown Doha, in a crowded area of the city—a location that can cause access problems for those seeking its services. Also, rural patients and families experience access problems due to extensive road work projects, traffic, and parking problems ongoing throughout the State of Qatar. Challenges in advertising and getting the word out about the services available at QCS have also arisen. QCS services are available to both cancer patients and their families, and to the general public; in essence, whoever seeks information about cancer.

Services offered by QCS, discussed in more detail below, center on promotion and prevention, support, and fund-raising.

8.6.1.1 Promotion and prevention information

The Supreme Council of Health (SCH), is the ministry that oversees all health matters in Qatar from policy, regulation, planning, and oversight of health care institutions. The Qatar

---

[1] This includes services and treatment not offered by the public hospitals.
[2] This objective refers to public health campaigns.
[3] Patients, both expatriates and locals, sometimes begin or complete treatment abroad.
[4] That is, assist with epidemiological and public health studies of cancer in Qatar.

National Cancer Strategy, published by the SCH in 2011, clearly recognizes the importance of information sources and education in preventative care: "That means increasing Education and Understanding of cancer through myth-busting campaigns, education in schools, cancer awareness events and a comprehensive Qatar-specific cancer information website" (SCH, 2011, p. 6).

As part of the Society's promotion and prevention services, the staff provides instruction on how to perform breast self-examination. The staff shared with the authors that this service has proven effective and has already saved lives.

In terms of information provision, the topics addressed by QCS, through the Internet, printed paper pamphlets, or through one-on-one discussions with a QCS health educator, include the following:

- where to find information about the disease;
- information provided on behalf of someone in their family;
- what families and patients can do to assist doctors and the healing process;
- how patients and families can deal with emotional stress and physical discomfort;
- how others can support their family member; and
- how the patient and patient's family can prepare for living with cancer.

A section on the QCS Web site currently under construction entitled "The Academy: Scientific and academic information" will link to evidence-based technical medical information for users. The activities of QCS, a donation-based organization, have been hampered by lack of funding and human resources. In addition, there is a severe lack of competent Arab Web development services in the area due to late adoption of the Internet in the region and lack of local ICT-training programs. The QCS educators who were interviewed expressed a desire to collaborate more with libraries in Qatar—however, database licensure issues remain with Weill Cornell Medical College in Qatar's (WCMC-Q)'s DeLib (a virtual, distributed electronic library with some printed sources located in the WCMC-Q building). The HMC medical library located in the major public hospital is not extensive. WCMC-Q's and Hamad's are the only two medical libraries in Qatar outside of smaller collections at the University of Calgary, Qatar (nursing library), Qatar University (which offers a PhD in pharmacy program), and College of the North Atlantic in Qatar (offering allied health programs, such as respiratory therapy). The QCS does not have a direct connection with the WCMC-Q DeLib, the medical library supporting all WCMC-Q education and research programs, but the HMC library is available to QCS staff as their primary information source. The libraries at WCMC-Q and HMC offer on-site access to the general public. Access to the WCMC-Q DeLib would help QCS to achieve its organizational and educational goals. The Qatar National Library, currently under construction in Education City (Hamad bin Khalifa University) has a growing list of academic databases, which are based on national licenses and which could be employed by QCS in a number of ways, including planning, education, and resources for patients.

QCS's informational campaigns are informed by data from the Qatar Cancer Registry database at HMC. The Hamad Cancer Registry ranking is considered the officially correct ranking in Qatar. Using these data, QCS targets the most prevalent cancers in Qatar, which are breast, lung, and colorectal cancer. Statistics are not collected on a systematic basis by QCS, although the society is building a database to refer patients to specific cancer information.

Health information delivery outside the clinic in a developing nation 179

The QCS produces a number of English, Arabic, and dual Arabic/English public information pamphlets that are distributed in clinics, hospitals, schools, and public places such as malls and at charity events. QCS educators describe these pamphlets as "evidence-based" and relying on the latest peer-reviewed medical information on cancer, but references are omitted in the pamphlets to make them more readable for the public. Figure 8.1 shows the interior panels of one of QCS's dual language pamphlets on colorectal cancer, demonstrating the efforts that are taken to keep text to a minimum and in plain language, and to use explanatory diagrams.

The Society also plays an instrumental role for the SCH in the National Cancer Strategy. In addition to recognizing the importance of information sources and education in preventative care, the SCH acknowledges the need to address prevalent myths about cancer. According to QCS educators, two common myths in Qatar that could be addressed in educational campaigns are that cancer is contagious and that it is always a fatal disease.

Figure 8.1 Dual language pamphlet on colorectal cancer.

The SCH recommends further that additional online information sources be created, as well as mobile apps, and QCS has been designated by the SCH as the obvious candidate for expanding its current Web site to provide more information to cancer stakeholders. The site, which includes plans for smart phone and tablet accessibility, will be Qatar-specific and provide patient-oriented information in Arabic and English to help cancer sufferers and their families find more information about their disease, that is, complementary information not designed to substitute for professional diagnosis and care (SCH, 2011). To address access and digital divide issues, the SCH recommends that public Internet kiosks be located in malls and primary health centers.

The SCH has even further plans for the QCS. Although QCS already produces a number of informational pamphlets, the SCH recommends that QCS develop "cancer Information Prescriptions," by SCH as "comprehensive information sheets for patients about their type of cancer including potential treatments and possible side-effects" (SCH, 2011, p. 27). Thus, the current role of the QCS is a broad and growing one in providing cancer information support to the people of Qatar.

8.6.2 Outreach to the community

QCS sponsors events for the public, such as World Cancer Day—with tents for Q/A with experts, demonstrations, puzzles, and games for kids. QCS also partners with other Qatar cancer events, such as the internationally known Terry Fox charity cancer run to raise funds for cancer research, held by a Canadian technical college in Qatar, the College of the North Atlantic ("Four Seasons Hotel Doha", 2014).

In addition, the National Cancer Strategy recommends a yearly schedule of public cancer awareness events, and QCS is carrying out this suggestion through its activities, "aligned with international cancer days/months such as 4 February (World Cancer Day), 31st May (World No Tobacco Day) and October (Breast Cancer Month). Other days, weeks or months for specific cancers should be designated within Qatar." The SCH further commends the volunteer sector of society, including QCS, for taking the lead on cancer events and education—such as the Think Pink Qatar cancer walk in October and the Hayat Cancer Support group's talks in schools and businesses—and suggests a more formalized leadership role for QCS. With these additional responsibilities, the SCH suggests that QCS could be provided with "additional support and administration provided by the SCH and other public sector bodies" (SCH, 2011, p. 14).

Planned future outreach projects of QCS include a mobile health clinic to reach areas outside of Doha. QCS also plans to convert one of their offices in Barwa tower to a lecture theater to deliver public lectures about cancer. The Ministry of Social Affairs will partner with QCS to build a new cancer center that will include "an education room for students and the general public; support groups for patients and families of the diagnosed; and a financial help system for Qatari and expat residents who have difficulty paying for life-saving treatments such as chemotherapy" (Khatri, 2014). The national telecommunications carrier Ooredoo has pledged 30 million QAR (8.2 million USD) to build and support this center. The challenge for the QCS will be to coordinate their internal plans and activities with the evolution of external entities that affect QCS's operations. However, given the array of current activities and the future plans for QCS, it clearly plays a central role in Qatar's national cancer strategy.

8.6.2.1 Visits to local schools

The main activities of the Society, in addition to providing ongoing phone support to cancer patients and financial assistance to low-income patients, are outreach workshops in schools, corporations, and other institutions in Qatar. QCS is working hard to introduce factual information about cancer to Qatar's adult population to reduce the misconceptions about cancer. In the school system, however, QCS is reaching out to Qatar's future generation. The Society is particularly active in primary and secondary schools in introducing the topic of cancer to young school children. Information delivered includes basic understanding of what cancer is, and its diagnosis, prognosis, and treatment. The objective is to open up a dialogue about cancer based on accurate information at an early age and reduce the risk of misconception about the disease later on. The hope is also that the children will discuss the topic with adults in their homes. Due to the Society's focus on preventive medicine, lectures often involve healthy eating practices. Hands-on activities for children include preparing portions of a healthy meal containing balanced nutrients, antioxidants, and fresh ingredients.

8.6.2.2 Targeting high-risk groups

One of the society's ongoing cancer campaigns will target 40-year olds since this is the age group often presenting with certain risk factors for cancer and individuals should become more vigilant about common types of cancer, such as breast and prostate cancer.

8.6.3 Patient and family support

Part of the emphasis on cancer information and education in the National Cancer Strategy stems from the growing recognition worldwide in various medical fields, and supported by the WHO, that the best care takes place in teams of specialists called multi-disciplinary teams. Professionals such as grief counselors, medical educators, psychologists, medical ethics experts, family liaisons, and family support specialists—previously seen as ancillary to medical care—are now being recognized as key partners in the healing process. QCS must ensure that their services are integrated with other entities involved with cancer care. Due to the long-term nature of cancer treatment, which often necessitates family involvement, its complexity, and the variety of treatment options available, cancer exemplifies a disease which requires more than simple clinical diagnosis and treatment following a set of predefined procedures. Creating a supportive and positive framework for the patient in which they are fully informed about the healing process at all times can be greatly facilitated by the services provided by nonclinical organizations such as QCS. Up until now, QCS has exercised some control over their activities. As the health care environment in Qatar evolves, however, QCS will need to expand both its plans and its staffing, which will require additional funding to facilitate. To that end, QCS must be prepared to evolve as well to respond to the changes in its external environment.

8.6.4 Support groups

The QCS also coordinates and assists with cancer support groups. According to the National Cancer Strategy, these groups should be encouraged and expanded:

> Three cancer support groups currently exist in Qatar – the Hayat Cancer Support group (which has links with the Qatar National Cancer Society), the Ladies of Harley cancer support group and a support group affiliated to Doha Mums. Cancer support groups need not be focused solely on patients – the members of the Doha Mums group are partners of cancer patients who have found it helpful to support each other in their experiences. More cancer support groups should be encouraged to develop in Qatar, especially a group that could cater for men.
>
> SCH (2011), p. 39.

Clearly QCS is already providing important support group services to cancer patients, a service that the QCS staff who were interviewed appeared to value highly.

QCS staff also reported on the direct impact on people with early diagnosis thanks to information provided through QCS's promotional activities. The Society is, however, working diligently to involve cancer survivors in their promotional activities, in their continued effort to break down the misconceptions and stigma that cancer still has in the society. That said, QCS staff has built their existing services on their own and are likely to succeed in finding appropriate ways to highlight the experiences of cancer survivors.

8.6.5 Utilizing cancer survivors for support services

QCS is currently encouraging cancer survivors to share their stories with others, providing comfort and assurance that one can survive cancer and lead a productive life. This form of public awareness is viewed as a necessity by QCS staff. The SCH has similar future plans: volunteers can also play a large role outside the clinic in supporting cancer awareness programs and aiding in support efforts for current cancer patients: "With some training, cancer survivors could also play a role in supporting newly diagnosed patients by drawing on their own experiences to, for instance, help people in understanding the reason for different tests and in coping with the side effects of treatment. There might even be potential to create a helpline hosted by the Qatar National Cancer Society and staffed by volunteers" (SCH, 2011, p. 39).

Cancer survivors sharing their stories publicly is a new phenomenon in Qatar, and is therefore approached cautiously. In their efforts to highlight cancer survivors in their public campaign, research findings indicate that the QCS should consider how to adequately provide information support to cancer survivors. Training programs for cancer survivors on how to effectively find relevant health information is an excellent avenue for collaboration with medical libraries in Qatar.

8.6.6 Financial support and fund-raising

Non-Qatari, low-income patients are supported by the QCS through grants and provided with follow-up telephone or face-to-face support that is not always available through the public health system. In addition to their existing services, financial assurance during an otherwise stressful time can provide cancer sufferers with the peace of mind they need to focus on getting well.

8.7 Cancer information delivery outside the clinical setting

Reaching out to cancer survivors is an important role for QCS, and information support is an important facet of that work, as borne out by studies of information seeking in the context of a cancer diagnosis. Hesse, Arora, Beckjord, and Finney Rutten (2008) found that 54.3% of participants reported that they preferred to get their cancer information from a health care provider, with 29.9% saying they preferred the Internet. However, when those who had reported that they had been looking for cancer information were asked to report where they looked *first*, the pattern switched: 46.7% reported that they searched on the Internet first, and only 23% had gone to their health care provider first. Preferring to go to a health care provider but in fact using the Internet first confirms findings from similar studies since 2003, that cancer survivors overwhelmingly prefer the Internet as their source of health information (Finney Rutten et al., 2015).

Hesse et al. (2008) also report that while cancer survivors continue to seek cancer information from a variety of sources over time, they also expressed frustration with the information-seeking process, reflecting a need for providing varied sources and formats of information as well as assistance with how to effectively retrieve it. The authors also point out: "[S]olving the information dilemma for patients and providers will offer one of the most important contributions to population health in the coming decades" (p. 2538).

Education and information provision go hand in hand. H.H. Sheikha Moza bint Nasser, consort of the former Emir of the State of Qatar, Sheikh Hamad bin Khalifa Al Thani, has been instrumental in developing Qatar's National Vision, reforming and advancing education at all levels in Qatar, and is serving as chairperson of Qatar Foundation for Education, Science and Community Development. Her position and role in Qatar society allow her to assert influence on topics that she addresses in public. In the *Qatar National Cancer Strategy*, H.H. Sheikha Moza bint Nasser emphasized the importance of education and increasing knowledge and awareness about cancers, a role that can be easily fulfilled by both libraries and institutions such as QCS:

> *The importance of education may be less apparent, but increasing the understanding of cancer amongst the people of Qatar is vitally important. Doing*

this will remove any misplaced stigma associated with cancer, will help people do what they can to prevent cancer and will encourage people to attend screening. Effective education to reduce the cancer burden is as important as excellent healthcare for those with cancer.

SCH, 2011, p. 1.

Cancer is not a topic that is generally discussed in Qatar. According to QCS staff, patients in Qatar frequently interpret a cancer diagnosis from their doctor as a death sentence. In addition, cancer carries a stigma in the society. The reason for this is that culturally, the norm in the Arabian Gulf Region is not to talk about bad news or misfortune. Complaining and becoming angry about one's negative health status could call one's religious faith into doubt, or one's trust in God's plan for the world. A cancer diagnosis can therefore provoke a wide range of conflicting emotions in patients in the Arabian Gulf, including fear, doubt, and anger. Additionally, due to the strong role of the family in the region, cancer patients also become excessively worried about being able to continue their roles as parents, husbands, and wives.

In addition to stigma, myths and misconceptions can decrease understanding of cancer. The health educators interviewed for this study reported a common question asked at QCS workshops, lectures, and discussions: Could cancer be "caught" from someone else? That is, is cancer an infectious disease? Should cancer patients be avoided?

According to QCS staff, although people have difficulty speaking about cancer in Qatar, the situation has improved over the last decade, from not being spoken about at all, to being dealt with at the medical and personal levels. QCS staff also reported a different attitude toward cancer in Doha than in the outlying areas, where many myths and stigma about the disease still persist among people, who do not regularly use health care services or interact with trained medical personnel. To reach people outside of Doha, QCS staff will first approach parents, teachers, and children in schools, and health care staff in clinics to introduce QCS, its staff, services, and mission before they talk directly to the public. Given the misconceptions and cultural norms about cancer and ill health in Qatar, QCS plays a pivotal role in providing up-to-date, factual and supportive information about cancer outside the clinical setting.

The concerns described above also hold true for young people, who expect to have a long and productive life ahead of them. While it is widely assumed today that young adults, for example, are computer savvy and know how to find suitable health information online, that may not always be the case. Although young Qataris are avid users of mobile technology, this technical expertise is not necessarily equivalent to having strong information-seeking skills. Combined with a lack of relevant health information available online, young people in Qatar find themselves in a similar situation to their peers in other countries (Dobransky & Hargittai, 2012; Jimenez-Pernett, Olry de Labry-Lima, Bermudez-Tamayo, Garcia-Gutierrez, & del Carmen Salcedo-Sanchez, 2010). Young people rely on their family, friends, and health care professionals for health information, despite their technical expertise. The work of the QSC, then, can begin to bridge this information gap among young people in Qatar. Collaboration with libraries is a logical route toward developing health information literacy skills, particularly in regards to Internet-based information, among young people.

According to the QCS staff interviewed, women are the primary users of QCS services and most often for concerns about breast cancer. This is compatible with findings in a study of the National Cancer Institute's Cancer Information Service of more than 83,000 information requests, where 79.4% of the callers were women aged 40 years or older with at least high school education. The study also revealed that women were more likely to ask for general cancer site information and psychosocial issues while men requested more specific treatment information. The study also found that callers with higher levels of education were less likely to ask for support services and more likely to ask about specific treatment information, referral to medical services and organizations (Finney Rutten, Squiers, & Treiman, 2006). Women's information-seeking behavior, then, resonates across cultures from the US to Qatar.

Considering the cultural environment the QCS is working within, tangible results from their outreach work is now becoming apparent. Staff believe the Qatari public is better informed and is addressing cancer both at a medical and personal level. This is a powerful testament of the power of information. QCS staff has found it effective to approach the public in outlying areas in Qatar in a stepwise fashion through public institutions. Cross-cultural similarities can be noted. Similar to reports from the US, the QCS is also finding that the majority of those who seek information and assistance from them are women. Overall, the QCS has made good progress by following H.H. Sheikha Moza's lead to address the misconceptions and stigma about cancer by educating the people of Qatar about the disease.

8.7.1 Role of medical libraries

To ensure its population is well informed, the *National Cancer Strategy* proposes various methods of delivering health information to patients outside the clinical setting, information that is typically available in medical libraries. The report calls on QCS to develop a Web site that serves as the official cancer-related information site in Qatar, to educate and inform the Qatari population about all the different types of cancer. The report also proposes to provide informational materials to Imams for dissemination (pp. 13–14) and to train community pharmacists to disseminate patient education materials provided to them (p. 45). In practice, however, El Hajj and Hamid (2011) reported that pharmacists need to be trained to fill this role; they also needed protected time and appropriate patient education materials. No studies in Qatar have included a medical library as a collaborator in disseminating health information outside the clinical setting.

Qatar Foundation institutions, including WCMC-Q, are referred to as partners in the National Cancer Strategy. WCMC-Q's DeLib possesses strong biomedical information resources and infrastructure. Licenses to proprietary content that are negotiated in the US prohibit resource-sharing activities. Therefore, information resources at WCMC-Q DeLib are available to the public on-site only. WCMC-Q librarians, however, have developed LibGuides—subject guides designed to orient the general searcher to specific topics—on various health subjects relevant to Qatar, including cancer, and aimed at the consumer.

The Qatar National Library, currently under rapid development and construction, will provide national licenses to a wide spectrum of information resources, including health information, to the Qatari public. Thus all residents of Qatar—both health care professionals and the general public—will have free or low-cost access to a wide variety of information resources which are currently restricted in most countries to institutions of higher learning and professional schools.

While medical libraries currently are not active participants in providing health information outside the clinical setting in Qatar, they are well positioned to assume this role. This situation is expected to change as the country continues to evolve, and also as the QCS expands its Web site and informational support.

8.8 Conclusions

This chapter focused on the role that a representative nonclinical health organization in Qatar could play with respect to health information provision to the general public outside of the hospital setting. The challenges to health information provision are considerable. Qatar has undergone tremendous change and economic development in the past few decades. The country is currently building a sophisticated infrastructure in medical education and health care, including a national cancer strategy and a national cancer research strategy. H.H. Sheikha Moza bint Nasser has remarked that to battle cancer in Qatar, the focus should be on education, to address any stigma and misconception people may have about cancer. This is important because culturally speaking, ill health is not generally discussed and cancer has been viewed as a death sentence. In addition, there is only a limited amount of health information available in Arabic, and even less information available online. In the Arabian Gulf region, people customarily rely on family, friends, and health care providers for health-related information, and do not turn to librarians for assistance with their medical questions.

The QCS is the only nonprofit organization in Qatar providing preventive cancer information and support services to the public. The organization fills an important need for this type of support in Qatar, serving as the primary source for cancer information and support in the country.

In response to the authors' questions, QCS staff indicated that most of those approaching them for assistance are women, which aligns with similar findings in the US. Similarly, they have also noticed a change in attitudes toward cancer in the last decade, with people being more willing to undergo cancer testing and publicly address cancer, including cancer survivors sharing their stories.

Future plans for the Society should include expanding their Web site, developing a program for young people and collaborating with a medical library to provide health literacy instruction. While there currently is no library involvement in providing health information outside the clinical setting, this is likely to change as the Qatar National Library acquires medical information resources and national licenses. WCMC-Q DeLib has developed LibGuides on consumer health topics, including cancer, and provides on-site information support to the general public. As the national cancer strategy is implemented, QCS is likely to expand their programs and collaborate with medical information specialists.

Appendix 1: Questionnaire

How many members does the Qatar Cancer Society (QCS) have?
What services does the QCS offer?
What services do members use?
Does QCS host any events? What is the purpose of these events?
What are the sources of any financial support?
Why do people come to QCS?
Where do most people in Qatar get their information about cancer?
What kind of information do members seek about cancer?
Does QCS collaborate with a medical library in Doha?
Where does QCS staff locate the information on cancer on the QCS Web site?
How many hits are there on the QCS Web site annually?
Does QCS produce any written materials or pamphlets?
What are the informational challenges facing QCS (language, culture, and religious issues) in Qatar's extremely diverse society?
What are the attitudes toward cancer in Qatar?
Do people have difficulty speaking about cancer in Qatar?

References

Al-Ayed, I. H. (2010). Mothers' knowledge of child health matters: are we doing enough? *Journal of Family Community Medicine, 17*(1), 22–28.

AlGhamdi, K. M., & Almohedib, M. A. (2011). Internet use by dermatology outpatients to search for health information. *International Journal of Dermatology, 50*, 292–299.

Al-Ghareeb, A. A. A. (2009). The role of health information resources in forming the health awareness of Saudi women: applied study in Riyadh. *Journal of the Social Sciences, 37*(2), 45–88.

Al-Harbi, A. (2011). Healthcare providers' perceptions towards health information applications at King Abdul-Aziz medical city, Saudi Arabia. *International Journal of Advanced Computer Science and Applications (IJACSA), 2*(10), 14–22.

Al-Meer, F. M., Aseel, M. T., Al-Khalaf, J., Al-Kuwari, M. G., & Ismail, M. F. (2011). Knowledge, attitude and practices regarding cervical cancer and screening among women visiting primary health care in Qatar. *Eastern Mediterranean Health Journal, 17*(11), 855–861.

Alnasir, F. A., & Skerman, J. H. (2004). Schoolteachers' knowledge of common health problems in Bahrain. *Eastern Mediterranean Health Journal, 10*(4–5), 537–546.

Al-Rowais, N., Al-Faris, E., Mohammad, A. G., Al-Rukban, M., & Abdulghani, H. M. (2010). Traditional healers in Riyadh region: reasons and health problems for seeking their advice. A household survey. *Journal of Alternative & Complementary Medicine, 16*(2), 199–204.

Al-Sane, M., Bourisly, N., Almulla, T., & Andersson, L. (2011). Laypeoples' preferred sources of health information on the emergency management of tooth avulsion. *Dental Traumatology, 27*(6), 432–437.

Al-Sughayr, A. M., Al-Abdulwahhab, B. M., & Al-Yemeni, M. R. (2010). Primary health care physicians' knowledge, use, and attitude towards online continuous medical education in Saudi Arabia. *Saudi Medical Journal, 31*(9), 1049–1053.

Alsughayr, A. (2013). King abdullah bin Abdulaziz arabic health encyclopedia (www.kaahe.org): a reliable source for health information in Arabic in the internet. *Saudi Journal of Medicine & Medical Sciences, 1*(1), 53–54.

Altuwaijri, M. (2011). Health information technology strategic planning alignment in Saudi hospitals: a historical perspective. *Journal of Health Informatics in Developing Countries*, *5*(2), 338–355.

Bener, A., Abdul Rahman, Y. S., Abdel Aleem, E. Y., & Khalid, M. K. (2012). Trends and characteristics of injuries in the State of Qatar: hospital-based study. *International Journal of Injury Control and Safety Promotion*, *19*(4), 368–372.

Bener, A., El Ayoubi, H. R., Moore, M. A., Basha, B., Joseph, S., & Chouchane, L. (2009). Do we need to maximise the breast cancer screening awareness? Experience with an endogamous society with high fertility. *Asian Pacific Journal of Cancer Prevention*, *10*(4), 599–604.

Dobransky, K., & Hargittai, E. (2012). Inquiring minds acquiring wellness: uses of online and offline sources for health information. *Health Communication*, *27*, 331–343.

Donnelly, T. T., Al Khater, A. H., Al-Bader, S. B., Al Kuwari, M. G., Al-Meer, N., Malik, M., et al. (2013). Beliefs and attitudes about breast cancer and screening practices among Arab women living in Qatar: a cross-sectional study. *BMC Women's Health*, *13*, 49.

El Hajj, M. S., & Hamid, Y. (2011). Breast cancer health promotion in Qatar: a survey of community pharmacists' interests and needs. *International Journal of Clinical Pharmacy*, *33*, 70–79.

Finney Rutten, L. J., Agunwamba, A. A., Wilson, P., Chawla, N., Vieux, S., Blanch-Hartigan, D., et al. (February 26, 2015). Cancer-related information seeking among cancer survivors: trends over a decade (2003–2013). *Journal of Cancer Education* (Epub ahead of print).

Finney Rutten, L. J., Squiers, L., & Treiman, K. (2006). Requests for information by family and friends of cancer patients calling the National Cancer Institute's cancer information service. *Psycho-Oncology*, *15*, 664–672.

Four Seasons Hotel Doha co-sponsors Terry Fox Marathon of Hope run 2014 in conjunction with national sport day. (February 6, 2014). Retrieved from http://press.fourseasons.com/doha/hotel-news/2014/four-seasons-hotel-doha-co-sponsors-terry-fox-marathon-of-hope-run-2014-in-conjunction-with-national-sport-day/.

General Secretariat for Development Planning (GSDP). (2008). *Qatar national vision 2030*. Doha: GSDP.

Gotting, F. (2006). *Healing hands of Qatar*. Doha: The Author.

Hamad Medical Corporation (HMC). (2008). *Annual health report 2008*. Doha: Hamad Medical Corporation.

Hesse, B. W., Arora, N. K., Beckjord, E. B., & Finney Rutten, L. J. (2008). Information support for cancer survivors. *Cancer*, *112*(11 Suppl.), 2529–2540.

International Energy Agency (IEA). (2013). *Key World Energy Statistics 2013*. Paris: IEA.

Jimenez-Pernett, J., Olry de Labry-Lima, A., Bermudez-Tamayo, C., Garcia-Gutierrez, J. F., & del Carmen Salcedo-Sanchez, M. (2010). Use of internet as a source of health information by Spanish adolescents. *BMC Medical Informatics and Decision Making*, *10*, e104.

Khatri, S. S. (March 16, 2014). *Ooredoo pledges QR30 million toward new cancer facility in Qatar*. Doha News, Retrieved from http://dohanews.co/ooredoo-pledges-qr30-million-toward-new-cancer-facility-qatar/.

Kronfol, Z., Ghuloum, S., & Weber, A. (2013). Country in focus: Qatar. *Asian Journal of Psychiatry*, *6*, 275–277.

Moawed, S. A., & Saeed, A. A. (2000). Knowledge and practices of mothers about infants' diarrheal episodes. *Saudi Medical Journal*, *21*(12), 1147–1151.

Qatar Cancer Society. (n.d.). *Qatar Cancer society in words*. Doha: QCS.

Rodríguez del Pozo, P., Fins, J. J., Helmy, I., El Chaki, R., El Shazly, T., Wafadari, D., et al. (2012). Truth-telling and cancer diagnoses: physician attitudes and practices in Qatar. *The Oncologist*, *17*(11), 1469–1474.

Saleh, N., Bener, A., Khenyab, N., Al-Mansori, Z., & Al Muraikhi, A. (2005). Prevalence, awareness and determinants of health care-seeking behaviour for urinary incontinence in Qatari women: a neglected problem? *Maturitas, 50*(1), 58–65.

Supreme Council of Health (SCH). (2011). *National cancer strategy: The path to excellence*. Doha: Supreme Council of Health.

Supreme Council of Health (SCH). (2012). *National cancer research strategy*. Doha: Supreme Council of Health.

Weber, A. S. (2011). Folk medicine in Oman. *International Journal of Arts and Science, 4*(23), 237–274.

Weber, A. S. (2013). Youth unemployment and sustainable development: case study of Qatar. *Revista de Asistență Socială, 12*(1), 47–57.

Weber, A. S., Verjee, M., Rahman, Z. H., Ameerudeen, F., Al-Baz, N. (2015). Typology and credibility of Internet health websites originating from Gulf Cooperation Council countries. *Eastern Mediterranean Health Journal, 20*(12), 804–811.

World Health Organization (WHO). (2010). *Framework for action on interprofessional education and collaborative practice*. Geneva: WHO.

Health information and older adults

Kay Hogan Smith
University of Alabama at Birmingham (UAB), Lister Hill Library of the Health Sciences, Birmingham, Alabama, United States

9.1 Introduction

This chapter presents evidence-based considerations and techniques for presenting health information targeted to the older adult population, through multiple formats and settings, by various information providers. While much has been written for clinical health information providers on communicating with elderly patients, this chapter's focus is on presenting information in a nonclinical environment, whether in person or via other formats. The reader is encouraged to consult the references provided at the end of the chapter for additional information about this growing area of research.

9.2 Background

Like most other segments of the population, older adults are characterized by a vast array of cultural backgrounds, life experiences, socioeconomic circumstances, abilities, and disabilities. According to a recent longitudinal study of older adults, "[I]n terms of change and development, there are more differences among older people than among younger people" (National Institute on Aging, 2010, p. 1). Gerontologists—engaged in the "scientific study of the process of aging and the problems of aged persons" (Meiner, 2011, p. 4)—have long recognized the wide variation among older adults. To address these differences demographically, older adults are typically grouped into subpopulations of "young-old" (aged 55–75) and "old-old" (formerly aged over 75, now generally those aged over 85) (Hooyman & Kiyak, 2011; Neugarten, 1974). Even within these divisions, there are substantial differences among individuals. While aging is correlated with an increase in health issues, it is not inevitably so (Lowsky, Olshansky, Bhattacharya, & Goldman, 2014). Indeed, there is a body of literature distinguishing aging from disease.

What is undeniable, however, is that the older adult proportion of the population is increasing. According to the US Administration on Aging (2013), Americans over 65 accounted for almost 14% of the total population in 2012. In 1900, their share was only 4.1% of the population. By 2050, those over 65 will make up 20% of the US population (Lowsky et al., 2014).

Along with the increasing share of older adults in the population, the United States, like other countries, is experiencing a rapid increase in the numbers of informal or family caregivers. These people may or may not be elderly themselves, although it is more often

Meeting Health Information Needs Outside of Healthcare.
Copyright © 2015 by K.H. Smith. Published by Elsevier Ltd. All rights reserved.

a middle-aged or "young-old" child of the older adult serving as caregiver, particularly a daughter. Current estimates indicate that approximately 42 million Americans serve as informal caregivers, providing unpaid care to support the formal health-care structure. This represents hours of physically and mentally demanding work (Gillick, 2013).

The "graying of America" means that health information providers of all types— for example, librarians, nurses, doctors, health educators, journalists, online health information designers, and content providers—must themselves be informed about the special needs, barriers, and circumstances of the older adult information seeker. At the same time, they need to remain alert to any stereotyping or ageist assumptions that may come up during the information transaction (Henry & Henry, 2004). They will also have to consider the inclusion of family caregivers as information audiences, particularly in situations where both older adult and caregiver might be the recipient of information.

What are the special needs and circumstances that may affect health information provision directed to older individuals? Older adults commonly experience sensory declines in vision and hearing. It is estimated that about 14% of older adults have trouble seeing without corrective lenses. That percentage rises to about 23% of those over 85. Hearing loss is more prevalent, with 31% of older women and 46% of older men reporting difficulty hearing without a hearing aid (Federal Interagency Forum on Aging-Related Statistics, 2012). In addition, there may be changes in thinking processes and cognition associated with illness, sometimes referred to as "the three D's" of depression, delirium, and dementia. There is also frequently a subtle slowing of cognitive processing and working memory with aging that does not necessarily progress to dementia (Gerontological Society of America, 2012; Meiner, 2011). The US Medical Expenditure Panel Survey (MEPS) reported that 5% of adults aged 65 and over, were affected by diagnosed cognition disorders in 2007; that percentage increased to over 18% for those over 85 (Stagnitti, 2011). Physical pain and illness can also create a barrier to comprehension of health information in older adults as well as other populations (Hochhauser, 2012).

The good news from these statistics may be found in the apparent majorities of older adults who do not experience these physical and mental limitations. In addition, it might be argued that regardless of health status, older adults, like other populations, need reliable health information to make the best decisions for their own health and on behalf of others (Manafo & Wong, 2012).

While there has been limited research into general health information seeking among older adults, there have been a few targeted studies, including a recent qualitative inquiry among British older adults (Hurst, Wilson, & Dickinson, 2013). There are differences between older and younger adults' health information seeking habits. Hurst et al. found that motivators for health information seeking included symptomatic triggers, such as unusual heartburn or pain, either in themselves or in a loved one. However, past health experiences, as well as societal attitudes toward aging also affected individuals' motivation for seeking health information and care. Older adults who perceived their health issue as a "natural" part of the aging process, or who anticipated that their doctors would view it through that lens, were less likely to follow up with research or care. Utilization of particular sources of health information for older adults in this study was influenced by what was most conveniently available to them, particularly social networks and prescription medication leaflets. Some older adults did utilize Internet sources, however. Other studies have noted the increasing use of the Internet by older adults, with health information consistently among the top motivators for going online (Hallows, 2013). Recent Pew Research Center data indicate that of the 59% of older Americans who use the Internet, finding health information is important for many of them (Zickuhr, 2014).

However, while older adults increasingly embrace computers and online technology, there are significant differences among those older adults who adopt technology in their daily lives and those who do not. Nonusers of technology tend to be older, poorer, and sicker (Smith, 2014); this is an important consideration for readers of this chapter. The Internet is still of limited use in reaching older adults, and those older adults willing to learn about new technologies may need a substantial amount of assistance to use it (Zickuhr, 2014). On the other hand, it is important to note that younger caregivers are typically more comfortable using technology to find health information and support (Pew Research Internet Project, 2015).

9.3 Settings: where do older adults go for information?

Before describing approaches to help information providers overcome potential physical or cognitive barriers to comprehension with older adult clients, it is important to consider the settings and means by which seniors typically receive such information. It is also necessary to consider the cultural context of medical and health information in the twenty-first century. In addition to the technological demands noted above, today's older adults are dealing with a societal shift in the nature of their encounters with the health-care system. This has changed from a top-down approach, in which the physician or nurse dispensed information on a need-to-know basis, to a patient-centered approach, in which it is assumed that the patient should be and wants to be involved in all health-care decisions. The patient-centered approach necessitates increased information accumulation (Centers for Disease Control and Prevention, 2009a). Information is still dispensed by health-care providers in clinical settings, but it is also disseminated by the traditional media, including television, radio, newspapers, books, and magazines. Older adults, especially non-Hispanic whites, are an important market for traditional media (Korzenny & Korzenny, 2007). Brochures remain a common method of dispensing health information, although these are increasingly published online only with downloading capability available. Online "eHealth" tools (such as patient portal sites) are more and more popular. Social networking patient communities such as PatientsLikeMe.com and health-care rating sites add to the bewildering array of Internet-based information available to older adults (Centers for Disease Control and Prevention, 2009a).

Nonclinical settings where older adults receive health information are numerous. They include senior nutrition and activity centers, community college courses, libraries, pharmacies, health fairs, farmers markets, churches and other places of worship, shopping malls, volunteer organizations, assisted living facilities and adult day care centers, and homes (California Department of Social Services, 2012). Meeting older adults in settings "where they are" is a commonsense approach to information dissemination, but may require advance collaboration with the institutional gatekeeper (National Council on Aging, 2012). While in-person delivery of health information avoids potential technological barriers to communication, these settings can provide teachable moments in terms of demonstrating online resources for older adults by utilizing tablets or laptops.

For older adults who are computer-literate, connecting online with others who have similar health issues can provide important psychosocial support as well as tips for coping with a condition (Fox, 2011). Friends and family can also be important sources of health information (Rainie, 2013). This exchange of information can be personal or through social media.

9.4 Health information format considerations

Beyond the dichotomy of online and offline formats, there are multiple presentation options for providing health information to older adults—each of which that can help or hinder their understanding. Possible formats include print (e.g., pamphlet, book, workbook, magazine articles); video and/or audio messages (including recorded telephone messages); charts and diagrams; in-person presentations (one-to-one or in a group setting); Internet-based (Web sites, e-mail, and social media); physical models (e.g., food models to demonstrate nutritional portion sizes), games, puzzles and toys; or some combination of any or all of these. Evidence-based guides to optimal presentation formats for health information communication with older adults are limited. However, some tips for format selection may be gleaned from the general health literacy and geriatric health education literature (Hainsworth, 2003).

9.4.1 Print format

With the increasing focus on health literacy issues in recent years, print health information has come under particular scrutiny. Up-to-date, reliable, easy-to-read health information in print, especially when it supplements verbal or recorded instructions, is both needed and desired by older adults with health concerns, even those with low literacy (Lorig, 2001). Besides educational pamphlets typically provided in clinical settings and pharmacies, print information is also accessed in popular magazines, newspapers, newsletters, and books. Some workshops and health education programs provide workbooks for participants as reinforcement of instructions or as self-guided instruction (Peterson et al., 2014). An obvious advantage to providing health information in print is its accessibility, particularly for older adults without access to or experience with computers. Printed health information materials should incorporate good design elements for older adults to be useful. Suggestions for formatting and design of print materials addressing potential visual, cognitive, and literacy issues among older adults are presented later in this chapter.

9.4.2 Audiovisual formats

There has been increased attention to audiovisual methods of transmitting health information in recent years. Television and radio reporting of health information has been shown to affect utilization of health-care services (Grilli, Ramsay, & Minozzi, 2002). Other traditional audiovisual means of transmitting information include PowerPoint or Prezi slides as enhancements to in-person or online presentations (Hainsworth, 2003). Examples of audiovisual information presentation formats include tailored health messages via DVD and online video, methods increasingly employed to address both educational and cultural barriers to comprehension (Houston et al., 2011; Lapane, Goldman, Quilliam, Hume, & Eaton, 2012). Recorded messages via MP3 and telephone are also useful formats for transmitting health information (Austin, Landis, & Hanger, 2012; Eames, Hoffmann, Worrall & Read, 2011; Stiles, 2011). Many audiovisual formats allow for self-pacing, whether for individual or group audiences, with telephone recordings timed for optimal reception by the older adult (Hong, Nguyen, & Prose, 2013). The formats can incorporate interactive voice or keypad response design to tailor the messages (Austin et al., 2012), and can integrate video and voice messages via smartphones.

Globally, "mHealth" (or mobile health) innovations show promise for disseminating public health information via smartphone or mobile devices to remote underserved areas. Although smartphone ownership among older adults lags behind that of other adult populations, it is increasing. For example, Scientific Animations without Borders (http://sawbo-illinois4.org/) uses animation in more than 50 languages to deliver health education about topics including Ebola and clubfoot—and could be considered for health information outreach to older adults in rural areas in the United States as well as Africa (Zickuhr, 2014).

9.4.3 Visual materials

This presentation format category includes displays, posters, charts, and other visual materials to supplement learning. The applicability of each type of material depends upon the setting and content of health information being presented. Employing such materials where appropriate as a means of encouraging participation can improve an audience's retention of the information beyond that generally witnessed in purely didactic presentations (Lee, Hoti, Hughes, & Emmerton, 2014).

9.4.4 In-person presentations

Evidence indicates that older adults are very receptive to information provided in person, particularly in small interactive groups with other seniors (Haber, 2003). Learning with others who are experiencing similar health issues can provide social support as well as validation and peer modeling of specific health behaviors (McAlister, Perry, & Parcel, 2008); thus, peer groups of older adults can be useful ways to present information about specific diseases and conditions. Community health advisors in global settings have devised creative methods to communicate health messages, for example, performing songs or short dramatic sketches with target audiences to promote healthy behaviors (Aja, Umahi, & Allen-Alebiosu, 2011). In-person presentations of health information have the advantage of being easily tailored to the individual or group recipient (Jolles, Clark, & Braam, 2012). Specific in-person communication tips for older adults are presented later in this chapter.

9.4.5 Online formats

Health information on the Internet is of primary interest to many providers and users in the twenty-first century, including older adults (Fox, 2013). There are many advantages to providing health information online via Web page, Facebook, Twitter, YouTube, or other social networking sites; online support groups; or blogs. The medium is relatively inexpensive for developers, and easy to edit and update. Users can easily—if not always skillfully—search for health information online (Lee et al., 2014). The drawbacks to the Internet as a health information vehicle, particularly for older users, include technology access barriers, inexperience with its use, distrust, and simple information overload (Smith, 2014). Older adult seekers of online health information should also be informed of the importance of critically evaluating the content they retrieve from the Internet, and educated about the best ways to do so (Hong et al., 2013).

9.4.6 Demonstration models and materials

Physical models and representations are frequently employed by health educators in presentations to enhance communication of health information and encourage active participation by the audience (Hainsworth, 2003). Models can take the form of anatomical parts, or simulate foods, for example, bags of sugar to demonstrate the sugar content in soft drinks, These materials can be produced by the presenter but may also be purchased from such commercial suppliers as ETR Associates (see http://pub.etr.org/) and Nasco "Hands on Health" (see http://www.enasco.com/healtheducation/). For an example of products targeting older adults as learners, see the "Senior Activities" (http://www.enasco.com/senioractivities/) online catalog.

9.4.7 Games and puzzles

In addition to the demonstration and visual materials categories discussed above, games, puzzles, and toys can engage the older adult audience, thereby strengthening their retention of the health information presented as well as influencing healthcare decisions. Health-related computer and video games, generally thought to be the domain of the young, are increasingly being tested by researchers on older adult audiences. One study demonstrated improvements in understanding of health-related quality-of-life factors involved in prostate cancer treatment decisions through the use of a computer-based interactive game (Reichlin et al., 2011). "Serious" video games have also been shown to impact health education outcomes (Primack et al., 2012). Other innovative games to reinforce health information employ simpler, paper-based media. Examples of the latter include "Jeopardy" simulations and trivia type games (Kennedy, 2006; Walker, Stevens, & Persaud, 2010). There are also Web-based quiz game educational products, for example, FlipQuiz, available at http://flipquiz.me/.

9.5 Format summary

Each of these health information presentation formats has unique advantages and disadvantages when used with older adult audiences. Print-based media and materials are familiar to older adults, relatively inexpensive and easily accessible; however, poorly designed print materials are ineffective (Haber, 2003). Audiovisual media can help address physical or cognitive barriers to comprehending health information, as well as heighten interest in the information presented. However, there may be barriers to the distribution of audiovisual materials among older adults as well as barriers to acceptance of audiovisual health information materials on their own, as opposed to supporting a broader health information presentation (Hong et al., 2013). Online health information resources are plentiful; the Web platform is fairly easy and inexpensive for information producers. However, there may be barriers for older adults in terms of accessibility and familiarity with the technology. Demonstration materials and models can enhance participation in health information outreach activities, but can be expensive to purchase. Finally, online or physical games and puzzles can reinforce comprehension of health information among older adult audiences as well as younger, but they are limited in their current availability. In summary, it is best to consider carefully the scope and nature of the health information topic as well as the particular composition of specific older adult audiences when designing or selecting presentation formats. It is advisable to consider combinations of presentation formats when appropriate (Higgins & Barkley, 2004). Older adults, like others, absorb more information when the information is demonstrated as well as told to them. If they are allowed to practice what they learn, retention is even better (Hunt, 2005).

9.6 Health information comprehension among older adults: barriers and solutions

The following section reviews specific physical and mental barriers to health information comprehension among older adults, and suggests ways in which information providers can surmount those barriers.

9.6.1 Vision problems

Approximately 14% of older adults self-identified as having trouble seeing even with glasses or contacts in 2010 (Federal Interagency Forum on Aging-Related Statistics, 2012). Visual problems include age-related declines, such as difficulty reading small print (*presbyopia*), dry eyes and floaters, or specks across one's visual field. More serious issues include age-related macular degeneration, cataracts, glaucoma, and eye conditions related to diabetes or stroke. Macular degeneration, which has some hereditary components, affects close vision and is most common in white women (Friedman, 2011). Cataracts are a common diagnosis among older adults and treatable through surgery. Cataracts cause a clouding of the lens in the eye such that visual sharpness is diminished. Additionally, yellowing of the lens causes color distortion (Friedman, 2011). Glaucoma can occur at any age, but adults over 60 have a higher risk of developing the disease. Glaucoma typically affects the peripheral vision first and can lead to total blindness if left untreated. Treatment for glaucoma includes both medical and surgical options (Friedman, 2011). Finally, diabetic complications affecting the eyes are an important issue, because of the high prevalence of older adults with diabetes. Diabetes mellitus was present in over 21% of people aged 65–74, and almost 20% of those over 75 in 2012 (Blackwell, Lucas, & Clarke, 2014); with almost 30% of diabetic older adults (over 65) in the United States are affected by diabetic retinopathy (Zhang et al., 2010). Prevention is the preferred approach to this condition. Zhang et al. (2010) note that improved screening among people with diabetes may be reducing rates of diabetic retinopathy. As the incidence of stroke increases with age (Ovbiagele & Nguyen-Huynh, 2011), so do the visual complications of stroke, which often affect one side of the visual field—from 20% to 57% of people who have experienced a stroke. Interventions for stroke-related visual defects are limited (Pollock et al., 2011).

Some studies have linked visual deficits with lower health literacy and lower self-management skills. However, it has also been shown that older adults with visual impairments can be provided with health information in a way that promotes their engagement in their care thus potentially improving their health outcomes (Harrison, Mackert, & Watkins, 2010; Press, Shapiro, Mayo, Meltzer, & Arora, 2013).

9.6.1.1 Communication tips

In a one-to-one interaction, the information provider can clarify the existence of vision problems by simply asking the older adult if vision is an issue. When a person identifies macular degeneration as a vision problem, ask him or her about any preference for left or right peripheral vision, as direct visual acuity is often diminished with this condition and with stroke-related visual field defects.

In a group setting, asking older adults to self-identify visual problems may raise privacy issues. It is best to use a "universal precautions" approach, wherein the information provider assumes that at least some of the audience members have visual difficulties and adjusts the information presentation accordingly (Lubinski, 2010).

9.6.1.2 Environment tips

Communicating health information to older adults with vision problems in person may be helped by simple environmental adjustments. For example, it is important to pay attention to lighting. Glare is sometimes an issue, especially for those with cataracts, but insufficient lighting for reading is also an obstacle. Task lighting may be better than overhead lighting for individual work—for example, filling out a questionnaire.

When turning lights on and off during a presentation, allow a little extra time for older adult eyes to refocus. Individuals who use glasses or contacts or magnifying devices for print materials should be encouraged to use them if these assistive devices are handy (Jinks & Baker, 1987).

9.6.1.3 Formatting tips

For printed materials, the font size for older adults should be at least 14 points. For both printed and visual presentations, avoid elaborate styles of print, such as script, and always use fonts with "feet" or serif bases, such as Times New Roman, in the body of the message. Organizing the material in "chunks" and using headings can also help make the information easier to read. Avoid using block capital letters. Use design elements such as boldface to highlight parts of the text, but avoid underlining and italics. Use clear, relevant images—photographs or cartoons—to demonstrate points, and label them with captions.

White space is also important! The Centers for Disease Control's (CDC) *Simply Put* (2009b) guide suggests at least 10% white space per printed page. Bullet points are also useful in breaking up blocks of text. In addition, the use of contrasting colors in signs or presentation slides—for example, dark text on a pale, but not stark white, background—can help older adults see them better (Jinks & Baker, 1987). In general, however, avoid using blue, green, or violet lettering (Thomas, 2011).

Recorded narration of print information may be useful for older adults with vision problems. Current computer technology makes it easy to provide voice-over narration of print. Screen readers can help on the user's end (see AccessibleTech.org, 2014, for an example) but information producers can make it easier for older adults by incorporating narration into an online presentation. For example, PowerPoint provides the option of recording an accompanying narration to a slideshow. An example of a health information resource making good use of narration technology is Healthy Roads Media (see http://www.healthyroadsmedia.org/index.htm). The health information topics on this site all offer easy access to narrated videos as well as print transcripts and simple audio narration.

9.6.2 Hearing problems

Hearing loss is fairly common among older adults, especially men. Data from 2010 indicated that 46% of men over 65 had trouble hearing, while 31% of older women reported hearing loss (Federal Interagency Forum on Aging-Related Statistics, 2012, p. 28). The most common form of hearing loss is a sensorineural type called presbycusis, also known as "age-related hearing loss." This typically affects both ears, resulting in difficulty hearing conversational tones and high-pitched sounds without a hearing aid. Older adults with hearing loss may also experience increased difficulty in situations with loud background noise.

Tinnitus, or ringing in the ears, also becomes more prevalent with age, with 14.3% of US adults between 60 and 69 reporting frequent episodes of tinnitus (Shargorodsky, Curhan, & Farwell, 2010). Persons affected by hearing loss may be advised to use hearing aids, although the expense may be prohibitive for some (Donahue, Dubno, & Beck, 2014). In severe cases, cochlear implant surgery may be recommended. Coping strategies also include communication tips and, in the case of tinnitus, desensitization and other therapies (Friedman, 2011). Loss of hearing can affect the quality of an older adult's life, particularly affecting social engagement (Health in Aging Foundation, 2012; Smith, 2012).

9.6.2.1 Communication tips

As discussed in the section on vision problems, it may be necessary to ask the individual directly if he or she is hard of hearing or uses a hearing aid. If so, the information provider may suggest that the individual use the hearing aid during the information presentation. Studies have identified numerous reasons for nonuse of prescribed hearing aids among those with hearing difficulties, ranging from cost to effectiveness and comfort of the hearing aid to psychological resistance (McCormack & Fortnum, 2013). For one-to-one information delivery, it is helpful to face the individual directly on the same level to help him or her read your lips (Gerontological Society of America, 2012). If the information presentation is to a group of older adults, the presenter can use the same universal precautions approach described earlier, that is, to assume that at least some present have a degree of hearing loss and adjust the presentation accordingly. Using a microphone can ease the listening concentration task for older adults with hearing loss; so can enunciating clearly, speaking at a moderate pace, and using a lower-pitched vocal tone (Jinks & Baker, 1987). It will also help to use simple, plain language, and check for understanding on occasion. Augmenting vocal instructions with printed handouts or transcripts of the presentation may improve understanding and reduce frustration on the part of the older adult audience.

These same universal precautions in communicating with persons with hearing problems apply to other information transactions besides in-person interactions. For example, communication can be enhanced through amplified audio by the user, clear enunciation by the speaker, and moderate pacing of speech in recorded presentations and auxiliary print information. For example, the NIH SeniorHealth online information videos available to consumers at http://nihseniorhealth.gov/videolist.html incorporate clear captioning and downloadable transcripts as well as the ability to increase the volume for each video posted (NIA, 2013a).

9.6.2.2 Environment tips

Background noise may interfere with message comprehension in the older adult with hearing loss. Thus, diminishing ambient noise whenever possible is vital for in-person as well as recorded information presentations. If necessary, ask the individual to turn down a television or radio playing in the background during an in-person information transaction (Gerontological Society of America, 2012). It may also be helpful for in-person presenters to ensure that they are in a well-lighted area, to help those who compensate for hearing loss by reading lips (Barbaro & Noyes, 1984). Repetition of unamplified comments or questions from an audience will help those who have not disclosed a hearing difficulty (Jinks & Baker, 1987).

9.6.2.3 Formatting tips

Since older adults with hearing loss depend more on their vision to receive information, it is generally advisable to incorporate print and/or other visual aids into any audio-recorded or in-person presentation of health information (Gerontological Society of America, 2012; Viggiani, 2003). However, providing auditory information to people with hearing loss should not be automatically dismissed, particularly if the audio is clear and the volume can be adjusted by the user. Online communication, such as captioned videos or online newsletters, is particularly appropriate for this population if they have a minimal familiarity with the technology (Gonsalves & Pichora-Fuller, 2008). Print materials for older adults with hearing loss, particularly those with accompanying visual decline, should follow the same formatting and design principles noted earlier in this chapter.

9.6.3 Cognitive issues

Older adults frequently worry about forgetfulness as they age, but minor memory lapses are generally not cause for concern. However, as noted in the background section of this chapter, changes in cognitive processing can occur with aging and illness, most frequently depression, delirium, and dementia. The causes range from medication to physical diseases (such as heart disease) to fundamental mood disorder to dementia and Alzheimer's disease (Kazer, 2011). In addition to the MEPS data on diagnosed cognitive disorders cited earlier (Stagnitti, 2011), the Centers for Disease Control and Prevention (CDC) reported that almost 13% of adults over the age of 60 experienced some degree of confusion or memory loss (CDC, 2013). Although depression is less prevalent among older adults than other populations, in 2010 it was estimated that over one million community-dwelling older adults experienced major depression (Substance Abuse and Mental Health Services Administration, 2012). The prevalence increases with institutionalization (Aziz & Steffens, 2013). Once diagnosed, depression in the elderly may be treated with drugs, exercise, or psychotherapy options (Kane, Ouslander, Abrass, & Resnick, 2013).

The National Institute on Aging estimates that about five million older adults have Alzheimer's disease, the most common cause of dementia, a serious cognitive disorder (National Institute on Aging, 2013b). The *Diagnostic and Statistical Manual of Mental Disorders*, fifth edition, defines "neurocognitive disorders" as the "group of disorders in which the primary clinical deficit is in cognitive function, and that are acquired rather than developmental" (American Psychiatric Association, 2013). The risk for dementia increases steadily with age. Some causes of dementia are reversible, but many are not. Treatment for Alzheimer's and other dementias typically focuses on moderating the symptoms of the disease, generally through the use of medications and behavioral approaches (Sink & Yaffe, 2014). An early form of dementia, mild cognitive impairment, has recently been identified as a potential precursor to Alzheimer's disease (Breitner, 2014). Older adults with early stage dementia can learn, and even those with later stage cognitive decline can respond to information, particularly if it is backed up with "props" such as calendars or notebooks (Buettner & Fitzsimmons, 2009; Davis, 2005).

Delirium is sometimes associated with the use of multiple medications as well as acute illness and surgery in the elderly, and is particularly prevalent among those in critical care hospital units (up to 80% of intensive care unit patients) (Vasilevskis, Han, Hughes, & Ely, 2012). The nonclinical information provider will not likely have cause to promote health information directly to the patient with delirium, although the patient's family may seek information about the issue.

9.6.3.1 Communication tips

Information overload is to be avoided when communicating with older adults. Thus, it is advisable to focus on the most important issues when designing the presentation. Simplifying sentence structure and using positive, active-voice, concrete language is also recommended in order to prevent confusion. A normal speaking rate with clear enunciation helps the older adult listener better understand a verbal presentation of information. It may be necessary to repeat important points (Gerontological Society of America, 2012). Whether or not the older adult information recipient has cognitive issues, it is important to allow him or her time to process the information (Speros, 2009). Older adults with dementia may be particularly sensitive to impatience or other negative attitudes on the part of the presenter. Empathy and a gentle tone might help avoid agitation in the older adult with cognitive disorder. Incorporating storytelling into new material has been demonstrated as an effective learning technique for older adults (Cangelosi & Sorrell, 2008). For instance, individuals or members of an audience can be asked to think about their personal experiences with a health issue and then relate those experiences to new material. Asking questions requiring simple "yes" or "no" responses is considered best for those with more severe cognitive decline, but open-ended questions that do not overly tax the older adult's short-term memory are generally appropriate (Gerontological Society of America, 2012).

9.6.3.2 Environment tips

Experts recommend that talking to small groups of older adults with cognitive decline will avoid overstimulation and promote learning (Day, Carreon, & Stump, 2000). Therefore, the health information provider who speaks to an older adult audience that may include individuals with cognitive decline should limit the size of the audience. In addition, environmental considerations should follow those described earlier for older adults with vision and hearing problems—that is, providing good lighting without glare and sound amplification if necessary. Ambient noise should be minimized, and enough space should be available to minimize the potential for agitation and to allow audience members to safely navigate the venue on foot or with assistive devices (Davis, 2005).

9.6.3.3 Formatting tips

There is limited research into the effects of various information formats on those with cognitive issues, as distinct from literature addressing older adults in general.

However, the literature supports some considerations in format with regard to older adults with cognitive decline, in addition to the in-person presentation guidelines provided above.

In general, any methods of presenting information in any format that reduce the cognitive burden on the older adult learner are more likely to be successful. For printed health information, for example, the text and design guidelines described under Section 9.6.1 above are also relevant for primary or supplemental materials provided to older adults with cognitive issues. The audience for such materials might also benefit from a suggestion to keep the information in a place they are likely to see it at home, for example, taped to the refrigerator.

Audiovisual materials should also follow the guidelines presented earlier. It is important to keep in mind the limitations of those with cognitive problems in deciphering complex messages; experts suggest keeping information transactions brief, with no more than 3–5 key messages for all formats. Suggesting health behaviors tied to daily cues, such as brushing one's teeth, is another strategy employed by nurse educators in working with older adults (Speros, 2009).

9.7 Comprehension summary

With the increase in the aging population, health information providers should consider the potential challenges as well as the rewards of preparing and presenting information to older adults. The older adult population is a diverse group in terms of socioeconomic background, education, ethnicity and culture, language as well as physical and mental health status. The focus of this chapter has been on communication techniques, setting, environmental and information format considerations.

In addition to the suggestions provided above, general guidelines stress the importance of treating the older adult respectfully and patiently. It is advisable for the health information provider to reflect carefully on any tendency toward ageist attitudes when preparing to present health information to an older adult individual or audience, no matter what presentation format will be used. Ageism may manifest in one's tone of voice, or a tendency to infantilize the older adult—for instance, referring to him or her as "sweetheart" or adopting a patronizing tone in the information presentation. Such ageism may be reflected in an assumption that the older adult is incapable of learning or self-care (Williams, Kemper, & Hummert, 2005). As noted above, even older adults with cognitive impairment can learn, when information presented to them does not place too many demands on their short-term memory and verbal capacity (Speros, 2009). In particular, the presenters should take care to avoid ignoring the older adult while addressing the information to an accompanying caregiver. When caregivers are present, the presentation should be inclusive of both parties (Gerontological Society of America, 2012).

In addition, information providers should take into account the cultural and ethnic makeup of the older adult audience, incorporating accompanying images, or even cultural attitudes into the materials which are reflective of the particular group (CDC, 2009b; Price et al., 2011).

In summary, health information providers of all professions and settings are encouraged to remember that their task—promoting health-related learning—is even more important when addressing learning in the older adult audience. Not only is this population at higher risk of health concerns that might be alleviated by the information provided, but the learning itself can help the older adult maintain health by promoting self-efficacy and confidence. The social interaction with other adult learners itself can have a positive impact.

Moreover, evidence suggests that maintaining life-long learning can help to postpone the onset of dementia and disability in old age. The involvement of older adults in learning activities can also help refute societal stereotypes about old age.

References

Accessible Technology.org. (2014). *What is a screen reader?*. Retrieved from http://accessibletech.org/assist_articles/webinfo/screenReaders_what_is.php.

Administration on Aging. (2013). *Profile of older Americans: The older population*. Retrieved from http://www.aoa.gov/AoARoot/Aging_Statistics/Profile/2013/3.aspx.

Aja, G. N., Umahi, E. N., & Allen-Alebiosu, O. I. (2011). Developing culturally-oriented strategies for communicating women's health issues: a church-based intervention. *Education for Health*, *24*(1). Retrieved from http://www.educationforhealth.net/text.asp?2011/24/1/398/101463.

American Psychiatric Association. (2013). Neurocognitive disorders. In *Diagnostic and statistical manual* (5th ed.). American Psychiatric Association. Available from. http://dsm.psychiatryonline.org.

Austin, L. S., Landis, C. O., & Hanger, K. H. (2012). Extending the continuum of care in congestive heart failure: an interactive technology self-management solution. *The Journal of Nursing Administration*, *42*(9), 442–446.

Aziz, R., & Steffens, D. C. (2013). What are the causes of late-life depression? *The Psychiatric Clinics of North America*, *36*(4), 497–516.

Barbaro, E. L., & Noyes, L. E. (1984). A wellness program for a life care community. *The Gerontologist*, *24*(6), 568–571.

Blackwell, D. L., Lucas, J. W., & Clarke, T. C. (2014). Summary health statistics for U.S. adults: National health interview survey 2012. *Vital and Health Statistics*, *10*(260). Retrieved from http://www.cdc.gov/nchs/data/series/sr_10/sr10_260.pdf.

Breitner, J. C. S. (2014). Mild cognitive impairment and progression to dementia: new findings. *Neurology*, *82*, e-34–e35.

Buettner, L. L., & Fitzsimmons, S. (2009). Promoting health in early-stage dementia. *Journal of Gerontological Nursing*, *35*(3), 39–49.

California Department of Social Services. (2012). Chapter three: how to reach older adults in your community. In *CalFresh outreach basics handbook* Retrieved from http://www.cdss.ca.gov/calfreshoutreach/res/Toolkit/Handbook-OlderAdults/OlderAdultsHandbook_CH3_HowtoReachOlderAdultsinYourCommunity.pdf.

Cangelosi, P. R., & Sorrell, J. M. (2008). Storytelling as an educational strategy for older adults with chronic illness. *Journal of Psychosocial Nursing*, *46*(7), 19–22.

Centers for Disease Control and Prevention. (2009a). *Improving health literacy of older adults: Expert panel report 2009*. Retrieved from http://www.cdc.gov/healthliteracy/pdf/olderadults.pdf.

Centers for Disease Control and Prevention. (2009b). *Simply put: A guide for creating easy-to-understand materials* (3rd ed.). Retrieved from http://www.cdc.gov/healthliteracy/pdf/simply_put.pdf.

Centers for Disease Control and Prevention. (May 10, 2013). Self-reported increased confusion or memory loss and associated functional difficulties among adults aged ≥ 60 years - 21 States, 2011. *MMWR: Morbidity & Mortality Weekly Report, 62*(18), 347–350.

Davis, L. A. (2005). Educating individuals with dementia: perspectives for rehabilitation professionals. *Topics in Geriatric Rehabilitation, 21*(4), 304–314.

Day, K., Carreon, D., & Stump, C. (2000). The therapeutic design of environments for people with dementia: a review of the empirical research. *The Gerontologist, 40*(4), 397–416.

Donahue, A. M., Dubno, J. R., & Beck, L. B. (2014). NIDCD research working group on accessible and affordable hearing health care. In *IOM (Institute of Medicine) and NRC (National Research Council), Hearing loss and healthy aging: Workshop summary.* Washington, DC: The National Academies Press. Retrieved from http://www.ncbi.nlm.nih.gov/books/NBK202191/.

Eames, S., Hoffman, T., Worrall, L., & Read, S. (2011). Delivery styles and formats for different stroke information topics: patient and carer preferences. Retrieved from Elsevier ClinicalKey *Patient Education and Counseling, 84,* e18–e23. http://dx.doi.org/10.1016/j.pec.2010.07.007.

Federal Interagency Forum on Aging-Related Statistics. (2012). *Older Americans 2012: Key indicators of well-being.* Retrieved from http://www.agingstats.gov/agingstatsdotnet/Main_Site/Data/2012_Documents/Docs/EntireChartbook.pdf.

Fox, S. (2011). *Peer-to-peer health care.* Retrieved from http://www.pewinternet.org/2011/02/28/peer-to-peer-health-care-2/.

Fox, S. (2013). *Health and technology in the U.S.* Retrieved from http://www.pewinternet.org/2013/12/04/health-and-technology-in-the-u-s/.

Friedman, S. (2011). Sensory function (chapter 31). In S. E. Meiner (Ed.), *Gerontologic nursing* (4th ed.). St. Louis, MO: Elsevier Mosby.

Gerontological Society of America. (2012). *Communicating with older adults: An evidence-based review of what really works.* Retrieved from http://www.agingresources.com/cms/wp-content/uploads/2012/10/GSA_Communicating-with-Older-Adults-low-Final.pdf.

Gillick, M. R. (2013). The critical role of caregivers in achieving patient-centered care. *JAMA, 310*(6), 575–576.

Gonsalves, C., & Pichora-Fuller, M. K. (2008). The effect of hearing loss and hearing aids on the use of information and communication technologies by community-living older adults. *Canadian Journal on Aging, 27*(2), 145–157.

Grilli, R., Ramsay, C., & Minozzi, S. (2002). Mass media interventions: effects on health services utilisation. *Cochrane Database of Systematic Reviews, 2002*(1), CD000389. Retrieved from The Cochrane Library.

Haber, D. (2003). *Health promotion and aging: Practical applications for health professionals* (3rd ed.). New York: Springer.

Hainsworth, D. S. (2003). Instructional materials (chapter 12). Retrieved from. In S. B. Bastable (Ed.), *Nurse as educator: Principles of teaching and learning for nursing practice.* EBSCOHost eBook Collection.

Hallows, K. M. (2013). Health information literacy and the elderly: has the internet had an impact? *The Serials Librarian, 65,* 39–55.

Harrison, T. C., Mackert, M., & Watkins, C. (2010). Health literacy issues among women with visual impairments. *Research in Gerontological Nursing, 3*(1), 49–60.

Health in Aging Foundation. (2012). *Hearing loss: Basic facts & information*. Retrieved from http://www.healthinaging.org/aging-and-health-a-to-z/topic:hearing-loss.
Henry, J. D., & Henry, L. S. (2004). Avoiding elder stereotyping: nurses expanding options for self and others. *The Oklahoma Nurse, 49*(4), 17.
Higgins, M. M., & Barkley, M. C. (2004). Improving effectiveness of nutrition education resources for older adults. *Journal of Nutrition for the Elderly, 23*(3), 19–54.
Hochhauser, M. (2012). Can sick patients understand informed consent? *SoCRA Source, 74*, 72–74.
Hong, J., Nguyen, T. V., & Prose, N. S. (2013). Compassionate care: enhancing physician-patient communication and education in dermatology. Part II: patient education. *Journal of the American Academy of Dermatology, 68*(3), e1–10. Retrieved from Elsevier ClinicalKey.
Hooyman, N. R., & Kiyak, H. A. (2011). The growth of social gerontology. In N. R. Hooyman, & H. S. Kiyak (Eds.), *Social gerontology: A multidisciplinary perspective* (9th ed.) (pp. 6–7). Boston, MA: Allyn & Bacon.
Houston, T. K., Allison, J. J., Sussman, M., Horn, W., Holt, C. L., Trobaugh, J., et al. (2011). Culturally appropriate storytelling to improve blood pressure: a randomized trial. *Annals of Internal Medicine, 154*(2), 77–84.
Hunt, R. (2005). Client teaching (chapter 6). In R. Hunt (Ed.), *Introduction to community-based nursing* (3rd ed.). Philadelphia, PA: Lippincott Williams & Wilkins.
Hurst, G., Wilson, P., & Dickinson, A. (2013). Older people: how do they find out about their health? A pilot study. *British Journal of Community Nursing, 18*(1), 34–39.
Jinks, M. J., & Baker, D. E. (1987). Addressing audiences of older adults. *Journal of Geriatric Drug Therapy, 1*(3), 89–100.
Jolles, E. P., Clark, A. M., & Braam, B. (2012). Getting the message across: opportunities and obstacles in effective communication in hypertension care. *Journal of Hypertension, 30*(8), 1500–1510.
Kane, R. L., Ouslander, J. G., Abrass, I. B., & Resnick, B. (2013). Chapter 7: diagnosis and management of depression. In R. L. Kane, J. G. Ouslander, I. B. Abrass, & B. Resnick (Eds.), *Essentials of clinical geriatrics, 7e*. Retrieved from http://accessmedicine.mhmedical.com/content.aspx?bookid=678&Sectionid=44833885.
Kazer, M. W. (2011). Cognitive and neurologic function (chapter 29). In S. E. Meiner (Ed.), *Gerontologic nursing* (4th ed.). St. Louis, MO: Elsevier Mosby.
Kennedy, L. (2006). PD trivia: making learning fun. *The CANNT Journal, 16*(3), 46–48.
Korzenny, F., & Korzenny, B. A. (2007). *Old and new media use: The multicultural marketing equation study 2007: Report #1*. Retrieved from http://hmc.comm.fsu.edu/files/2012/02/2007-Multicultural-Study-on-Old-and-New-Media-Use.pdf.
Lapane, K. L., Goldman, R. E., Quilliam, B. J., Hume, A. L., & Eaton, C. B. (2012). Tailored DVDs: a novel strategy for educating racially and ethnically diverse older adults about their medicines. *International Journal of Medical Informatics, 81*, 852–860.
Lee, K., Hoti, K., Hughes, J. D., & Emmerton, L. M. (2014). Interventions to assist health consumers to find reliable online health information: a comprehensive review. *PLOS One, 9*(4). Retrieved from http://www.ncbi.nlm.nih.gov/pmc/articles/PMC3978031/.
Lorig, K. (2001). Arthritis self-management (chapter 3). In E. A. Swanson, T. Tripp-Reimer, & K. Buckwalter (Eds.), *Health promotion and disease prevention in the older adult*. New York: Springer Publishing Company.
Lowsky, D. J., Olshansky, S. J., Bhattacharya, J., & Goldman, D. P. (2014). Heterogeneity in healthy aging. *Journals of Gerontology. Series A, Biological Sciences and Medical Sciences, 69*(6), 640–649.
Lubinski, R. (2010). Communicating effectively with elders and their families. *The ASHA Leader*. Retrieved from http://www.asha.org/publications/leader/2010/100316/communicatingeffectivelywithelders/.

Manafo, E., & Wong, S. (2012). Exploring older adults' health information seeking behaviors. *Journal of Nutrition Education and Behavior*, *44*(1), 85–89.
McAlister, A. L., Perry, C. L., & Parcel, G. S. (2008). How individuals, environments, and health behaviors interact: social cognitive theory (chapter 8). In K. Glanz, B. K. Rimer, & K. Viswanath (Eds.), *Health behavior and health education: Theory, research, and practice* (4th ed.). San Francisco, CA: Jossey-Bass.
McCormack, A., & Fortnum, H. (2013). Why do people fitted with hearing aids not wear them? *International Journal of Audiology*, *52*, 360–368.
Meiner, S. E. (2011). Overview of gerontologic nursing (chapter 1). In S. E. Meiner (Ed.), *Gerontologic nursing* (4th ed.). St. Louis, MO: Elsevier Mosby.
National Council on Aging. (2012). *Issue brief: Evidence-based health promotion programs for older adults: Key factors and strategies contributing to program sustainability*. Retrieved from http://www.ncoa.org/improve-health/NCOA-Health-Promo-Issue-Brief.pdf.
National Institute on Aging. (2010). *Healthy aging: Lessons from the Baltimore longitudinal study of aging*. Retrieved from http://www.nia.nih.gov/sites/default/files/healthy_aging_lessons_from_the_baltimore_longitudinal_study_of_aging.pdf.
National Institute on Aging. (2013a). *AgePage: hearing loss*. Retrieved May 5, 2015 from http://www.nia.nih.gov/health/publication/hearing-loss.
National Institute on Aging. (2013b). *The dementias: Hope through research*. Retrieved August 29, 2014 from http://www.nia.nih.gov/alzheimers/publication/dementias/introduction.
Neugarten, B. (1974). Age groups in American society and the rise of the young-old. *Annals of the American Academy of Political and Social Science*, *415*, 187–198.
Ovbiagele, B., & Nguyen-Huynh, M. N. (2011). Stroke epidemiology: advancing our understanding of disease mechanism and therapy. *Neurotherapeutics*, *8*, 319–329.
Peterson, J. C., Link, A. R., Jobe, J. B., Winston, G. J., Klimasiewfski, E. M., & Allegrante, J. P. (2014). Developing self-management education in coronary artery disease. *Heart & Lung*, *43*, 133–139.
Pew Research Internet Project. (2015). *Health fact sheet*. Retrieved from http://www.pewinternet.org/fact-sheets/health-fact-sheet/.
Pollock, A., Hazelton, C., Henderson, C. A., Angilley, J., Dhillon, B., Langhorne, P., et al. (2011). Interventions for visual field defects in patients with stroke. *Cochrane Database Systems Review*, *5*(10), CD008388.
Press, V. G., Shapiro, M. I., Mayo, A. M., Meltzer, D. O., & Arora, V. M. (2013). More than meets the eye: relationship between low health literacy and poor vision in hospitalized patients. *Journal of Health Communication*, *18*, 197–204.
Price, A. E., Corwin, S. J., Friedman, D. B., Laditka, S. B., Colabianchi, N., & Montgomery, K. M. (2011). Older adults' perceptions of physical activity and cognitive health: implications for health communication. *Health Education & Behavior*, *38*(1), 15–24.
Primack, B. A., Carroll, M. V., McNamara, M., Klem, M. L., King, B., Rich, M., et al. (2012). Role of video games in improving health-related outcomes: a systematic review. *American Journal of Preventive Medicine*, *42*(6), 630–638.
Rainie, L. (2013). *E-patients and their hunt for health information*. [Slideshare slides]. Retrieved from http://www.pewinternet.org/2013/07/26/e-patients-and-their-hunt-for-health-information-2/.
Reichlin, L., Mani, N., McArthur, K., Harris, A. M., Rajan, N., & Dacso, C. C. (2011). Assessing the acceptability and usability of an interactive serious game in aiding treatment decisions for patients with localized prostate cancer. *Journal of Medical Internet Research*, *13*(1). Retrieved from http://www.jmir.org/2011/1/e4/.
Shargorodsky, J., Curhan, G. C., & Farwell, W. R. (2010). Prevalence and characteristics of tinnitus among US adults. *American Journal of Medicine*, *123*(8), 711–718.

Sink, K. M., & Yaffe, K. (2014). Cognitive impairment & dementia. In B. A. Williams, A. Chang, C. Ahalt, H. Chen, R. Conant, C. Landefeld, et al. (Eds.), *Current diagnosis & treatment: Geriatrics* (2nd ed.). Retrieved from http://accessmedicine.mhmedical.com/content.aspx?bookid=953&Sectionid=53375646.

Smith, J. M. (2012). Toward a better understanding of loneliness in community-dwelling older adults. *The Journal of Psychology, 146*(3), 293–311.

Smith, A. (2014). *Older adults and technology use: Main findings.* Retrieved from http://www.pewinternet.org/2014/04/03/older-adults-and-technology-use/.

Speros, C. L. (2009). More than words: promoting health literacy in older adults. *Online Journal of Issues in Nursing, 14*(3). Retrieved from CINAHL Plus with Full Text.

Stagnitti, M. N. (2011). Statistical brief #310: person characteristics of the elderly reporting one or more cognitive disorders, 2007. *MEPS: Medical Expenditure Panel Survey.* Retrieved from http://meps.ahrq.gov/mepsweb/data_files/publications/st310/stat310.shtml.

Stiles, E. (2011). Promoting health literacy in patients with diabetes. *Nursing Standard, 26*(8), 35–40.

Substance Abuse and Mental Health Services Administration. (2012). *Mental health, United States, 2010.* [HHS Publication No. (SMA) 12-4681] Rockville, MD: Substance Abuse and Mental Health Services Administration. Retrieved from http://www.samhsa.gov/data/2k12/MHUS2010/MHUS-2010.pdf.

Thomas, M. H. (2011). Health care delivery settings and older adults (chapter 9). In S. E. Meiner (Ed.), *Gerontologic nursing* (4th ed.). St. Louis, MO: Elsevier Mosby.

Vasilevskis, E. E., Han, J. H., Hughes, C. G., & Ely, E. W. (2012). Epidemiology and risk factors for delirium across hospital settings. *Best Practice & Research Clinical Anaesthesiology, 26*, 277–287.

Viggiani, K. (2003). Special populations (chapter 9). Retrieved from. In S. B. Bastable (Ed.), *Nurse as educator: Principles of teaching and learning for nursing practice.* EBSCOHost eBook Collection.

Walker, E. A., Stevens, K. A., & Persaud, S. (2010). Promoting diabetes self-management among African Americans: an educational intervention. *Journal of Health Care for the Poor and Underserved, 21*, 169–186.

Williams, K., Kemper, S., & Hummert, M. L. (2005). Enhancing communication with older adults: overcoming elderspeak. *Journal of Psychosocial Nursing, 43*(5), 12–16.

Zhang, X., Saaddine, J. B., Chou, C. F., Cotch, M. F., Cheng, Y. J., Geiss, L. S., et al. (2010). Prevalence of diabetic retinopathy in the United States, 2005–2008. *JAMA, 304*(6), 649–656.

Zickuhr, K. (2014). *Older adults and technology.* [Slideshare slides]. Retrieved from http://www.pewinternet.org/2014/04/29/older-adults-and-technology/.

Re-envisioning the health information-seeking conversation: insights from a community center

10

Prudence W. Dalrymple, Lisl Zach
College of Computing & Informatics, Drexel University, Philadelphia

10.1 Introduction

The growing importance of health literacy in meeting the US national health goals has captured the interest not only of the public health and health-care communities, but also of the information professional community. The Medical Library Association released a policy statement addressing the role of the librarian in consumer health information and patient education almost 20 years ago (Medical Library Association, 1996). However, the ways in which information professionals can contribute to the improvement of health literacy are still developing. Information professionals have considerable experience in facilitating access to relevant information. Applying this experience to the challenge of seeking health information in the twenty-first century is a natural role for information professionals.

In considering health information seeking, it may seem obvious that "health information" is "information about health." However, such a simple concept may vary along several dimensions. Is health information really "disease information," for example, the ways in which Lyme disease is contracted; or is it "treatment information," for example, treatments for Lyme disease? Is "health information" more concerned with acquiring and maintaining a state of optimal health? Or all of the above? Who is doing the "seeking" for health information—the worried well? The patient? The consumer? The caregiver for the patient/consumer? The health provider? Depending on which of these dimensions one chooses, the theoretical framework and research tradition may be quite different.

In this chapter, we take a broad approach to health information seeking that is not limited to disease or treatment information; our focus is on the individual, regardless of his or her personal relationship to the health-care system. In keeping with the theme of this edited volume, we do not focus on information that is provided prescriptively to the patient in a clinical encounter, or upon discharge from a health-care facility. Nor do we focus on information that is "pushed out" to change behavior, as in the targeted communications developed for health promotion. Rather, we focus on information that is being sought by an individual in the context of his or her daily life and interactions with health issues. Thus, we view health information seeking from a perspective that acknowledges that information is not only presented to and received by individuals; it is also sought by individuals, from health providers, from friends and family, through the media or, increasingly, on the Internet.

In this chapter we describe key concepts of information behavior drawn from information science theory and research and review some key studies from that literature. We then summarize previous health information work we have done in a low-income, primarily African-American community in Philadelphia. We present the findings from a series of focus group sessions during which members of that community completed a brief survey about their level of Internet use and responded to open-ended questions aimed at elucidating their experiences while looking for information related to their health situations outside of a clinical encounter. The focus group questions addressed their preferred sources of information (e.g., health providers, friends or family, the Internet) and their level of satisfaction with the information that they found. Participants were encouraged to be candid in their descriptions of their current information seeking—or lack of it—and their perceptions of the value of the information sources that are available on the Internet. The goal of this exploratory research was to gain insights into the information behaviors of this specific population, which in turn could be used to formulate a broader research agenda in the area of health information seeking—an important component of health literacy as well as part of the mission of health sciences libraries.

10.2 Understanding information behaviors

Information is everywhere; we live in an "information age" in which the amount of information doubles each year, and new terms such as "zettabyte" and "exabyte" describe the vast amounts of data generated in today's technology-driven society. At the same time, this rapid expansion in the amount of data does not necessarily create a more informed society. Rather, the ubiquity of data may result in misinformation or "disinformation," prompting interest in examining how people find and use information, how they evaluate it, and how they apply it to solve problems. Understanding how individuals seek and use information has long been a central focus of information science. Health information seeking is a subset of general information seeking, which is in turn one of a cluster of concepts, together with information need and information use, which are known in the information science literature as "information behaviors" (Case, 2012; Fisher, Erdelez, & McKechnie, 2005; Spink & Cole, 2001). While these behaviors are recognized as components of health literacy by theorists in the fields of public health, nursing, and medicine (Anker, Reinhart, & Feeley, 2011; Nutbeam, 2000; Ormandy, 2011), they have also been explored in depth by library and information science (LIS) researchers. Information scientists study the ways people ask questions, identify and evaluate information sources, and apply the information they find to life situations. Despite the apparent clarity of terms such as "information needs," "information seeking," and "information use," these areas still lack common conceptual frameworks among the disciplines. In this chapter, we offer definitions of these terms derived from leading theorists in the LIS field. For a thorough review of information behavior theory and research, the reader is referred to Donald Case's excellent survey *Looking for Information* (Case, 2012) and to Johnson & Case, *Health Information Seeking* for information seeking behaviour specific to health (Johnson & Case, 2012). A selection of the most influential theories is presented here.

10.2.1 Information needs

Robert Taylor's theory of information needs posits that an **information need** evolves in four stages: *visceral, conscious, formalized*, and *compromised* (Taylor, 1962). At the point where the visceral becomes conscious, the need can be stated formally as a query that can be the basis for an information search.

For example, suppose that an individual decides that listening to music while running is a desirable goal. In order to accomplish that goal, it will be necessary for her/him to obtain access to some songs, select a few she/he likes, download them, and save them on a device. "Finding downloadable music suitable for use while running" is this individual's conscious information need. Some runners don't want to "listen to music while running," and so they will not experience that information need. Experienced runners may only rarely have that need as they already have their playlist, and don't feel any "need" to update or expand it. Some runners who want to incorporate music in their routine but do not have a playlist may have an even more complex information need; they may need instruction on how to access digital music and add it to a device. Still others may need information on how to select a device that is appropriate for use while running.

Runners who already have their playlist setup on a device have only to grab it as they go out the door for a run. Their information need, which may have been conscious at one point, has now become a latent or "visceral" information need. That is, they may recognize the need only when they forget their music, and experience their run as less satisfactory. Only then do they realize the effect that decision, made long ago, has on them in the here-and-now.

It may be that runners who listen to music take longer, more frequent runs and report that they enjoy them more, but if the person who is struggling to incorporate running into her/his exercise routine may not even realize the potential positive effect of listening to music. This person, too, may have what Taylor (1991) calls a "latent" information need. That is, she may have noticed that some days you have a better run than others, and it is only upon reflection that you realize that when you are listening to music you seem to have a better run.

Taylor also notes that not all information needs lead to taking action. Some information needs may be noted, then modified or compromised—even abandoned. In the example under discussion, the benefit of adding music to an exercise routine may not be seen as worthwhile, and the need is therefore simply ignored.

10.2.2 Information seeking

As can be seen in the foregoing example, having an information need frequently leads to taking action to address it. To the information scientist, the action that is of most interest is **information seeking**. The work of Brenda Dervin spans both the information science and the communications fields, and the concepts of information needs and information seeking. Her 1986 review of information needs and uses, co-authored with Michael Nilan, introduced her approach to the information science community where it has been highly cited (Dervin & Nilan, 1986). Her theory, known as "sense-making," posits that individuals experience information "gaps" in their lives, and that these gaps are that which drive information seeking behavior.

In the previous example, an individual who is engaged in an exercise program has a need for information about downloadable music to help optimize her/his routine. The information need is affected both by what the runner is feeling, and what the individual believes is the optimal situation. If there is a "gap" between the individual's experience and her/his desired goal—a pleasant, satisfying run—the individual is said to be experiencing an information need. Dervin calls the actions that the individual takes to fill that gap "making sense" of a situation.

In the example of the runner, the gap ("My run is not optimal") is filled by seeking information about adding music to a running routine ("Will adding music to my run improve it and if so, how do I go about doing that?"). Sense-making theory emphasizes the subjective nature of an information need, rather than an objective set of action steps. That is, the runner's information gap may be subjective, arising from a need for reassurance or support—even companionship. Dervin's theory allows for the possibility that "emotions are at least as important as cognitions in 'gappy' situations: searchers may be intent upon reducing their anxiety as much as their uncertainty" (Case, 2012, p. 85).

For Dervin, information needs are situation-based and highly contextual; it follows, then, that the information that fills that gap may vary widely. The music that one individual chooses to search for, select, and download may be quite different from someone else's; indeed, even the presence or absence of music while running may vary according to the individual and the situation. One individual might find that music is motivational, or helps establish a rhythm and so uses specific criteria to build a playlist. Others may find that on certain days, music helps, while on other days—in other situations—music is distracting or even dangerous, if it interferes with the runner's ability to be aware of her/his surroundings while running. The sense-making approach contends that information is actively constructed by individual users in their unique situations and must be understood through their eyes (Dervin, 1997). Dervin's large body of work is nicely summarized in Glazier & Powell (1992).

In addition to Taylor's and Dervin's very influential and quite different views of information needs and information seeking, other information scientists have contributed to an understanding of these aspects of information behavior in various spheres. For example, Carol Kuhlthau's model emphasizes the affective or feeling aspects of information seeking. Kuhlthau notes that various feelings may arise during a search for information, and that these feelings may change over time as new information is encountered (1991). By becoming aware of how these feelings change, not only information seekers, but those who want to facilitate information seeking—librarians, teachers, and information professionals—can help individuals adapt and improve their information-seeking skills.

There is a rich literature on information-seeking habits of various demographic populations as well as members of various occupational and social groups, such as information seeking and use by health providers, when the audience for the information is the provider himself or herself. Of specific relevance to this chapter are studies on the information-seeking habits of low-income and marginalized populations (Chatman, 1999; Harris & Dewdney, 1994; Harris, Wathen, & Fear, 2006; Westbrook, 2009).

It should be noted that any single individual is a member of many different groups simultaneously. For example, an individual information user may be described as a scientist, an independent voter, and a middle-income, African-American female. She may engage in multiple, diverse information behaviors at any given time, depending on her situation. The value of this approach to looking at information seeking is that it recognizes the context of the information need and situates it within a specific type of user.

A potential limitation of the information seeking theoretical approach is that it can create stereotypes rather than recognizing that membership in a specific group may change over time. The value of this approach to looking at information seeking is that it recognizes the context of the information need and situates it within a specific type of user. In contrast to these studies, and more closely aligned with Dervin's approach, researchers such as Reijo Savolainen and others have proposed models of information seeking that focus on everyday needs. These models propose that seeking information to solve problems of daily life is a richer approach than information sought as part of work or professional life (Savolainen, 1995).

10.2.3 Information use

The third behavior is **information use**. In a classic article, "Use, users, uses," Zweizig & Dervin (1977) were among the earliest proponents of examining the ways in which individuals make use of information, questioning what impact information has on daily life. They advocated for looking at whether and how individuals use information in a particular context, rather than looking at specific occupational or social groups (Zweizig & Dervin, 1977). That is, they proposed starting with the act of using information—applying information to a specific situation or problem—rather than looking at demographic characteristics.

In the example above, the information user is a middle-income African-American female who is not affiliated with a specific political party. These attributes may be of less significance in studying her information use than the fact that she needs information to complete a specific task. In attempting to facilitate information use for this individual, information professionals might therefore concentrate on identifying the information needs and preferences of other individuals who are performing similar tasks, rather than on the information needs of African-American women.

More recently, Karen Pettigrew-Fisher developed the concept of "information grounds," which captures the contextual aspects of information use. She has stated that "the study of information use cannot be considered in terms of an isolated individual or outside a specific social context," thus linking her work to Dervin's situation-based, sense-making approach (Fisher, Durrance, & Hinton, 2004; Pettigrew, 2000). In the "listen to music while running" example presented earlier, examining the "use" of the information sought would reveal whether the "need" for adding music to a running routine actually resulted in the user's taking steps to identify and acquire the music, and then play it on a device while running. Ultimately, of course, the "use" of greatest interest to most people is determining whether listening to music during a run increases enjoyment and leads to more frequent, enjoyable exercise and better health. However, examining the impact or outcome of information use more properly belongs to the realm of communications, placing it beyond the scope of most information science research. This distinction is especially important with regard to health information where information provision and use is frequently associated with behavior change, for example, a health promotion campaign aimed at weight loss.

10.2.4 Why information scientists study information behaviors

The preceding discussion of information behaviors is an attempt to deconstruct the process of looking for and using information. While this discussion has been general in scope, it is useful to consider how and whether information behavior changes when health is the topic of concern.

While information science and communications researchers often look at similar problems, their focus is different. The information scientist, and particularly the information professional, stops short of studying the effect of information on behavior—asking whether it made a difference—preferring to leave the assessment of impact to the individual. Communications researchers, especially those working in the health field, are often determining how to maximize the effect of information on a person, investigating how and whether it will lead to behavior change. As noted above, information science typically focuses on the processes rather than the outcomes of information behaviors, but it is also important that we understand how it can contribute to changing behaviors, such as creating more enjoyable exercise routines that result in better health outcomes.

10.2.5 Current studies of health-related information behavior

It is against this backdrop of information science theory that we examine current research about health information seeking and use. Although our review is selective, we cast a wide net, searching both LIS and biomedical literature databases and following up on citations to the work of the scientists mentioned previously. We were guided in part by the fact that the population from which the participants in our focus groups were recruited is largely low-income minority, primarily African-American, and is known to have a significant burden of disease, yet has limited access to established health-care services. Many of the participants receive primary care in a nurse-managed, neighborhood care center that provides services without regard to ability to pay. For complex or acute medical needs, this population seeks care at any of the many hospitals in the city. Thus, overall, our focus group participants can be characterized as belonging to a health disparities population; their health profile is "closely linked with social, economic, and/or environmental disadvantage" (U.S. Department of Health & Human Services, 2011). For this reason, our literature review focuses on studies of similar groups of people.

Much health information-seeking research examines information seeking in the context of a specific disease, such as cancer or multiple sclerosis, or a health-related condition, such as pregnancy. These studies can be very enlightening when the goal is to design information services or communication campaigns to disseminate information to specific patient populations. The focus of this chapter is on information seeking that occurs outside the clinical context, so we have omitted these studies from review. Furthermore, our focus groups were recruited without regard to whether these individuals had been diagnosed with any specific disease; rather, we were interested in learning more about the factors that affect the residents of this particular public housing neighborhood, regardless of their health status.

In an early investigation of information seeking by disadvantaged populations, Chatman described an information seeking environment as a "small world" (Chatman, 1996; Yu, 2010). In a "small world," members of a specific community determine which types of information are important to them and which information sources they believe they can trust. While the use of personal contacts for information is surely not unique to such communities, it is the strength of this community bond, often resulting in the exclusion of external, mainstream information sources, that Chatman highlighted. Chatman defined those living in such environments as the "information poor"—people who perceive themselves as lacking (or having limited) sources of information that might help them. She found that most people in "small world" communities depended on local contacts to identify and authenticate trustworthy sources of information; often these information sources did not include mainstream media, libraries, or other sources of "authoritative" information. In the small world, information seeking and sharing behaviors reflect the social norms and attitudes particular to the social environment (Burnett, Jaeger, & Thompson, 2008; Chatman, 1999).

These findings are confirmed in work done by Sligo and Jameson in (2000); they suggest that in tightly knit communities, information seeking and sharing occurs within the group according to cultural and social norms, but when care is sought, group members may be more likely to trust a health-care provider (Sligo & Jameson, 2000). Spink & Cole, too, found in 2001 that the lower-income, mostly African-American residents of a public housing project in Dallas, Texas did not tend to seek information from those outside of their community; their information needs related to problems of daily survival, and most of their information was gathered from informal channels close to home. An important exception, however, was health information needs, which were addressed by consulting a family physician more often than a family or community member (Spink & Cole, 2001). It should be noted that at the time these studies were done, channels of information seeking were largely based on print materials, broadcast media, and interpersonal communication.

In the past 15 years, much has changed in terms of the affordances available to individuals seeking health information in everyday life. The use of the Internet for health information seeking has been studied extensively by the Pew Center for the Study of the Internet in American Life (Pew Research Center, Internet & American Life Project). Pew reports that as of 2014, 87% of the US population has Internet access at home and, according to 2012 data, 72% of these Internet users reported looking online for health information (Pew Research Center, Internet & American Life Project, 2012). The ways in which people search vary, but most Internet users (77%) begin their search for information by consulting a general search engine, such as Google, and most search for a specific disease, treatment, or procedure. Searching the Internet for health information requires Internet access, and increasingly, that access is achieved through mobile devices. The demographic distribution of smartphone usage indicates that 90% of the American population owns a cell phone and 58% owns a smartphone, with the highest rate of ownership occurring among Hispanics (61%). Smartphone ownership is also concentrated in the young; it is most common among adults under 50, with the highest prevalence (83%) occurring among 18- to 29-year olds (Fox & Rainie, 2014). The growth in ownership of mobile devices with ever-increasing capabilities is expected to continue.

While these data indicate that the Internet is widely accessible across the American population, other evidence reported by Pew indicates that broadband access and computer use still lag behind mobile access. In 2006, Lorence, Park, and Fox studied a stratified sample of the Pew data and confirmed the persistence of digitally underserved groups (2006). More recent work indicates that using the Internet for health-related activities (referred to as "eHealth" or "mHealth") has continued, and possibly reinforced, these patterns. Between 2010 and 2013, broadband access increased slightly from 66% to 70%; lack of broadband access tends to be associated with lower socioeconomic status (SES) and educational attainment. This is important because broadband access provides greater flexibility in using resources, not simply accessing them (Lustria, Smith, & Hinnant, 2011).

The growing diversity of Internet access has prompted health information-seeking research with a focus on the Internet. In a pilot study with a convenience sample of low-SES, largely African-American participants, we found that only 21% of those with access to the Internet reported using it to look for health information, and then only when they had a specific question in mind (Zach, Dalrymple, Rogers, & Williver-Farr, 2012). In a much larger study ($n=270$) conducted in 2012/2013, we surveyed convenience samples from several areas of a metropolitan area, and found that although computer-based or mobile access to the Internet was almost universal among the population being studied, only 36% of respondents reported using the Internet to look for health information (Rogers et al., 2013). In a reanalysis of data collected in 2005 and 2007, Malone, While, & Roberts (2014) concluded that in addition to the digital divide related to Internet access, a preferential divide exists, based on differing perceptions of information availability and utility, as experienced locally. This preferential divide is described in terms similar to those of Chatman's small world. Neter & Brainin (2014) observed that the discourse on the digital divide has shifted to examining the factors that affect patterns of access and effective use, rather than simply access to technology.

This suggests that attitudes and habits differ across some populations when it comes to seeking health information. Morey's study of an African-American community in Buffalo, NY, revealed that most respondents relied on health providers, but reported following up with an Internet search (2007). Recently, Mitchell, Godov, Shabaxx, & Horn (2014) studied a sample of African-American parents who were active users of the Internet but did not use it for seeking health information. Despite their lack of interest in searching, they expressed interest in passively receiving health information via mobile phones, either through text messaging or Internet access (Mitchell et al., 2014).

The digital divide is no longer about access alone. The most recent Pew data show that 87% of the US population has access to the Internet (Pew, 2014). However, groups outside the mainstream are often of low SES and experience a narrower "world" of information compared to those with access to more resources, both as individuals and as residents of a community. Such groups do not appear to search for health information on their own, but although exactly what the factors are that contribute to this phenomenon remain elusive. A deeper understanding of this phenomenon remains elusive.

Despite the growth in the Internet as a source of health information, many studies indicate that health professionals remain a valued source of information. The focus of this chapter is on health information behaviors that take place outside the clinical encounter. However, it is important to keep in mind that information about health issues is still strongly associated with health-care professionals, either positively or negatively. For example, recent work by Veinot and colleagues explored African-American youth attitudes toward an informatics intervention aimed at HIV/STD prevention and found that focus group participants expressed distrust of "expert" information (Veinot, Campbell, Kruger, & Grodzinski, 2013). Consultation with friends and family is quite likely a universal phenomenon, regardless of SES and racial and ethnic affiliation; differences do exist in the extent to which external sources—such as the media and libraries—are regarded as trusted sources. Respondents in Veinot's study displayed varying degrees of reluctance to using the Internet for health information seeking, primarily because of perceived lack of trust in the results.

In a study of 90 low-income minority men, Song, Cramer, and McRoy (2014) found that participants relied most on health professionals, family, and friends for issues of general health and the Internet was least relied upon as a source of health information. Zahradnik (2011) found that middle-aged African-American women had marked preference for obtaining asthma information from health providers rather than from friends, family, or other sources (Zahradnik, 2011).

This review suggests that despite the insights emerging from research, understanding of health information behaviors in marginalized communities is limited and findings often contradictory. These studies are important because they help to explain the persistence of attitudes and beliefs that may be impervious to outside influence of any sort. They also suggest that one of the information behaviors—information seeking—may differ from one community to another.

10.3 Health information seeking in a local context

10.3.1 An "on-ramp" to health information

The exploratory research described in this chapter took place in the context of a larger research program reported elsewhere (Dalrymple, Rogers, Zach, Turner, & Green, 2013; Zach et al., 2012). This larger program was designed to test a simple, replicable intervention designed to encourage the use of reliable information sources on the Internet. Since 2007, we have been conducting research involving members of a neighborhood community center in North Philadelphia. We have established a high level of trust and cooperation with its leadership, including its advisory board, composed of people living in the local public housing complex. The center offers an array of "healthy living" activities, including cooking, gardening, and exercise classes, to residents of an area that has been designated as "medically underserved" by the U.S. Department of Health and Human Services. According to information provided by the center, the unemployment rate among adult residents of the community is approximately 49% (one of the highest in Philadelphia), and the median family income is approximately $13,000 (one of the lowest in Philadelphia). The population in the area is approximately 90% African-American and 6% Latino. Users of the center's services are predominantly women between the ages of 25 and 64 years, many of whom are mothers or grandmothers of young children.

Conspicuously absent from the services being offered to the community was any formal training in using technology to find relevant information to support the "healthy living" activities. We felt that this lack might be addressed by offering formal instruction in effective information seeking as part of the services available at the center, and began a series of discussions aimed at determining a future direction. Based on these discussions with staff at the center and on input from the community advisory board, we developed and tested an intervention using short text messages containing a tiny hypertext link to credible and authoritative sources of health information. We referred to this as an "on-ramp" to health information seeking on the Internet because it provided curated, authoritative resources that were relevant to specific health topics. We also collaborated with the health sciences librarians at our institution to select materials for use in the study. By providing this high-quality material, we hoped to encourage health information seeking and thereby improve health literacy.

Women in the test group reported finding the text messages useful and supportive. Most also accessed the Web sites and many shared their experience with the other participants in their group activities (Dalrymple et al., 2013).

10.3.2 Survey on health information seeking

In addition to implementing the text-messaging intervention, we sought to understand more about the community's attitudes toward and experience with health information seeking in general. To gain further understanding of the community's information behaviors, we conducted a series of focus groups at the center, during which participants completed a brief paper-and-pencil survey on the ways in which they accessed and used the Internet. The survey (see Appendix A) was distributed at the beginning of each focus group session and collected at its completion. Not all focus group participants returned a completed survey.

10.3.3 Findings from the survey on Internet access and information sources

Seventy-nine percent (19/24) of the focus group participants reported having access to the Internet; 46% accessed the Internet primarily from their home computers and 29% from their cell phones. This is consistent with the Pew data for the same time period and with our earlier research with similar populations (Pew Research Center, 2014; Zach et al., 2012). Sixteen out of the 19 participants who reported having access to the Internet also reported that they used the Internet to look for various types of information. Of these 16, 12 looked specifically for health-related information.

However, the Internet did not appear to be the information source of choice for most participants. Table 10.1 shows the frequency of preferred information sources for health-related information reported by all focus group participants. Participants were asked to indicate their first and second choices for sources. The majority of participants (20/24) preferred to use a person (doctor, nurse, family member, or friend) as their primary source of health-related information; only two participants listed the Internet as their *preferred source* of health-related information.

Of the 12 participants who used the Internet to look for health-related information, five reported using the Internet daily, six reported using the Internet at least once a month, and one reported using the Internet "not very often." The perceived success that the participants experienced in seeking health-related information, as measured by their self-reported ability to find the information they were looking for and their confidence in the information they found, also varied (see Table 10.2).

Table 10.1 Frequency of preferred information sources

	Preferred source	Secondary source
Doctor	21%	29%
Nurse	17%	21%
Family/friend	46%	
Internet	8%	12%
Library		8%
No response	8%	29%

Table 10.2 **Frequency of perceived success of information seeking**

	Always	Mostly	Usually	Sometimes	Never
If you currently use the Internet to look for health information, do you find what you want?	1/12	6/12	1/12	4/12	–
If you currently find health information on the Internet, are you confident that it is accurate?	2/12	3/12	2/12	5/12	–

10.3.4 Focus groups on health information seeking

Consistent with our practice of involving the community in the design of any interventions proposed for the center, we chose to conduct a series of informal focus groups as a way of gauging interest, increasing buy-in, and soliciting ideas about the issues and problems most meaningful to community members. We hoped to engender a candid exchange of opinions and perceptions about the perceived value of information services. The goal of these focus groups was to understand the information needs of a minority population, largely of low socioeconomic and educational levels, and to elicit their preferences and opinions regarding the types of information sources that they consulted.

Five focus group sessions were held in March through May 2011; each focus group consisted of four to five participants and at least one member of the research team, who facilitated the discussion using a semistructured protocol. A total of 24 participants attended the five sessions, each of which lasted approximately 45 min. Participants for the focus groups were recruited from existing users of the various healthy living programs provided by the center, using a flyer approved by the Institutional Review Board (IRB) at Drexel University in September 2010. No focus group participant was a participant during any previous phase of the study.

The focus group sessions were audio-recorded so that a complete record of the discussions was available; the audio recordings were professionally transcribed and used in addition to the facilitators' notes as a basis for content analysis. The focus group facilitators used a semistructured protocol consisting of five basic questions to elicit perceptions and attitudes about health information available on the Internet. The questions were open-ended, with follow-up probes included in the protocol. The facilitators encouraged the participants to elaborate on their individual experiences in looking for health information.

The basic questions that guided the focus group sessions were the following:

Q1. Where do you go when you need to find specific health information?
Q2. Have you looked for specific health information on the Internet before?
Q3. If you use the Internet to look for health information, do you find what you want?

Q4. If you currently find health information on the Internet, are you confident that it is accurate?
Q5. What would help you the most to find the health information you need?

Although each focus group was asked the same basic questions, the character of the responses differed somewhat as a result of the individual group's dynamics. As facilitators, we guided the conversation to cover each of the basic questions, but otherwise allowed the group to follow its own natural progression, so that some focus groups spent more time on one question than on another.

Participants in the focus groups represent a convenience sample of community members who were making use of various healthy living activities offered by the center, such as cooking and exercise classes. Recruitment flyers were posted at the center and recruitment was limited to users of the center's existing services. The five focus groups were offered on different days of the week and at different times of the day to provide as many options as possible for potential participants to choose. Focus groups were not limited in size. Focus group participants were predominately female (75%) and ranged in age from 21 to 75 years, with 46% of the participants falling between the ages of 45 and 60 years. Four participants had never completed high school, 11 participants had only high school diplomas, and eight had at least some college. (One person did not disclose his/her age.) All were African-American. The sample recruited, while not statistically representative of all users of the center, does represent a cross section of those using the center during the time period covered.

10.3.5 Findings from the focus groups

The transcriptions from the five focus groups were analyzed by two members of the research team; content analysis techniques were used to identify patterns, themes, and categories emerging from the participants' own voices. Each team member coded the transcriptions separately, and the results were compared to ensure intercoder reliability. In keeping with the restrictions imposed by the IRB protocol, no identifying data were collected about individual focus group participants, so the transcriptions differentiate only between the facilitator's comments and those made by the participants; the facilitators' notes augmented the transcriptions and provided an indication of the concepts that were raised and discussed by several participants. The facilitators' notes were used to confirm that the responses in the transcriptions were not attributable to a single person, but rather represented themes of common interest among the group.

10.3.5.1 Typology of users

Analysis of the focus group transcripts identified three composite profiles of information seekers that typify the opinions expressed by the focus group participants. These are presented here to help the reader gain a richer sense of the ways in which the participants responded to the questions and the probes. Classic typologies such as the Myers-Briggs Type Indicator have long been used to understand individuals' reactions to and interactions with information, especially when related to decision-making. While our typology does not include multiple dimensions for each personality type, it does suggest different ways in which the participants think or feel about looking for health-related information and assessing its relevance in their lives. Since the comments were collected anonymously, we are unable to identify individual participants; the same participant may have provided opinions that were coded as responses from different types of information seekers. No attempt is made to assign an individual participant to a single category; rather, the typology emerged as a way of presenting the findings from all of the focus groups in a coherent way. All participant comments were included in the analysis regardless of whether the participant used the Internet for information seeking or not. The three categories of information seekers that were created are the "Caregiver," the "Sufferer," and the "Surfer." Examples of responses in each category of information seeking are described below in the participants' own words.

10.3.5.2 The caregiver

The *Caregiver* type is characterized by a focus on needing information to respond to the health situation of family members, typically aging parents or young children. Information seekers of this type tend to have specific information needs and are looking for the best sources that they can find to answer their questions, whether those sources are family, friends, professionals, or the Internet.

> *Participant: I've got a son who at one point in time...he's three now, but for a whole year we couldn't get his fever to drop below 102F. We went to four specialists...and we couldn't get his fever to drop below 102F. The only thing they come up with it's just something that happens in children. They never find the actual source of what it actually is, it's just something that happens. They still haven't determined exactly what it is. Just one day it went away, and it was okay. For a whole year, from the time he was a year and a half until he was about two and a half years old, before he turned three, we couldn't get his fever drop below 102F.*

> *Participant: I've got to get with this computer situation. I've got to get online...I learned a lot as far as my dad has got low blood pressure. He got rushed to the hospital last week. We didn't know what to do...I've got to get some computer lessons and stuff like that, and I think I would be good after that. My son and my daughter has got eczema and I don't know what to do...*

10.3.5.3 The sufferer

The *Sufferer* type is an expert in her or his own disease. This type of information seeker looks for anything that will address her or his own specific health issues.

> *Participant: I looked for myself like back pain or something like that...like if I'm having, if I'm sort of going through pains, and I don't have anything, if I go to the doctor and they tell me one thing, I will look it up before I take the medication for it, and I will look up the different medications that they give you. I look up the side effects. I look over everything and see if this is something I'm supposed to be taking or just something that I shouldn't take.*

> *Participant: I know before I was pregnant, I was taking Lovenox and heparin, so before I was taking it, there was a website. Before I took it, it was okay, but later on there was a lot of webpages sending information to my email. I looked it up and seen it was like a recall on heparin was saying that you weren't supposed to use it, but I had a blood clot in order to lessen it or in order to help the blood clot dissolve they kept me on heparin...I had gestational diabetes and all the different things they told me I had when I was pregnant, so I was looking up all this stuff to see what I can do, so I won't have to take medicine.*

10.3.5.4 The surfer

The *Surfer* is characterized by a general interest in health-related information, which may or may not relate to a specific issue. Information seekers of this type typically use the Internet as a convenient source of information, but may also rely on family, friends, or professionals

> *Participant: I think most of the answers on the Internet are all right. I stay home. I'm disabled. I'll be on the Internet all day. I learn so much stuff...*

> *Participant: Also if you go to Google...stuff will come up. It might be a forum from another grandmother that had the same problem that was describing what she did, who she sought out, what sites that she found. I think online forums and communities are a lot better sometimes, because they can give you a different perspective from personal experience as well as legitimate sites that they have already used and found that they were helpful.*

> *Participant: [I'm searching] all the time. Even if it's not I'm not going to necessarily say that Sally Sue had the same experience I'm not going to take hers as gospel, but I'm basically searching for it...maybe ask my doctor about something. Once you have a name for something it definitely makes you feel a lot better about it. Then you know exactly how to research it.*

10.3.6 Patterns and themes

Early in the analysis process, one critical element of the conversation became clear: participants wanted to tell their own stories about their health issues and how they had tried to resolve them—sometimes by using traditional information-seeking techniques (for example, asking a professional or an expert) but more often by a hit-or-miss approach, finding useful information about one topic while looking for something different. Their stories often focused on their success and failures during information seeking.

> *Participant: A lot of times if you're looking for one thing, I will just say a cough, it brings you to all the other things that has nothing to do with what it is that you initially put in. Then you get distracted. You start looking for or seeing something*

else, and then you are looking at something else, but not the original thing that you were actually looking for…You'll find a lot of stuff, but then you've got to be careful because there will be other websites that will be about other things. It will be other things that pop up like pop up blockers on the side like staying well. It could be like if you had a side effect, and it's like lawyers, different law groups I notice pop up. It will be like 'if you took this medicine…' or 'if this is the medicine you're taking…' and you click on it and then it might be somebody saying, 'well if you took this, then you are eligible for this law suit. Click on here if you want to hear more information.' I'm like oh that's the wrong thing, but it will say the medicine.

A second critical element that became clear was that most participants in the focus groups did not articulate a specific "information need" when starting to look for information. Rather, they approached information seeking as a "journey"—to use Dervin's term—leading them to encounter various types of information that might or might not address their questions. Furthermore, they appeared to be reluctant to look for health information on their own, preferring instead a more passive role, receiving their information from a doctor or nurse. These participants sometimes viewed looking for information on the Internet as an early or preliminary step in the process before referring their question to a more authoritative source.

Participant: You know children have all sorts of symptoms, and you don't know whether they are important or not. You have to go [to the Internet] and you say she's coughing or she's not sleeping or she's sleeping too much or whatever then you have to look at all the different things that could be and you think about if it's important enough and whether it's time to go to the doctor. There's only so much the Internet will provide for you…

The concept of "information need," as discussed above, has been a central focus of LIS research and discussion for several decades, often implies an articulated need for information to solve a problem or support a decision; in a population such as the one present in our focus groups, a perceived lack of individual empowerment can stand in the way of formulating a specific question that requires a specific answer.

Participant: I was thinking I was having an asthma attack, but I kept taking the inhaler, and it wouldn't work. I was taking the albuterol, and it wouldn't work. I was taking Advair, but it wouldn't work. [The doctor] said well it was pneumonia. I didn't get my charts at the time, so he gave me a bunch of prescriptions and discharged me. They gave me treatments while I was in there, but I still wasn't feeling right. Once I got the prescription filled, I still wasn't feeling right…I couldn't do nothing…

Throughout discussions with the focus group participants, we found the conversation returning to the desire for relevant, authoritative information. One participant commented, "I don't mess around. I just go straight to the doctor. I don't need no side bars." Other participants tended to agree with this attitude: "When you know it's coming from a medical professional, you have much more confidence in the information." However, there was also a sense of frustration and/or lack of trust in the "experts" who either did not know the answers, or could not communicate the answers effectively to the information seekers.

> *Participant: We were looking it up [on the Internet], because we went to so many specialists that nobody couldn't tell me anything. They all said the same thing. It's like somebody is to have to be able to tell me something.*

These insights into the participants' experiences with looking for information highlight the disconnect between the perceived needs of the participants and their ability to find appropriate information to satisfy those needs. The major finding from the focus group sessions was that the Internet has not replaced personal connections—whether with doctor, nurse, or family member—as a preferred source of health-related information, even though the professional sources do not always provide the desired information. Participants described an iterative process: looking for information from a variety of sources, such as family members, clinicians, and the Internet, to build a consensus about what the "right" answer to a particular question might be.

> *Participant: Sometimes for me the information that I am getting if it matches up with maybe a family member, or friend, or someone may have the same situation, a doctor may get the information from, if it correlates to what I've heard or what I've read then that works for me. Sometimes they may have other associates drugs to go to. That's helpful for me.*

10.3.7 Discussion

Traditionally, information seeking has been linked to a perceived information need on the part of the individual doing the seeking. Information need has variously been defined in the LIS literature as a gap: that which enables a person to move forward toward making a decision or taking an action. As noted earlier, this definition has worked well in the past and yielded an entire genre of studies looking at the information-seeking habits of scientists, physicians, nurses, teachers, and students in the pre-Internet age. Their purpose was to support the design of tools, services, and programs for those who are assumed to need information to make informed decisions. Such studies focused, in large measure, on how these groups interacted with printed sources or personally known and trusted individuals.

Today's world is more diverse in terms of its information sources and the ways in which they can be accessed. There are now numerous methods of accessing the abundance of information on the Internet—there are curated, authoritative sources on almost any topic, and a universe of social media "conversations" in which personal—albeit often anonymous—connections provide answers to almost any question. Disparities in access to information still exist, but the patterns and the reasons for these disparities are not always clear. Although Chatman's "small world" (Chatman, 1996) has changed dramatically, her insights are still worth considering. She observed that members of "small world" communities frequently are not active seekers of information outside of their immediate environments. She explained this phenomenon by suggesting that outside sources were perceived as unresponsive to these individuals' concerns, and judged quite likely irrelevant to the concrete realities of their situations. She went on to say that members of such communities determine which types of information are important to them and which information sources they believe they can trust based on their shared experiences.

This type of information behavior was also seen in our focus group participants, even though their access to resources available from the Internet provides them with opportunities to reach well beyond the limits of a traditional (geographically limited) "small world" to satisfy their information needs.

In thinking about this behavior, it is useful to revisit Dervin's definition of an information need in terms of the "gaps" that are experienced by individuals trying to make sense of their world (Dervin & Nilan, 1986). The gap or need is filled by information, which is either sought actively by the individual or provided to her/him passively, often by an authoritative or trusted source. As noted in our findings, most participants in the focus groups did not think of themselves as having a specific "information need" that they could resolve on their own, but rather thought of themselves as being in a situation that required some external solution, typically from a doctor or nurse. The participants' perception limits the degree to which they feel empowered to look for information independently. While not limited by geography or lack of access to information resources, this lack of personal empowerment creates a different type of "small world" phenomenon: one in which individuals are limited by a preferential divide based on their perceptions of information availability and utility. One role that information professionals can take on is that of a guide, to assist potential information seekers to become more successful in finding information that they perceive to be useful in their own situations.

As we consider the typology of users suggested by the findings from the focus groups, it is important to think about how personality typologies can be used to characterize individual preferences and behaviors. Like the Briggs-Myers personality types, our typology provides a short-hand way of referring to different types of information seekers. Because information seeking is situational in nature, the same person may at one time exhibit the characteristics of a "caregiver" but at other times the characteristics of a "sufferer." Information professionals must be prepared to enter into a dialog with potential information seekers and to be sensitive to the situation in which each user finds herself/himself; it cannot be a "one size fits all" response. In a 2010 lecture, Dervin made an impassioned plea to health information professionals for an approach to information provision that is dialogic and conversational, in contrast to one in which information is a commodity that is transferred from one entity to another (Dervin, 2010). As researchers in information behavior, our interest in information seeking accords with this desire. Information seeking is an active process of identifying a need and interacting with sources to meet that need; it is not passive receptivity to messages that are communicated from outside the situation.

The work of the researchers and theoreticians discussed in this chapter, as well as our own previous research, helps us to understand the challenges faced by the practitioners in our field—librarians and other information professionals—when facilitating access to health information. Our work with the community center taught us that many of the traditional LIS preconceptions about information seeking need to be reenvisioned. Information seeking in low-income, low-education health disparity populations like those at the community center appears less structured than the process presumed by theoretical models based on the information behaviors of groups defined by specific attributes such as occupation. But this information seeking still describes the same goal—the desire to find relevant information to help understand/solve a problem or make a decision. Current research highlights the lack of successful experiences that many disadvantaged populations have with traditional information seeking methods. Their work suggests that such populations may be unlikely to adopt traditional library-like search techniques because of their preferences for personal sources such as family members or friends in their social sphere. Rather, they will triangulate for information—looking to personal sources to confirm what they have been told by "experts" or heard over the media. When an answer is found consistently from multiple sources, that answer is often deemed to be a "good" answer. Authority in terms of the quality of information found often relies heavily on the source—the information seeker wants to know if it derives from someone or someplace that is personally trusted. This view of an authoritative source may or may not include published materials such as those found in libraries or on "reliable" Web sites—the sources typically recommended by librarians.

As researchers within the LIS tradition, our bias also may be to find ways of applying traditional library-like tools to support the information seeking of all groups. However, it seems that a more socially and culturally response will be to develop a deeper understanding of the ways in which people from diverse backgrounds and situations currently look for and find—or don't find—the information that they feel is valuable. All of the professionals who strive to facilitate successful health information seeking, whether librarians, journalists, teachers, or social workers, need to reenvision the conversation about health information provided to the public. Rather than being prescriptive about the "best" ways to look for and evaluate information, particularly online information, we should be investigating better ways of making health information seeking a successful and rewarding experience. Creating an "on ramp" such as the text-messaging intervention, a tool that can be used effectively by individuals to meet their own personal health information needs, is likely to be more productive than trying to train these same individuals to conduct formal information searches. Only by understanding and embracing the multiple ways in which information is looked for by the public, and building on those approaches, can we truly change the ways in which information is found and used.

10.4 Conclusions

Information seeking is of growing importance to the health status of citizens. More and more, individuals are faced with decisions about how to spend their time and resources to ensure the highest quality of life. Asking questions, seeking information, and understanding the answers are activities associated with improved quality of life; research and theory from information science can contribute much to examine and explicate the complexities of ensuring that all sectors of society have access to resources that will help them.

Our experience and the insights afforded to us through our work with a specific community indicate that the study of health information seeking, especially by marginalized populations, is a complex and multifaceted problem. Given that Internet access appears to be widespread, it is not clear why this and similar populations are not more active seekers of health information. Our data are limited, but our experience leads us to hypothesize: there is a tension between the desire for self-empowerment and a preference for a more passive role. All our research with this population suggests that lack of access to the Internet is not the barrier; rather, a complex mix of factors inhibits individuals from seeking health information. Therefore, greater emphasis must be placed on developing ways to improve health information.

The focus of our contribution has been to illustrate the diversity in how individuals from a specific community seek and use information; how information impacts their lives, or fails to impact their lives. Information professionals and others seeking to enter into conversation with these populations must first understand more about their information-seeking behaviors and source preferences, and then design interventions that are culturally in tune with their potential users.

Appendix A: Participant Survey (To Be Completed before the Workshop)

Hello. Drexel University is conducting workshops to show you how to find health information on the Internet. Following the workshop, we will have a short focus group to discuss how you currently look for health information. What you tell us will help us find better ways of communicating health information to you. The workshop will take approximately 45 min, and the focus group will take approximately another 45 min. The focus group will be recorded, but we will not know your name and everything that you say will remain confidential. If you complete the focus group, you will receive a gift card worth $20. *If you have attended one of these sessions before, you are not eligible to stay for the focus group.*

Answering the survey questions will indicate your consent to participate in this study. To ensure that your responses remain confidential, we will not ask for your name in this study and no identifying information will be collected. If you feel uncomfortable answering any questions and wish to discontinue this survey, you may do so at any time. There are no wrong answers to these questions. They will help us to understand how you look for health information.

About you:

Q1. Are you a patient of the 11th Street Health Center? ☐ Yes ☐ No

You must be a patient of the Center to participate in this study.

Q2. What is your age? _____ *You must be between the ages of 18 and 65 to participate in this study.*

Q3. What is your gender? ☐ Female ☐ Male

Q4. What is the highest level of education that you've completed?

☐ Middle school ☐ Some high school ☐ High school diploma ☐ Some college ☐ College degree

☐ Other: _____

About you and the Internet:

Q5. If you currently go online to use the Internet, how do you get there the most?

☐ Home computer ☐ Cell phone ☐ Mobile device ☐ Other: _____

☐ Don't use the Internet (please skip to Q10)

Q6. If you currently use the Internet, what do you use it for the most? (Please check just one choice.)

☐ Email ☐ Entertainment (downloading music, games, etc.) ☐ Facebook/My Space, etc.

☐ Looking for information ☐ Other (please describe) _____

Q7. Do you ever use the Internet to look for information? ☐ Yes ☐ No (if no, please skip to Q10)

Q8. If you currently use the Internet to look for information, what do you want to know about?

(Please check the 3 topics you look for most often.)

☐ Banking ☐ Directions ☐ Education ☐ Finding people ☐ General information

☐ Health information ☐ Job searching ☐ News/Weather ☐ Shopping

☐ Other (please describe) _____

Q9. If you currently use the Internet to look for information, do you find what you want?

☐ Never ☐ Sometimes ☐ Usually ☐ Mostly ☐ Always ☐ Don't know

About you and health information:

Q10. When you have a question about your health or the health of somebody you provide care for, where do you go to find information? (Please check the 3 sources you use most often.)

☐ I ask somebody in my family

☐ I ask a good friend

☐ I ask a nurse at the Center

☐ I ask a doctor

☐ I go to the library

☐ I use the Internet

☐ Other: _____

Q11. If you currently use the Internet to look for health information, how often do you use it?

☐ Not very often ☐ Once a month ☐ 2 or 3 times a month ☐ Once a week ☐ Daily

☐ Don't use the Internet

Q12. If you currently use the Internet to look for health information, do you find what you want?

☐ Never ☐ Sometimes ☐ Usually ☐ Mostly ☐ Always ☐ Don't know

☐ Don't use the Internet

Q13. If you currently find health information on the Internet, are you confident that it is accurate?

☐ Never ☐ Sometimes ☐ Usually ☐ Mostly ☐ Always ☐ Don't know

☐ Don't use the Internet

Q14. Are you (or were you) a member of a Centering Pregnancy group that received cell phone messages with links to Internet websites? ☐ Yes ☐ No

Q15. If the 11th Street Health Center sends you information about your health via the Internet, would you prefer to receive it on your:

☐ Home computer ☐ Cell phone ☐ Mobile device ☐ Other: _____

Please read the behavior descriptions and the examples below and then give yourself an honest rating between 1 and 5.

Rating: 1 = Never 2 = Sometimes 3 = Usually 4 = Mostly 5 = Always.

Behavior	Example	Rate yourself
I know that there are people who know a lot about a particular topic so I ask them rather than ask just anybody	If I'm not feeling well, I will ask an expert, like a nurse	
I don't go to the same place for the answers to all my questions. I look for the best place	Sometimes I go to a family member or friend, but sometimes I need to go to an expert	
If I don't find my answers right away, I keep on asking questions until I find the information I need	If my friends don't know the answer, I will go to the Internet or the library to look for information	
I understand most of the information that I am given or find on the Internet	If I get or find information about my medical condition, I usually know what it means	
I pay attention to the quality of the information I get or find	I can tell if the information that I have found is good or useful	
I make decisions based on the information that I get or find	Having good information helps me make better choices	

References

Anker, A. E., Reinhart, A. M., & Feeley, T. H. (2011). Health information seeking: a review of measures and methods. *Patient Education and Counseling, 82*, 346–354.

Burnett, G., Jaeger, P. T., & Thompson, K. M. (2008). Normative behaviour and information: the social aspects of information access. *Library and Information Science Research, 30*, 56–66.

Case, D. O. (2012). *Looking for information: A survey of research on information seeking, needs and behavior* (3rd ed.). Bingley, UK: Emerald Group.

Chatman, E. A. (1996). The impoverished life-world of outsiders. *Journal of the American Society for Information Science, 47*, 193–206.

Chatman, E. A. (1999). A theory of life in the round. *Journal of the American Society for Information Science, 50*, 207–217.

Consumer and Patient Health Information Section (CAPHIS/MLA). (1996). The librarian's role in the provision of consumer health information and patient education. Medical Library Association. *Bulletin of the Medical Library Association, 84*(2), 238–239. Retrieved from http://caphis.mlanet.org/chis/librarian.html.

Dalrymple, P. W., Rogers, M., Zach, L., Turner, K., & Green, M. (2013). Collaborating to develop and test and enhanced text messaging system to encourage health information seeking. *Journal of the Medical Library Association, 101*(3), 224–227.

Dervin, B. (1997). Given a context by any other name: methodological tools for taming the unruly beast. In P. Vakkari, R. Savolainen, & B. Dervin (Eds.), *Information seeking in context* (pp. 13–38). London, UK: Taylor Graham.

Dervin, B., & Nilan, M. (1986). Information needs and uses. *Annual Review of Information Science and Technology, 21*, 3–33.

Dervin, B. L. (2010). Clear...unclear? Accurate...inaccurate? Objective...subjective? Research...practice? Why polarities impede the research, practice and design of information systems and how Sense-Making Methodology attempts to bridge the gaps. Part 1. *Journal of Evaluation in Clinical Practice, 16*(5), 994–997.

Fisher, K. E., Durrance, J. C., & Hinton, M. B. (2004). Information grounds and the use of need-based services by immigrants in Queens, New York: a context-based, outcome evaluation approach. *Journal of the American Society for Information Science & Technology, 55*(8), 754–766.

Fisher, K. E., Erdelez, S., & McKechnie, E. F. (Eds.). (2005). *Theories of information behavior.* Medford, NJ: Information Today.

Fox, S., & Duggan, M. (November 2012). *Mobile Health 2012.* Retrieved from http://www.pewinternet.org/2012/11/08/mobile-health-2012/.

Fox, S., & Rainie, L. (February 2014). *The Web at 25.* Retrieved from http://www.pewinternet.org/2014/02/25/the-web-at-25-in-the-u-s.

Glazier, J. D., & Powell, R. (Eds.). (1992). *Qualitative research in information management.* Englewood, CO: Libraries Unlimited.

Harris, R. M., & Dewdney, P. (1994). *Barriers to information: How formal help systems fail battered women* (No. 81). Westport, CT: Greenwood Press.

Harris, R. M., Wathen, C. N., & Fear, J. M. (2006). Searching for health information in rural Canada. Where do residents look for health information and what do they do when they find it? *Information Research, 12*(1). Retrieved from http://InformationR.net/it/12-1/paper274.html.

Johnson, J. D., & Case, D. O. (2012). *Health information seeking.* New York: Peter Lang.

Kuhlthau, C. C. (1991). Inside the search process: information seeking from the user's perspective. *Journal of the American Society for Information Science, 42*(5), 361–371.

Lorence, D. P., Park, H., & Fox, S. (2006). Racial disparities in health information access: resilience of the Digital Divide. *Journal of Medical Systems, 30*, 241–249.

Lustria, M. L., Smith, S. A., & Hinnant, C. C. (2011). Exploring digital divides: an examination of eHealth technology use in health information seeking, communication and personal health information management in the USA. *Health Informatics Journal, 17*, 224–243.

Malone, M., While, A., & Roberts, J. (2014). Parental health information seeking and re-exploration of the "digital divide". *Primary Health Care Research & Development, 15*, 202–212.

Mitchell, S. J., Godov, L., Shabaxx, K., & Horn, I. B. (2014). Internet and mobile technology use among urban African American parents: survey study of a clinical population. *Journal of Medical Internet Research, 16*(1), e9. http://dx.doi.org/10.2196/jmir.2673.

Morey, O. T. (2007). Health information ties: preliminary findings on the health information seeking behaviour of an African-American community. *Information Research-An International Electronic Journal, 12*(2).

Neter, E., & Brainin, E. (2014). eHealth literacy: extending the digital divide to the realm of health information. *Journal of Medical Internet Research, 14*(1), e19.

Nutbeam, D. (2000). Health literacy as a public health goal: a challenge for contemporary health education and communication strategies into the 21st century. *Health Promotion International, 15*, 259–267.

Ormandy, P. (2011). Defining information need in health:assimilating complex theories derived from information science. *Health Expect, 14*(1), 92–104.

Pettigrew, K. (2000). Lay information provision in community settings: how community health nurses disseminate human services information to the elderly. *Library Quarterly, 70*, 47–85.

Pew Research Center Internet & American Life Project. (2014). *2014 Health fact sheet.* http://www.pewinternet.org/fact-sheets/health-fact-sheet/.

Rogers, M., Dalrymple, P., Zach, L., Luberti, A., Abrams, D., & Jobs, J. (April 26, 2013). The year of mHealth in Philadelphia: drawing a portrait of technology use in an urban area, revisited. In *6th Annual healthcare informatics symposium, presented by the center for biomedical informatics (CBMi) at the children's hospital of Philadelphia.*

Savolainen, R. (1995). Everyday life information seeking: approaching information seeking in the context of "way of life." *Library and Information Science Research, 17*, 259–294.

Sligo, F. N., & Jameson, A. M. (2000). The knowledge-behavior gap in use of health information. *Journal of the American Society for Information Science, 51*(9), 858–869.

Song, H., Cramer, E. M., & McRoy, S. (2014). Information gathering and technology use among low-income minority men at risk for prostate cancer. *American Journal of Men's Health* pii: 1557988314539502. [Epub ahead of print].

Spink, A., & Cole, C. (2001). Information and poverty: information-seeking channels used by African American low-income households. *Library and Information Science Research, 23*, 45–65.

Taylor, R. S. (1962). The process of asking questions. *American Documentation, 13*(4), 391–396.

Taylor, R. S. (1991). Information use environments. In B. Dervin, & M. Voight (Eds.), *Progress in communication sciences* (Vol. 10). Norwood, NJ: Ablex.

U.S. Department of Health & Human Services. (2011). *Healthy people 2020.* Washington, DC: U.S. Department of Health & Human Services, Retrieved from http://www.healthypeople.gov/2020/default.aspx.

Veinot, T. C., Campbell, T. R., Kruger, D. J., & Grodzinski, A. (2013). A question of trust: user-centered design requirements for an informatics intervention to promote the sexual health of African-American youth. *Journal of the American Medical Informatics Association, 20*, 758–765.

Westbrook, L. (2009). Crisis information concerns: Information needs of domestic violence survivors. *Information Processing & Management, 45*, 98–114.

Yu, L. (2010). How poor informationally are the information poor? Evidence from an empirical study of daily and regular information practices of individuals. *Journal of Documentation, 66*(6), 906–933.

Zach, L., Dalrymple, P., Rogers, M., & Williver-Farr, H. (2012). Assessing Internet access and use in a medically underserved population: implications for providing enhanced health information services. *Health Information & Libraries Journal, 29*, 61–71.

Zahradnik, A. (2011). Asthma education information source preferences and their relationship to asthma knowledge. *Journal of Health and Human Services Administration, 34*, 325–351.

Zweizig, D. L., & Dervin, B. (1977). Public library use, users, uses: advances in knowledge of the characteristics and needs of the adult clientele of American public libraries. In M. Voight, & M. Harris (Eds.), *Advances in librarianship* (Vol. 7) (pp. 231–255). New York: Academic Press.

For the mutual benefit: health information provision in the science classroom

Albert Zeyer[1], Daniel M. Levin[2], Alla Keselman[3]
[1]Health Division, Bern University of Applied Sciences, Bern, Switzerland; Institute of Education, University of Zurich, Zurich, Switzerland; [2]Department of Teaching and Learning, Policy and Leadership, University of Maryland, College Park, Maryland, USA; [3]Division of Specialized Information Services, US National Library of Medicine (NLM), National Institutes of Health, Bethesda, Maryland, USA

11.1 Background

11.1.1 Opportunities and challenges of the era of participatory care

Today, individuals are expected to be active participants in their health care, collaborating with health professionals in choosing the course of action in care, treatment, and prevention. To encourage patients to ask questions, to enable shared decision-making, and thus to overcome "white-coat silence" is seen as an important focus of an ongoing effort in health care quality improvement (Judson, Detsky, & Press, 2013). The Society for Participatory Medicine defines participatory medicine as "a model of cooperative health care that seeks to achieve active involvement by patients, professionals, caregivers, and others across the continuum of care on all issues related to an individual's health." It is "an ethical approach to care that also holds promise to improve outcomes, reduce medical errors, [and] increase patient satisfaction" (Society for Participatory Medicine, 2014).

An emphasis on participatory medicine and shared decision-making requires the public to actively engage with health information, reviewing and contributing to health records, evaluating descriptions of clinical trials, and weighing benefits and risk factors of different treatment options. The information that individuals are expected to engage with is complex. For example, many health-care systems and health plans now provide their consumers with personal health records combining patient-entered data (e.g., over-the-counter medications, family history) with provider-entered data (e.g., physical exam findings, laboratory reports). In order to achieve "improve[d] outcomes" and "increase[d] patient satisfaction," individuals need to integrate these types of information, written at different granularity levels and using different terminology. Patients are also expected to participate in decision-making in situations that have no routine standard of care. In the US, ClinicalTrials.gov, currently the world's largest database of clinical trials, grew out of the 1997 congressional law that mandated building such registry for broad audiences, including "individuals with serious or life-threatening diseases or conditions, members of the public, health-care providers, and researchers" (Clinical Trials.gov, 2015). At the time this chapter was being written, ClinicalTrials.gov yielded 3278 open trials for a search on the term "diabetes mellitus," 2032 for "breast cancer," and 1224 for "arthritis." Navigating among trial entries involves evaluating descriptions of purposes, designs, primary and secondary outcome measures, and inclusion and exclusion criteria for eligibility.

Research suggests that lay individuals often have difficulty evaluating and comprehending complex health information created by health professionals. For example, an informed consent document required to participate in a clinical trial is notoriously challenging for patients to understand. In one study of cancer patients participating in a clinical trial, 63% did not fully understand the risks described in their consent form and 70% did not understand the treatment's unproven nature (Joffe, Cook, Cleary, Clark, & Weeks, 2001). Keselman et al. (2007) found that when patients review their medical records, many find physicians' and nurses' notes and radiology reports particularly difficult to understand (e.g., only 36% of the participants rated physicians' notes as "easy"). Individuals also have trouble understanding prescription drug labels and instructions (Davis et al., 2009).

A central concept providing context for understanding individuals' interaction with health information is health literacy. In the US, health literacy discourse typically draws on the definition developed by the National Library of Medicine and used in *Healthy People 2010*, a nationwide health promotion agenda led by the Department of Health and Human Services (Ratzan & Parker, 2000). According to this definition, health literacy is "the degree to which individuals have the capacity to obtain, process, and understand basic health information and services needed to make appropriate health decisions." A growing literature shows negative consequences of poor health literacy, mediated through three pathways: access and utilization of health care, patient–provider relationship, and self-care (Paasche-Orlow & Wolf, 2007). Low health literacy is an independent risk factor for poorer outcomes, including increased mortality, lower satisfaction with care, lower quality of care, poorer patient safety, and higher health care costs (Berkman, Sheridan, Donahue, Halpern, & Crotty, 2011).

11.1.2 Deconstructing health literacy

The *Healthy People 2010* definition cited above (Ratzan & Parker, 2000) does not enumerate specific skills that constitute health literacy. In fact, even delineating these skills as well as assessing health literacy in research studies (see the chapter by Logan in this volume) has been a challenge. First attempts to operationalize the definition have focused on basic academic skills. For example, in 1999, the American Medical Association defined health literacy as "a constellation of skills, including the ability to perform basic reading and numerical tasks required to function in the health care environment" (Ad Hoc Committee on Health Literacy for the Council on Scientific Affairs, 1999). Obviously, this type of definition, though very common, is very restricted, as it refers to what has been called "functional literacy." Functional literacy is a one-to-one transfer of literacy in its primary sense, that is, basic reading, writing, and numeric skills, into the health domain. Much empirical research that has been done on health literacy, particularly in medicine, is tied to this functional literacy definition. For example, in many studies, health literacy is measured by the test of functional health literacy in adults (Parker, Baker, Williams, & Nurss, 1995), which assesses reading comprehension and numerical ability with respect to prescription labels, clinical appointments, and prose passages (Schulz & Nakamoto, 2012).

Though this functional approach to health literacy is common, it has been criticized, and alternatives have been sought. Nutbeam (2000) expanded the construct by developing the concepts of interactive literacy and critical literacy. *Interactive* literacy is the ability to extract information and its meanings from different forms of communication. For example, for potential polyarthritis patients, understanding a video, which points out the importance of an early-onset treatment, asks for interactive literacy. Not only do the viewers need to understand the basic content of the video, but they also have to realize that the presented symptoms match their own situation, and calling a doctor might be wise.

Critical literacy refers to the ability to analyze the information and to use the results of this analysis in a directed manner in order to manage and be responsible for one's own health. This holds not only for health, but also for disease. For example, a polyarthritis patient, treated with cortisol, needs critical literacy in order to decide whether to believe her doctor—who recommends continuing with the treatment—or an Internet site that warns against doing so.

Nutbeam's approach is in line with the World Health Organization (WHO) definition extending far beyond the functional literacy approach: "Health literacy represents the cognitive and social skills which determine the motivation and ability of individuals to gain access to, understand, and use information in ways that promote and maintain good health" (World Health Organization, 1998, p. 10).

One of our main concerns in this chapter is that the role of biomedical knowledge, or the bases of health and disease, while implicitly present in all these definitions, has frequently been underestimated in the health literacy discussion (also cf. Zeyer, 2012). Keselman et al. (2007) found that insufficient conceptual biomedical knowledge was the second most prevalent barrier to successful use of records after problems with records access—more common than problems with medical terminology and difficulty understanding records' structure and organization. Participants' comments suggested that the concepts related to laboratory tests—understanding test purposes and interpreting results—were among the most difficult for them to understand. In another study, Keselman and Smith (2012) developed a classification of errors in lay comprehension of medical documents, a study from ClinicalTrials.gov and a typical office visit note. Once again, misunderstanding biomedical concepts, such as the function of hormones and the role of insulin in metabolism, was among the most common identified comprehension barriers. There is much that health professionals and librarians can do to support the public's ability to deal with health information, but foundational biology, chemistry, and physics concepts are usually taught in schools, and thus schools and teachers must play a role in building health literacy.

11.1.3 Health literacy in context: the case of human papillomavirus vaccination

Consider the following case: Human papillomaviruses (HPVs) are a group of more than 150 related viruses, certain types of which can cause cervical cancer. Vaccines have been developed that are highly effective in preventing infection with carcinogenic types of HPV. HPV vaccination has the potential to reduce cervical cancer deaths around the world by as much as two-thirds. The vaccines are approved in most parts of the world. For example, the U.S. Food and Drug Administration has approved one of them, Cervarix, for use in females ages 9–25 years (National Cancer Institute, 2011).

In Switzerland, where the empirical studies described later in this chapter took place, the HPV vaccination was introduced in 2007. Since then, all girls between the ages of 11 and 26 years have been offered the vaccination for free through a school and public health campaign organized by the government. Consequently, every Swiss girl and young woman has to decide whether to take the vaccination or not, which is a typical decision drawing on an individual's health literacy.

To understand the dimensions of an expanded conceptualization of health literacy, consider the following information on HPV vaccination that was selected and included in an official Swiss guide for girls and young women (Kanton Luzern: Gesundheits- und Sozialdepartement, 2012). The original document is a two-sided flyer obviously designed to appeal to young women—the basic color is pink, and the titles appear in a font that imitates handwriting. The cover shows a portrait of a girl; the back shows a young couple holding hands. The flyer contains the following information[1]:

1. HPV is the cause of almost all cases of cervical cancer.
2. 70% of all girls and women are infected with HPV during their life.
3. There are more than 100 types of HPV. Some of these are responsible for cervical cancer as well as mucosal warts. The vaccination offers over 95% protection against some of these viruses, which are the cause of 70% of cervical cancer cases.
4. For comprehensive protection, girls aged between 11 and 14 require 2 doses and from the age of 15 years girls require 3 doses, which are administered within 6 months of each other.
5. The vaccinations are considered safe and generally well tolerated. Severe side effects are extremely rare.
6. The vaccination is free when part of a cantonal vaccination program.

These six points are part of the flyer, which also contains additional information on other topics such as cervical cancer and sexual transmission of HPV. Obviously, understanding all this information presupposes functional literacy. A client cannot understand point 4, for example, without being functionally literate; she must be able to read the phrase and understand the numbers used in it. In fact, point 4 will even be a challenge for many literate readers. Additionally, it has been shown that many people have trouble understanding quantitative data expressed as percentages; the phrasing in point 3, citing 95% protection against viruses that cause 70% of cases, is likely to be an excessive demand for many individuals (Paasche-Orlow, Schillinger, Greene, & Wagner, 2006).

Beyond functional health literacy, the brochure also calls for interactive literacy. In order to understand the message, a client must combine points 1–3 and relate them to her own situation. She has to understand that she will be most likely to be infected with HPV viruses, some of these viruses induce cervical cancer, the vaccination will offer protection against infection, and, therefore, against cervical cancer, and taking this vaccination will mean she will be protected against cervical cancer. Being able

---

[1] Translated in a paraphrase close to the original text by one of the authors.

to draw these inferences and make conclusions based on the information in this flyer, and apply them to one's own personal situation, is the central idea of interactive health literacy.

Critical literacy, in the sense of the WHO definition, expects even more. The critically literate client uses the information in the flyer to decide whether or not HPV vaccination is good for her health, and whether or not she should choose to be vaccinated. Making this decision requires the client to draw upon her background biomedical knowledge of things that cannot be addressed in a short guide, but that affect overall meaning construction and the client's interpretation of personal relevance of the information. For example, an understanding of the mechanism by which vaccine provides protection could help the client evaluate its safety (thus, linking points 3 and 5). As shown further in this paper, gaps in or misunderstandings of biomedical knowledge, affect response to the guide. Ultimately, the client must answer the questions of personal relevance and usefulness herself. They cannot be answered by others, not even by health professionals. This is a typical situation of shared decision-making, and represents a central challenge of health information provision.

11.2 The science classroom as a setting for health literacy

In view of the above discussion, we claim that the science classroom is a crucially important setting for supporting the development of health literacy (Keselman, Hundal, & Smith, 2012). First, we suggest that the science classroom is a well-fitting context for delivering the foundational background knowledge needed to deal with information about health and disease in a critically literate way. Content in science classrooms in secondary schools internationally is naturally connected to issues of personal and environmental health (Neulight et al., 2007). Biology content is especially well-connected and has the potential for applying science to health-related issues in the science classroom. Understanding of cellular and physiological processes can serve as a framework for learning about disease, and understandings of the flow of matter and energy in the ecosystem can support exploration of modern and historical environmental health issues, including those connected to climate change and resource distribution and disposal. The physical sciences also connect to health issues, as understanding of chemistry can potentially support learning about properties of water and toxic chemicals, and study of pressure and fluid dynamics connects to understanding of healthy and abnormal physiology, for example, issues of blood pressure and heart function. However, at present, science classrooms typically introduce students to the basic concepts, such as cellular processes or pressure and fluid dynamics, yet do not complete the final step of connecting them to personal and social issues of health and disease.

There is potential synergy between health and science education, but for decades, the two have been alienated through the distance between the two cultures. Only in recent times have first efforts been started in order to bridge this unnecessary gap and to underline common interests and mutual benefits of these fields (Zeyer & Dillon, 2014).

In the United States and in many European countries, health education is primarily taught as a separate class, in which underlying scientific principles and global health issues take a back seat to coverage of sex education, alcohol and drug awareness, and personal health and wellness (Jourdan, 2011). For example, in one school district in the US, the curriculum framework for health education focuses on seven topics: mental and emotional health, alcohol, tobacco, and other drugs, personal and consumer health, family life and human sexuality, safety and injury prevention, nutrition and fitness, and disease prevention and control. There is little expectation, nor any guidance for teachers, to connect with underlying scientific concepts and mechanisms (Montgomery County, n.d.). The health classroom emphasis is typically on wellness and prevention, rather than disease, and is framed in behavioral terms (recommended ways to do things), rather than conceptual terms (the underlying mechanisms). Although health education topics do sometimes naturally come up in science and social studies classes, these teachers rarely collaborate with health teachers in linking the content (Collins et al., 1995). A renewed science pedagogy for science education may reconceptualize the relationship between the fields of science and health education (Zeyer & Kyburz-Graber, 2012).

The second part of our claim is that the science classroom is a fitting context for fostering critical literacy, which is grounded in such critical thinking skills, as inferring, problem-solving, and evaluating, reconciling, and integrating information, all essential to the practice of science. Scholars and policy makers in both the US and Europe have long advocated the importance of critical thinking in health education (Education Commission of the States, 1981; Jourdan, 2011). Critical thinking, as it is described in this literature, refers to problem-solving and decision-making about health-related issues (Collins et al., 1995). It is not clear from these recommendations, however, how critical thinking connects to specific "practices" of science that students should engage in, such as analyzing and interpreting data, constructing explanations, and engaging in arguments from evidence. A focus on specific practices of science is promoted internationally by recent science education research in the US (National Research Council, 2011) and by recent policies in Europe (EACEA/Eurydice, 2011), and these practices should be highlighted in students' reasoning about health issues.

Generally, in health classes, students are expected to make decisions by directly absorbing the information they have been provided (smoking is harmful, therefore, one should not smoke), and not through practices of analysis and argumentation, which would support students' understanding of the scientific basis of health concepts, provide practice in authentic scientific pursuit, and promote interactive and critical health literacy. Throughout this chapter, we will be showing that the assumption of direct transmission–absorption of information does not work, and how such instructional practices (e.g., delivering short lectures, handing out pamphlets) do not change beliefs and behaviors. The science classroom, where analysis and argumentation are becoming more central, can serve as a natural context for pursuing knowledge and inquiry into health issues and a laboratory for the continued study of the relationships between students' scientific knowledge, inquiry, and health education.

Our argument that the science classroom is a good arena in which to foster health knowledge and promoting health literacy is in line with current science education research and policy beliefs about science education. For decades, scientists and science educators have expressed different views on what should be taught to children and teens in science classes. These views are generally tied to the beliefs about the role of science education in society. Recent science education reform in the US has focused on the development of ideas, reasoning, and scientific practices of all students (National Research Council, 2007, 2011). The contemporary focus of science education reform is in preparing the future science and engineering workforce and producing an educated general public, supportive of scientific pursuits, engaged in socioscientific discourse and advocacy, and capable of applying science to daily tasks. From this perspective, health constitutes an important subject for engaging both future scientists and nonscientists by teaching them about topics that they are likely to perceive as relevant to their lives.

11.2.1 From health information to health behavior: the role of knowledge and other factors

11.2.1.1 Impact of scientific knowledge on daily life and health

For the past several decades, science educators have routinely discussed the importance of teaching science to all students—future scientists and nonscientists alike—offering the applicability of science to daily life's decisions as one of the justifications (National Research Council, 2011). At the same time, few empirical studies have addressed the use of scientific knowledge in everyday life; those that produced results were difficult to interpret. Some of the most thought-provoking studies were conducted decades ago, without more recent follow-up. In a 1993 UK study interviewing elderly people planning their heating budgets, Layton, Jenkins, Macgill, and Davey (1993) concluded that participants rarely framed their problems as scientific, and were more likely to speak about the importance of "keeping up appearances" than about the mechanism of convection heating. Another study showed a relationship between lay theories of residential thermostat heat control expressed in interviews and recorded patterns of actual thermostat control (Kempton, 1986).

Conclusive demonstration that knowledge leads to better health decisions may be challenging, because humans are not entirely rational. Individuals frequently hold conflicting beliefs without resolving them. For example, in a cross-cultural study, Sivaramakrishnan and Patel (1993) asked college-educated mothers of East Indian descent living in Canada to explain their beliefs about causes of childhood diseases. They found that while the mothers mentioned many Western biomedical concepts in their explanations, they also rooted their causal explanations in the Indian Ayurvedic belief system that attributes illnesses to the imbalance of humors or bodily fluids. This resulted in piecemeal, internally contradicting explanations, often inconsistent from the Western medicine perspective (e.g., attributing malnutrition to liver problems), despite the fact that the explanations were expressed in largely Western biomedical terminology. In the US, Johnson and Piggliucci (2004) showed that college science majors are not less likely than business majors to believe pseudoscientific claims, such as the existence of telepathy, despite their better knowledge of science facts.

People also have difficulty using their scientific knowledge when assessing health myths. For example, Keselman et al. found that while biology majors were more likely to draw on their depth of biological knowledge when reasoning about common health beliefs, they were only marginally better than nonscience majors at evaluating the accuracy of those beliefs (Keselman, Hundal, Chentsova-Dutton, Bibi, & Edelman, 2015).

This review may suggest that conceptual knowledge does not affect real-life health decisions. We believe instead that these studies underscore the complex nature of the interaction between knowledge and behavior, and the challenge of cataloging situations when knowledge has positive impact. While direct evidence of the impact of conceptual knowledge on daily life is limited, indirect evidence comes from studies that document negative impact of poor health literacy on health behaviors and outcomes. For example, in a study of health literacy, behavior, and outcomes among HIV patients, Kalichman and Rompa (2000) showed that those with lower health literacy had lower CD4 cell counts and higher viral loads (CD4 cells, also called T cells, are white blood cells that initiate the body's immune response to infection. HIV attacks these cells, thus reducing their number and weakening the immune system). These patients were also less likely to comply with prescribed medications and were hospitalized more frequently. Baker et al. (2002) demonstrated a similar relationship between poor health literacy and number of hospitalizations among elderly adults. Schillinger et al. (2002) found that in patients with diabetes, inadequate health literacy was linked to poorer glycemic control and higher instances of retinopathy. Health literacy may also be related to patients' perception of their ability to be active participants in their health care (Smith, Dixon, Trevena, Nutbeam, & McCaffery, 2009).

Feinstein points out that developing an effective science curriculum for all students requires filling the "empirical vacuum": identifying specific real-life context and situations where scientific knowledge is likely to be beneficial (Feinstein, 2010, p. 169). Science education researchers and practitioners should continue collaborating on identifying such contexts, as well as develop methods to teach science in ways that help individuals make health decisions and deal with real-life health information beyond school. This chapter is our contribution to this discussion.

11.2.1.2 The role of background biomedical knowledge in information processing

While showing an unequivocal positive impact of scientific knowledge on behavior may be challenging, an empirical study conducted by one of the authors underscores how insufficient scientific knowledge can hinder a health information campaign. HPV vaccination was described earlier in this chapter. This is a subject of some controversy among the Swiss public. Zeyer and Sidler (2015) investigated how Swiss student teachers, women young enough to be recommended recipients of the HPV vaccine, responded to the official vaccination guide (described in *Health literacy in context: Considering HPV vaccination* section) and how the guide affected their personal decisions about vaccination. The study focused on teachers because the vaccination guide is often promoted in schools, where teachers' attitudes may influence that of students.

The majority of the participating student teachers continued to have a negative or ambivalent view of vaccination after studying the information guide. A detailed analysis of responses showed that most of them had superficial knowledge of vaccination that led to misunderstandings. In fact, even correct pieces of information gave rise to misunderstandings that clearly had an impact on student teachers' attitudes toward HPV vaccination.

Four main misunderstandings led to skepticism about HPV vaccination. The first was that the vaccine was made out of attenuated (reduced in virulence) HPV viruses, which sometimes might change back into the virulent wild variant and thus could cause infection and sometimes even cervical cancer. This misunderstanding may well be a result of historic narratives provided in Swiss schoolbooks. In fact, early vaccines contained attenuated pathogens, and some vaccines (like the measles vaccine) still do. However, recent developments in genetic engineering have resulted in new vaccines, the HPV vaccine being one of them, that contain only synthetic marker particles, which are not at all infectious.

The second misunderstanding was that the limited number of strains in the vaccine offers only incomplete protection and thus makes the vaccine fairly ineffective. This idea arose from a lack of knowledge: specifically, knowing that only four strains of the HPV can cause cervix carcinoma, and that therefore, not every infection caused by the papillomavirus will result in a malignant transformation of cervix cells.

The third misunderstanding was that physical examination can discover an HPV infection, so that in case of regular checkups with a gynecologist, there is no further need for vaccination. This third belief is correct up to a point; a regular Pap smear during annual checkups can be used to detect dysplasia, or abnormal precancerous change in cells. However, many people do not know that in this case there are no therapeutic options beyond a regular follow-up. If diseased cells remain, the next step is conization (in this procedure, a cone-shaped piece of the cervix containing the area with abnormal cells is removed by a scalpel or a laser. It is associated with complication risks in pregnancy, and is not sure to definitely cure the illness.).

The fourth and final misunderstanding was that the vaccination is only useful if administered before someone has sex for the first time. However, HPV vaccination is now recommended for females less than 26 years old, because it has been shown that the vast majority of women in this age are not yet infected with those HPV strains that cause cancer (Paavonen, 2008).

The misunderstandings held by the participants show the challenge of developing critical literacy. The biomedical knowledge needed for dealing with consumer-level health information can be considerable and demanding. In the case of HPV, it includes basic knowledge about infectious diseases, tumor biology, immunology, genetics, and risk management. The student teacher participants investigated in this study have all passed their Matura, the final exam that young adults (aged 18 or 19 years) take at the end of their secondary education (high school) in Switzerland. This means that they have taken biology courses at the senior high school level, because in Switzerland biology courses are compulsory for all types of higher secondary education. Yet, these courses obviously did not equip these future teachers with the necessary biomedical knowledge to interpret these official vaccination guides.

Public understanding of science is the field that focuses on how the public understands, assimilates, and judges science knowledge and information. It addresses the interaction of these literacies as the problem of "bounded understanding" of scientific contexts: that even the most well-educated members of the general public do not possess the knowledge of experts (Bromme & Goldmann, 2014, p. 60). There are roughly two attitudes, or orientations, toward this phenomenon. The learning orientation seeks to support "full understanding," appropriate for the learner's age and educational level. The communication orientation, on the other hand, concedes the boundedness of lay peoples' understanding and focuses on attitudes about science and trust in scientists (Bromme & Goldmann, 2014). This article supports and underlines learning orientation as the appropriate approach to deal with bounded understanding of science. However, we believe that, beyond questions of depth and breadth of knowledge, there are other important decision-making influences that mediate the impact of knowledge in health literacy.

11.2.1.3 Decision-making influences that mediate the impact of information and knowledge

As previously discussed, functional and interactive literacy concepts are based on the transmission model of learning, a model which assumes that knowledge can be directly communicated and learners can more or less directly understand and use knowledge depending on their level of literacy. Below, we present a research study (Di Rocco & Zeyer, 2013) that shows how much reality can differ from this assumption. It is again concerned with health literacy in the case of HPV vaccination.

Since 2007, the city of Zurich has been offering free vaccination to all female students between the ages of 11 and 19 years through a campaign organized by the local school doctors' service ("Schulärztlicher Dienst"). As previously noted, the HPV vaccination is controversial in Switzerland; many parents consider HPV vaccination to be unnecessary for their children. In a qualitative study, Di Rocco and Zeyer (2013) investigated the information process in four classes of two schools of lower secondary level (i.e., grades 6–8). The schools primarily served students from upper middle-class families. These schools were chosen for the study because in Switzerland this local school medical board reported particularly frequent occurrences of vaccination skepticism.

In the course of the study, the school doctor was accompanied during instruction of the female students, aged between 11 and 13 years. All the stakeholders—the local school doctors' service, the students, the parents, and teachers—were interviewed. The researchers' aim was to determine how these students decided for or against being vaccinated and which actors influenced their decision.[2]

---

[2] This formulation may raise questions about whether 11 to 13-year-old students in Switzerland are really deciding for themselves to be vaccinated or not. We chose it in this (and in all the following passages) in order to capture the fact that, in Switzerland, participatory approaches are very common, not only in politics, but also in school contexts. Of course, in a legal sense, in Switzerland, as everywhere, parents need to make medical decisions for underage students. However, the approach to this decision is participatory, i.e., it is the students, who are informed first about the vaccination procedure. They also receive an information sheet addressed to them and their parents. In the end, the parents must give their consent to the HPV vaccination by signing an agreement, but they are expected to respect the students' personal decision and not sign without their child's consent.

In accordance with the transmission model of learning, the information presented by the school doctor was organized in a two-lesson event for each of the two classes. The school doctor visited each class in their classroom. The female students were provided with information about the HPV vaccination, essentially similar to the content of the aforementioned information guide distributed among girls and young women and discussed above. The school doctor, a young female practitioner, started with a brief introduction of her role, in particular that of a student's confidante. This was followed by an overview of vaccinations in general, and then of the HPV vaccination in particular. The entire presentation lasted for 45 min.

The school doctor assumed that the event had been successful. However, the researchers' interviews with the students showed that the students generally could not reproduce the information with which they had been presented. Only about half of the interviewed students knew that HPV vaccination prevented cancer, and that HPV could be transmitted by sexual intercourse. Some confused HPV infection with hepatitis; some assumed that the vaccination was about breast cancer. Even more importantly, the presentation had no detectable influence on a student's decision to be vaccinated. Actually, it was the parents, particularly the mothers, who had the largest influence on this decision. Other actors influencing students' decisions were the media, the teachers, and their peers, though their influence seemed to be less important.

It was striking how many of these actors showed a negative position toward vaccinations and the visit of the school doctor. However, many of them were only partially open to the doctor concerning their skepticism. For that reason, she was not aware of the rather adverse atmosphere, and so she did not take it into account when conducting the presentation. During the interviews with teachers and a mother, it became clear that even some of the pediatricians and gynecologists in the school area were skeptical about HPV vaccination; these health-care professionals seemed to actively influence the girls' mothers in particular. As the study did not include interviews with these medical doctors, the reason for their skepticism is not known. However, it is an attitude not uncommon among European pediatricians, which they mostly justify by calling the vaccination unnecessary and questioning the official arguments for vaccination as biased by commercial interests (cf. Hirte, 2012).

As a result, less than 50% of students in the two schools, where the study took place, were vaccinated, although the vaccine was free. This failure clearly points out the problem with the transmission approach. The transmitted knowledge is often not understood; in only a few cases it is taken on-board and applied. Indeed, in our example, the health-related decision was hardly influenced by the transmitted knowledge. In contrast, other factors were central—factors that are often not taken into account.

In science education, problems with transmission approaches are well-known, which is why moderate constructivism has become standard. Constructivist views of science education view human knowledge as created via the personal cognitive construction process, undertaken by the individual in order to make sense of his or her social or natural environment (Duit & Treagust, 2003). In order to take this approach into account and to frame the results of the presented research on HPV vaccination, Zeyer has adapted and elaborated a moderate constructivist model originally developed in environmental education (Gräsel, 2000) (Figure 11.1) (see also Zeyer, 2012).

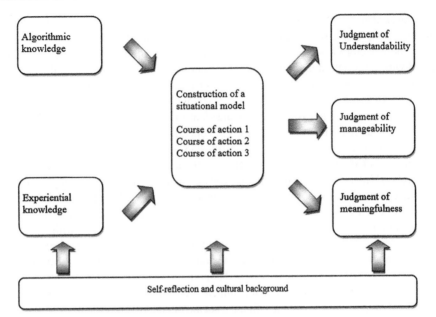

Figure 11.1 Framework model of health literacy.

The core idea of this model is that every person confronted with a certain health issue (e.g., HPV vaccination) makes their own and idiosyncratic situational construction, which includes various options of action (e.g., to vaccinate, not vaccinate, postpone vaccination). Different types of knowledge influence the situational construction, which entail different types of judgments. In the following, we explain the different factors included in the framework model.

The framework model uses a subclassification of knowledge into *algorithmic* knowledge and *experiential* knowledge, a formulation proposed by the Harvard economist Stephen Marglin (2008). It seems to be particularly appropriate for questions of health and disease. Algorithmic knowledge can be stored in textbooks or computers. Its application is guided by algorithms—hence its name—and rational inference. Experiential knowledge is embodied—that is, it is an integral part of a person—and its application depends on personal experience and intuition.

Scientific medicine, for example, mainly depends on algorithmic knowledge, expressed in textbooks, guidelines, and flow charts that can, in principle, be learned and applied by everybody with access to these materials. Alternative medicine, on the other hand, points out the importance of experiential knowledge, which may not be acquired solely by studying textbooks, but instead requires scaffolding, instruction, and support by a mentor, and learning by doing. When it is applied, the user needs a sort of embodied intelligence that uses analogies instead of inferences. For example, in homeopathy, the doctor chooses the therapy not because of algorithmic inference, but because of the core principle of similia similibus curantur (i.e., a disease is cured by those remedies that produce effects similar to the disease itself). The way this principle is applied depends very much on the user, their experience, and their idiosyncratic way to think and conclude. The framework model proposes that, generally, the construction of health and disease situations always involves both algorithmic and experiential knowledge. Celebrated clinicians, for example, are known for combining stupendous handling of algorithmic knowledge with apt intuition based on their vast experience.

The situational construction informs personal judgments, which provide the basis for decisions to act. Judgments can be further understood as judgments of understandability, manageability, and meaningfulness (Antonovsky, 1997). Understandability is the extent to which impacts from the external and internal environment are structured, explicable, and predictable. For example, prevention of high blood pressure is understandable, because high blood pressure causes damages on blood vessels and thus stroke problems in heart and brain in a foreseeable way. Manageability is the extent to which resources are available to a person to meet the demands posed by these stimuli. For a patient, who needs to treat high blood pressure, manageability of this treatment for him depends on factors including the price of the medication, their availability, and side effects. Meaningfulness is the extent to which these demands are challenges worthy of investment and engagement—for example, does this same patient consider the treatment of high blood pressure to be worth the trouble.

Finally, all the factors of this framework model are influenced by individual self-reflection and cultural background. Self-reflection, careful thought about one's behavior and beliefs, is necessary in considering and analyzing one's own actions (Gräsel, 2000). For example, if somebody is too shy to ask their medical doctor about reasons for a certain treatment, this phenomenon may be subject to critical self-reflection. Cultural background is important because it articulates the fact that human beings are embedded in cultural contexts, shared systems of meaning and symbols, in terms of which social interaction takes place (Geertz, 1973). From this point of view, scientific medicine may well be considered as a culture, a culture which embraces a scientific orientation, displayed not only in certain practices like clinical assessment or clinical reasoning, but also in dressing in white coats with all its consequences. Conversely, the culture of alternative medicine is decisively distinguishable for patients through reference to nonscientific theories or by different dress codes.

The research presented about HPV vaccination may provide interesting insights into the mechanisms of the framework model. Algorithmic knowledge, for example, can be misunderstood, as evidenced by Zeyer and Sidler's (2015) student teachers' belief that HPV vaccination used attenuated infectious particles. The situational construction resulting from this misunderstanding leads to the individual's negative judgment about the treatment's manageability, because the concern that vaccinated people risk contracting HPV through vaccination is too severe. Self-reflection—the analysis of one's values and circumstances—may also influence the decision. Some of the student teachers felt that they did not need vaccination because their partner was HPV-free and they would always live in a faithful partnership. They concluded, in their situational construction, that HPV vaccination, though reasonably sound, was not of any relevance to them. On the other hand, when the student teachers took into account their experiential knowledge—for example, family experience with cancer—this led them to make a judgment of meaningfulness: that one should do everything to avoid cancer. The role of the cultural background can be seen most strikingly in the study of the student teachers' decision-making about vaccination. Though the school doctor was very engaged and tried to motivate students to take their vaccination, she had no chance in the context of a strong culture of vaccination skepticism, a culture shared even by the pediatricians and some of the gynecologists in the neighborhood.

11.3 Challenges and opportunities for bringing health education into the science classroom

11.3.1 Making room for health

If health-related science knowledge is helpful in daily life and potentially helpful in sparking future nonscientists' interest in science, bringing health content into the science classroom should be easy. Unfortunately, this is not the case. Science educators are overwhelmed by time pressures, testing requirements, and multiple competing priorities based on their sense of accountability to administrators, colleagues, and students (Keselman et al., 2012; Levin, 2008). These pressures broadly restrict educators from making connections across the curriculum, innovating, and taking up much of the spirit of educational reform initiatives. In this particular case, there are also no widespread calls to amplify the role of learning about health in science education reform.

In the recently released U.S. Next Generation Science Standards (NGSS Lead States, 2013), health biology is not emphasized as a particular strand throughout the K-12 standards. Nevertheless, there are abundant opportunities for students to learn about health issues and to engage in practices of acquiring and analyzing health information embedded in the Standards. These Standards are organized along strands of *science and engineering practices, cross-cutting concepts, and disciplinary core ideas*. The disciplinary core ideas are laid out as a "learning progression" (Smith, Wiser, Anderson, & Krajcik, 2006) of research-based sequences that identify what aspects of the core idea should be developed within each age group. For the disciplinary core idea of *structure and function* in life science, for example, in grades K-2, students are expected to understand that "All organisms have external parts that they use to perform daily functions"; by grades 9–12 they are expected to understand that "systems of specialized cells within organisms help perform essential functions of life. Any one system in an organism is made up of numerous parts. Feedback mechanisms maintain an organism's internal conditions within certain limits and mediate behaviors" (NGSS Lead States, 2013). There are opportunities from one end of the age/grade spectrum to the other for students to learn about health. Very young children can begin to understand, for example, what it means for someone to have a broken limb, or an illness that inhibits their function. With secondary school students, a natural way to study the role of specialized systems and feedback mechanisms is to study cases of disease—understanding the structure and function of systems by understanding what happens when they do not work properly. Understanding the biology in these authentic health contexts can lead to more robust understanding of the science (Ratcliffe & Grace, 2003).

11.3.1.1 Health education in sciences outside biology: some examples

When talking about biomedical knowledge in science education, most people think about biology. However, interdisciplinary approaches are a key to health literacy in every school science discipline. Many topics of physics or chemistry can be treated in health contexts instead of only in the traditional contexts of science and technology. Another key is to take into account that health literacy is not only about health, in its core sense, but also about disease. Medicine is a treasure trove for interdisciplinary topics in science education.

Zeyer and Müller (2007) provide an example of such a bridge between medicine and a science field other than biology. In Swiss lower secondary school (8th–10th grade), students typically learn about the electromagnetic spectrum and, in particular, X-rays, and the relationship between X-rays and light waves. This content is normally taught without a health connection, but the topic obviously is health related. To show the connection while helping students grasp the principle of X-ray beams, Zeyer and Müller led them in experiments that involved casting shadows of skeletal bones onto paper screens using a video beamer. This teaching unit included also other aspects, like the history of X-rays and a prism experiment. By the end of the unit, students were able to give a basic anatomical interpretation of X-ray photographs. In fact, the teachers had invited them to bring personal X-rays to the science class, and many had done so and had asked their pediatrician for their own photographs.

This example of a teaching unit demonstrates the potential win–win situation between health and science education. The unit includes interdisciplinary (mostly physics related) science content like the electromagnetic spectrum and the principle of photographic imaging. It also gives students the opportunity to discuss many health-related issues like the interpretation of X-ray photographs, the anatomy of inner organs, and bone fractures. Students are also given a platform to share stories about diseases and accidents, which fosters health literacy as an (intended) "side effect."

When tested, this teaching unit proved to be very attractive to the students. They loved to present their own X-ray photographs and eagerly scrutinized them in order to find the fractures. Both boys and girls were fully engaged in creating their own X-ray "photographs," thereby drawing their own "fractures." In this medical context, they intensively discussed the electromagnetic wave-spectrum and its applications. An informal evaluation of the unit revealed the students' enthusiasm and their demand for more lessons of this kind (cf. Zeyer, 2012).

Many interdisciplinary science topics are conducive to including health aspects. Exploring this idea, Zeyer (2007) has developed more than a dozen health-related science units that were tested in Swiss schools. Here, we include three of them as specific instances of the general principle.

The percentile curves showing human growth and development in pediatric practice provide an opportunity to teach and learn about kinematics—the physics of bodies in motion—in a medical context. Usually, in physics classes, kinematics is illustrated by technical objects like cars, airplanes, or rockets. However, percentile curves about growth and growth velocity are kinematic graphs, too. Because they are position versus time curves and velocity versus time curves, they admit exactly the same kinematic calculus as physical objects. Kinematic calculus allows us to transform one curve into another, and to interpret position and velocity in terms of human biology and medicine. In this way, these variables can become meaningful to students, who are more interested in human contexts than in purely physical and technical topics.

Or take the example of fibroendoscopy, a technique that allows seeing into inner spaces of the body, such as the knee or the abdomen, through very small holes, for diagnostic or surgical reasons. Fibroendoscopy is based on the physical phenomenon of total inner reflection, a topic that usually is taught in every science class. It means that light under certain conditions can be totally reflected by a surface between two different types of glass. Total internal reflection enables light to be transmitted inside thin glass fibers. The light is internally reflected off the sides of the fiber, and, therefore, follows its path. Fibroendoscopy uses this effect to transmit pictures from bodily organs to an external viewer who, in this way, is able to have visual access to anatomical structures that normally can be seen only during an overt operation. This context combines the physics of optics with anatomy and medicine. Moreover, it has an important impact on health literacy, because many students will be future patients confronted with the technique of fibroendoscopy in the context of their disease.

Finally, consider the lever principle, applied to the human body. Usually, in physics education, the lever principle will be used for cranes or scales. Applied to the human body, it can explain why people should, when lifting heavy weights, take them as close to their body as possible. Indeed, to translate this context into levers and forces, for example, for the human spine, is a sophisticated task, and provides insight into the huge forces acting on a spine disc, when heavy weights are lifted improperly. In this way, physics can explain why disc herniation may result from inappropriate manipulation of heavy objects, reinforcing important injury prevention rules, in a very natural way.

These are only three examples in a list that could be much longer. Others include bone structure, which can be understood by analyzing physical forces; the chemistry of soap bubbles, which provides insight into the lung therapy needed by premature neonates (Zeyer & Welzel, 2005); the physics of energy, which helps in calculating weight loss by jogging. Blood sedimentation, a common marker of inflammation, can be understood and calculated in terms of free fall of red blood cells in blood plasma (Zeyer & Welzel, 2006). The human eye is a fascinating object for applying the optical theory of lenses. Electrocardiography is nothing more than electrodynamics applied to the heart physiology.

11.3.2 Structuring student opportunities for critical meaning construction

For fostering health literacy in science education, the use of interdisciplinary contexts is not enough. Moreover, as seen in the HPV-related studies presented in this chapter, health information transmission via lectures and brochures is insufficient. Therefore, in addition to introducing health-related contexts into science education, educational approaches are needed that emphasize critical thinking and support knowledge integration. If one is to take the presented framework model seriously, then students must be given opportunities to build their own individual situational construction of health issues and to then elaborate on their personal judgments. In this context, we point out two approaches that have become key themes in contemporary science education research. One is inquiry-based science education (IBSE) (Osborne & Dillon, 2008), and the other has become known as cultural border crossing (CBC) (Aikenhead & Jegede, 1999).

IBSE is an approach to science education characterized by four aspects (European Seventh Framework Programme, 2007):

1. IBSE is based on authentic, problem-based teaching and learning activities that are not primarily focused on producing the correct results.
2. IBSE uses a certain amount of hands-on science. In this context, hands-on science does not only include science experiments, but also acquiring information in libraries, searching the Internet, and interviewing experts.
3. IBSE contains elements of autonomous, self-organized learning.
4. IBSE promotes reasoning and communication with peers in groups and in the classroom, which is also referred to as "talking science" (Lemke, 1990).

Inquiry-based approaches to science teaching enable students to independently elaborate on scientific topics and acquire scientific knowledge through self-directed learning. IBSE attempts to find the middle ground between rote learning, where knowledge is transferred by the teacher only, and exploratory learning, in which students are completely left on their own when discovering scientific content. IBSE is committed to fostering students' scientific literacy and fits well with the framework for health literacy proposed above. Obviously, IBSE will be likely to embrace health issues as welcome contexts for science education. Furthermore, because it is not primarily focused on (scientifically) correct results but on students' active engagement with scientific questions, it opens up teaching to constructional processes and does not block students' reasoning for fear of mistakes. Moreover, IBSE asks for a certain amount of hands-on science. This fosters students' acquiring experiential knowledge and provides teachers with space for scaffolding their students, that is, for supporting and mentoring them on their way to scientific reasoning. Since IBSE contains elements of autonomous, self-organized learning, it endorses self-reflection, another important factor in the framework model.

IBSE's emphasis on discursive reasoning, that is, reasoning in active exchange with others, and communication, called "talking science" (Lemke, 1990), helps students to engage in meaning construction through argumentation, with a focus on evaluating and refuting evidence and claims. The focus on developing students' proficiency in scientific discourse, and in scientific inquiry in general, aligns with the scientific practices advanced by the Next Generation Science Standards. These standards describe particular aspects of inquiry or "practices" in which students must develop proficiency, including asking authentic questions, constructing explanations, and engaging in argumentation (NGSS Lead States, 2013). Talking science is also an ideal instrument for engaging students in cultural refection and for fostering cross-cultural exchange.

Cross-cultural understanding is the core issue in the second approach to science education that seems to be ideal for promoting health literacy in science contexts: the CBC approach. The basic idea of CBC is to interpret the frequent failure of teaching–learning processes in science education not as a reflection of bad teaching, nor of insufficient learning, but as a result of a cultural gap between science culture and the learners' reference culture (Aikenhead & Jegede, 1999). An influential study of interest in science (Haste, 2004) identified a group of girls as "alienated from science." These students were repelled by the frequent use of technical contexts in science classes, not because they did not understand machines, but because of cultural alienation. Such a "cultural clash" may prevent students from truly assimilating science content into their personal culture, in spite of their learning and often understanding this content on a formal level. In consequence, science knowledge will play no role in those students' lives, although they may be able to perfectly reproduce it on demand on exams.

The CBC concept seems to be essential for understanding learning processes in issues of health and disease. As has been shown above in the example of vaccination decisions, cultural considerations play a major role in people's decision-making. In Switzerland, more than 40% of people feel attracted by alternative medicine and its culture (comparis.ch, 2014). In this alternative medicine culture, scientific medicine is usually criticized as paternalistic, reductionist, and commercially biased. Thus, a purely transmissive approach by medical stakeholders, one focusing solely on medical (algorithmic) knowledge about HPV vaccination, and without taking cultural aspects into account, is likely to produce a "cultural clash" (Aikenhead, 2000). Public suspicions about scientific medicine are thus confirmed, and people may reject vaccination even more strongly than before. It seems likely that this happened with the school doctor's intervention in one of the research projects discussed above.

To make it clear, a CBC approach does not mean complete abandonment of teaching of algorithmic knowledge about vaccination—transmissive or otherwise. Instead, the strategy is complement teaching algorithmic knowledge by helping students to explicitly deal with cultural influences on their situational construction. The IBSE methods can be useful for doing so. Talking science, for example, may enable students to reflect on their skepticism toward scientific medicine. Self-organized learning provides the opportunity to search documents in the literature and in the media and to find out if these documents confirm or reject their personal assumptions about the culture of scientific medicine. The results of these investigations may be presented in talking science events.

In the CBC context, teachers and health professionals take a different role than they usually do in the learning–teaching process. As "culture brokers" (Aikenhead, 2000), they aim not to directly transmit scientific knowledge into students' heads, but to advise, guide, and encourage students as they are crossing the borders into scientific medical culture. The goal of CBC is autonomous acculturation. This principle leaves it up to each student what insights and techniques they want to "take home" from the visited medical culture, and how they prefer to connect these findings with their own everyday culture. For example, in the vaccination case, it may well be that a student, in spite of their substantial learning of biomedical knowledge, decides to adhere to alternative concepts of health and nature. Nevertheless, they may reevaluate their position on vaccination, or at least on some of the vaccinations they get to know during their learning process. In terms of CBC, such a "hybrid" world conception—stick to alternative health concepts, while accepting one or more vaccinations—would not be seen as a failure of the teaching process, but indeed as a success (Roth, 2008).

11.3.3 "Scientific practices" and "practice-based" teacher education

The IBSE and CBC approaches are only two of the many models that could be considered productive for structuring student opportunities to critically make meaning of health issues. Each of these aligns with the NGSS focus on scientific "practices."[3] This term, as described in the NGSS Front Matter (NGSS Lead States, 2013)

> Describe[s] the major practices that scientists employ as they investigate and build models and theories about the world...We use the term "practices" instead of a term such as "skills" to emphasize that engaging in scientific investigation requires not only skill but also knowledge that is specific to each practice...[P]art of our intent in articulating the practices...is to better specify what is meant by inquiry in science and the range of cognitive, social, and physical practices that it requires. (p. 2)

The intention of the change in wording from "inquiry" to "practices" is intended to address some of the confusion about the meaning of inquiry, and to refine and deepen its meaning by an explicit definition of its practices. The NGSS specifies eight science and engineering practices, including the following:

- Asking questions and defining problems
- Developing and using models
- Planning and carrying out investigations
- Analyzing and interpreting data
- Using mathematics and computational thinking
- Constructing explanations and designing solutions
- Engaging in argument from evidence
- Obtaining, evaluating, and communicating information

The IBSE focus on authentic problems, inquiry, and the acquisition and use of knowledge aligns with these practices; so does the CBC approach of helping students to gradually inhabit a scientific culture. For example, as students have opportunities to engage in argumentation, they develop their practices of critique (Ford & Forman, 2006)—the evaluation of knowledge claims that is central to participation in scientific culture. Both the IBSE and the CBC approaches emphasize the importance of students' participation in classroom discourse, an environment which provides the medium for students to engage in explanation and argumentation. Students need to learn not only to communicate explanations, but also to construct knowledge through talk, as scientists do; to argue, to collaboratively investigate, and to develop and use models.

---

[3] NGSS describes engineering practices as well. Discussion of those is beyond the scope of this chapter.

If the science classroom is to become an environment, where students may engage in scientific practices and discourse in considering health issues, and where students may have cross-cultural borders to see themselves as critical constructors and users of health knowledge, there are surely implications for teaching and teacher education. Consider, for example, the expectation that students will engage in scientific discourse, particularly argumentation. How will teachers manage these discussions? How will they know how to guide students in ways that will help them develop understandings of concepts of health and disease and engage in productive discussion and argumentation around issues of personal and environmental health? If teachers are to manage discussions about health-related issues, and to support students' construction and evaluation of explanations and arguments, teacher education must be organized to prepare teachers for these demands.

A growing approach in teacher education also uses the term "practices," in this case, focusing on the "core practices" of teaching that support productive student learning (Ball, Sleep, Boerst, & Bass, 2009; Grossman, Hammerness, & McDonald, 2009). "Leading discussion," is described as a core practice by several authors, for example; this presumes that there is expertise to be developed in learning to lead discussion and structure the discourse, and that the focus on learning to teach should be on developing expertise in this and other core practices (Ball & Forzani, 2009; Grossman & McDonald, 2008).

Central to the idea of the development of core practices is the notion that preservice teachers must *practice* to achieve proficiency. They must have the opportunities to enact leading discussion, for example, if they are to become more proficient at making some of the moment-to-moment decisions that a robust discussion around a health issue would likely require. Practice-based teacher education programs aim to create such opportunities through "representations of practice" and "approximations of practice" (Grossman et al., 2009). "Representations of practice" might include videos of classrooms, samples of student work, or sample lessons plans. A preservice science teacher education program attempting to support teacher candidates' development of practices of leading discussion might first have preservice teachers (or student teachers) have candidates analyze a representation; this might mean watching videos of students engaged in classroom argumentation around a health topic. Then, the program might gradually introduce students to situations in which they can "approximate practice." These approximations could include having preservice teachers try out leading a health-topic discussion in their teaching methods courses, then planning for and enacting a health-related discussion in their teaching practicum.

This extended example of the core practice of leading discussion is presented as a way to describe how efforts to engage students in scientific practices in exploring health issues depend on providing teachers with the practice they need to guide students in these pursuits. Teacher education should provide teachers themselves with opportunities to engage in these same kinds of scientific practices around health issues. Understanding how they themselves engage in scientific practices can provide teachers with a greater understanding of how their students will engage.

Furthermore, an important aspect of engaging both preservice and in-service teachers in their own inquiry around health-science issues, and in helping them develop pedagogies to support students' participation, is bringing the message of current ideas and research in science and health education to teachers and into the schools, where other systemic influences might discourage new and innovative approaches. This underscores the importance of the practice-based teacher education approach not just in preservice teacher education, but in ongoing teacher professional development.

11.3.4 Fostering teachers' knowledge development

We make the assumption that teachers' enactment of core practices will depend largely on their content knowledge and on their "pedagogical content knowledge (PCK)" (Schulman, 1986). Content knowledge, in this case, refers to teachers' understandings of the biological mechanisms underlying health and disease. PCK refers to a category of knowledge that integrates teachers' knowledge of the content with knowledge of pedagogy and of students. It may include understandings of common student misunderstandings of the content, or understandings of how to leverage students' experience to help them make meaning of the content.

If science teachers are to engage students in scientific practices around issues of health and disease, it is important that they have both sufficient content knowledge and pedagogical content knowledge. Zeyer and Sidler's (2015) work alone raises an alarm. If preservice teachers themselves do not understand the biological principles of vaccination, we cannot expect that their students will have access to such understanding. Furthermore, we cannot expect that these teachers will understand what their students do or don't understand correctly, and thus may have little understanding in how to help their students make conceptual progress.

Unfortunately, there is little work that unpacks the content and pedagogical content knowledge that teachers need to enact of core practices in health science, and little work in understanding how to better prepare science teachers to develop the content and pedagogical content knowledge necessary. Thus, this is clearly an important area of further research. In a study with students of a university of teacher education, Zeyer and Knierim (2009) investigated whether teacher students' conceptual biomedical knowledge could foster their critical reasoning about, and rejection of, vaccination myths. The students' attitudes toward vaccination were investigated by means of a questionnaire administered before and after an intervention. Results showed a significant change in the students' judgment regarding seven statements for and against vaccination in favor of the pro-vaccination statements.

Thus, it has been demonstrated that the teaching of conceptual biomedical knowledge can have an impact on the health literacy of teacher students. On the other hand, Van Driel, Verloop, and de Vos (1998) identify teaching experience as the most important factor in the development of science teachers' PCK. Thus, science teacher education and professional development can make efforts to augment this experience: by providing teachers opportunities of their own to engage in scientific practice around health issues, by presenting them with representations that highlight students thinking about health issues, and by engaging them in approximations of practice that allow them to try out, and gain experience in, supporting students in scientific practice around health issues. Recent case studies in science and mathematics teacher education suggest that such approaches can support beginning teachers learning about content, pedagogy, and students' thinking (Levin, 2014; Levin & Richards, 2010).

11.4 Conclusions and implications

11.4.1 Science education for better health: from facts-transmission to the development of understanding and critical thinking

The key message of this chapter is that health information provision with the aim of affecting behavior is a challenging task. Actions that we as educators or public health professionals view as desirable health behaviors can hardly be prompted by handing out pamphlets. Instead, health informing and educating for behavioral change should involve building critical health literacy. This, we believe, is achieved by developing a solid biomedical knowledge foundation and integrating it with experiential and cultural knowledge.

Historically, health and science education have been seen as disparate disciplines, but the science classroom is a well-fitting environment for the development of critical health literacy. The science classroom provides opportunities for students to learn about health issues, to draw on their experience with health and disease to learn about science, and to investigate health-related personal and socioscientific issues and consider action. This requires that teachers go beyond simply providing students with health information and create opportunities for them to problem-solve, investigate and develop understanding of health risks, engage in argumentation about causes of health issues, communicate with others about health issues, and design actionable plans to mitigate health risks. Teachers also need to act as "culture brokers," using the health domain to show how science is relevant to everyday experiences of all students, and helping students to find their own situational constructions of health issues and draw their own life world conclusions. As discussed above, bringing this kind of change to classrooms is fraught with challenges. A great part of this change will fall to teachers, and this has implications for teacher education and professional development. Innovative teacher education and professional development programs will have to deal with the challenges of incorporating health education into the complex world of the classroom, building teachers' own knowledge of health issues, and helping them recognize opportunities to help students construct meaning by drawing on their scientific knowledge and their everyday experience with health and disease.

Two of us have approached this problem by working on designing health curricula with teachers, drawing on their knowledge of the classroom context and helping to build their own content and pedagogical knowledge in creating opportunities for students to construct meaning (Hundal, Levin, & Keselman, 2014). There is not yet evidence about how such targeted professional development translates into teachers' classroom practice. There is considerable work to be done in understanding features of productive professional development with science teachers around health issues.

11.4.2 The collaborative nature of health education and information delivery

Effective health information provision in a science classroom should be a collaborative endeavor among science education researchers, teachers, administrators, school librarians, and developers of science education resources. It is up to science education researchers to identify scientific concepts that are most useful for making health-related decisions and outline teaching strategies for underscoring this applicability. Researchers also assume the task of identifying most effective teaching approaches and techniques. Science teachers are the ones implementing the instruction. They need help and support, in finding space in a tightly packed curriculum and in discovering connections with health educators and health professionals.

Teachers also need lesson plans and materials—textbooks and Web sites—that could help them connect scientific concepts to health. For example, a typical high school biology textbook may organize anatomy and physiology information into systems (e.g., digestive, nervous, etc.), largely focusing on the normal structure and function, and, perhaps, giving brief examples of disorders. These units typically do not provide in-depth information about illnesses and diseases, show impact on multiple systems and their interconnectedness, outline mechanisms and side effects of treatments, and discuss the impact on daily life. In general, the emphasis is more on health and wellness than on disease. At the same time, many authoritative federal organizations, national health associations, and academic institutions develop quality consumer health information resources that contain the information missing in the textbooks. Unfortunately, as these resources are created without classroom curriculum in mind, teachers find it difficult to make connections and integrate these resources into their busy lessons, let alone to find the time to explore and organize these resources. The information professions can be of help here. Librarians' outreach could focus on organizing consumer health information resources around classroom needs and promoting them to teachers. Librarians could also educate resource developers about best ways to structure and organize information to make classroom connections easier, and to collect and organize innovative resources.

Whatever the specific tasks in their job descriptions, professionals involved in delivering health education to the public will benefit from the understanding of the range of personal and social factors that mediate the effect of the information. Professionals interested in informing and educating young people toward better health will benefit from understanding how these factors interact with one another, and what approaches are likely to produce actionable knowledge. Health information is not a panacea; but providing it in a way that results in construction of deep understanding, taking into account cultural beliefs and social concerns, is likely to greatly increase its impact.

References

Ad Hoc Committee on Health Literacy for the Council on Scientific Affairs/American Medical Association Health Literacy. (1999). Report of the council on scientific affairs. *The Journal of the American Medical Association, 281*(6), 552–557.

Aikenhead, G. S. (2000). Renegotiating the culture of school science. The contribution of research. In R. Millar, J. Leach, & J. Osborne (Eds.), *Improving science education* (pp. 245–264). Philadelphia: Open University Press.

Aikenhead, G. S., & Jegede, O. J. (1999). Cross-cultural science education: a cognitive explanation of a cultural phenomenon. *Journal of Research in Science Teaching, 36*(3), 269–287.

Antonovsky, A. (1997). *Salutogenese. Zur Entmystifizierung der Gesundheit.* [Salutogenesis. To demystify health]. Tübingen: dgvt Verlag.

Baker, D. W., Gazmararian, J. A., Williams, M. V., Scott, T., Parker, R. M., Green, D., et al. (2002). Functional health literacy and the risk of hospital admission among medicare managed care enrollees. *American Journal of Public Health, 92*(8), 1278–1283.

Ball, D. L., & Forzani, F. M. (2009). The work of teaching and the challenge for teacher education. *Journal of Teacher Education, 60*(5), 497–511.

Ball, D. L., Sleep, L., Boerst, T. A., & Bass, H. (2009). Combining the development of practice and the practice of development in teacher education. *The Elementary School Journal, 109*(5), 458–474.

Berkman, N. D., Sheridan, S. L., Donahue, K. E., Halpern, D. J., & Crotty, K. (2011). Low health literacy and health outcomes: an updated systematic review. *Annals of Internal Medicine, 155*(2), 97–107.

Bromme, R., & Goldmann, S. R. (2014). The public's bounded understanding of science. *Educational Psychologist, 49*(2), 59–69.

Clinical Trials.gov. (2015). *History, policy, and laws.* Retrieved from http://clinicaltrials.gov/ct2/about-site/history#CongressPassesLawFDAMA.

Collins, J. L., Small, M. L., Kann, L., Pateman, B. C., Gold, R. S., & Kolbe, L. J. (1995). School health education. *Journal of School Health, 65*(8), 302–311.

comparis.ch. (2014). *Jeder Zweite interessiert sich für Komplementärmedizin.* [Every second person is interested in alternative medicine]. Retrieved from http://www.presseportal.ch/de/meldung/100580182.

Davis, T. C., Federman, A. D., Bass, P. F., 3rd, Jackson, R. H., Middlebrooks, M., Parker, R. M., et al. (2009). Improving patient understanding of prescription drug label instructions. *Journal of General Internal Medicine, 24*(1), 57–62.

Di Rocco, S., & Zeyer, A. (2013). Wie entscheiden sich Schülerinnen für oder gegen eine HPV-Impfung. [How do female students decide in favor or against HPV vaccination]. *Prävention und Gesundheitsförderung, 8*(1), 29–35.

Duit, R., & Treagust, D. F. (2003). Conceptual change: a powerful framework for improving science teaching and learning. *International Journal of Science Education, 25*(6), 671–688.

EACEA/Eurydice. (2011). *Science education in Europe: National policies, practices, and research.* Brussels: EACEA/Eurydice.

Education Commission of the States. (1981). *Recommendations for school health education: A handbook for state policy makers.* Denver, CO: Education Improvement Center, Education Programs Division, Education Commission of the States.

European Seventh Framework Programme. (2007). *Mind the gap: Learning, teaching, research and policy in inquiry-based science education.* Retrieved from http://cordis.europa.eu/publication/rcn/14996_en.html.

Feinstein, N. (2010). Salvaging science literacy. *Science Education, 95*(1), 168–185.

Ford, M. J., & Forman, E. A. (2006). Redefining disciplinary learning in classroom contexts. *Review of Research in Education, 30*, 1–32.

Geertz, C. (1973). *The interpretation of cultures: Selected essays.* New York: Basic Books.

Gräsel, C. (2000). *Ökologische Kompetenz: Analyse und Förderung.* [Ecological competency: Analysis and promotion]. München: Ludwig-Maximilians-Universität München.

Grossman, P., Hammerness, K., & McDonald, M. (2009). Redefining teaching, re–imagining teacher education. *Teachers and Teaching: Theory and practice, 15*(2), 273–289.

Grossman, P., & McDonald, M. (2008). Back to the future: directions for research in teaching and teacher education. *American Educational Research Journal, 45*(1), 184–205.

Haste, H. (2004). *Science in my future: A study of values and beliefs in relation to science, and technology amongst 11–21 year olds. Nestlé Social Research Programme, Report 1.* London: Nestlé Social Research Programme.

Hirte, M. (2012). *Impfen pro & contra*. [Vaccination: Pros and cons]. München: Knaur.

Hundal, S., Levin, D. M., & Keselman, A. (2014). Lessons of researcher-teacher co-design of an environmental health afterschool club curriculum. *International Journal of Science Education, 36*(9), 1–21.

Joffe, S., Cook, E. F., Cleary, P. D., Clark, J. W., & Weeks, J. C. (2001). Quality of informed consent in cancer clinical trials: a cross-sectional survey. *Lancet, 358*(9295), 1772–1777.

Johnson, M., & Piggiluci, M. (2004). Is knowledge of science associated with higher skepticism of pseudoscientific claims? *American Biology Teacher, 66*(8), 536–548.

Jourdan, D. (2011). *Health education in schools. The challenge of teacher training*. Saint-Denis: Inpes, Coll. Santé en action.

Judson, T. J., Detsky, A. S., & Press, M. J. (2013). Encouraging patients to ask questions: how to overcome "white-coat silence". *The Journal of the American Medical Association, 309*(22), 2325–2326.

Kalichman, S. C., & Rompa, D. (2000). Functional health literacy is associated with health status and health-related knowledge in people living with HIV-AIDS. *Journal of Acquired Immune Deficiency Syndromes, 25*(4), 237–244.

Kanton Luzern: Gesundheits- und Sozialdepartement. (2012). *Schütze Dich vor Gebärmutterhalskrebs*. [Protect yourself against cervical cancer]. Retrieved November 15, 2014, from www.hpv-impfung.lu.ch.

Kempton, W. (1986). Two theories of home heat control. *Cognitive Science, 10*(1), 75–90.

Keselman, A., Hundal, S., Chentsova-Dutton, Y., Bibi, R., & Edelman, J. A. (2015). Are biology classes and biological reasoning associated with students' acceptance or rejection of common misconceptions about health? *American Biology Teacher, 77*(3), 170–175.

Keselman, A., Hundal, S., & Smith, C. A. (2012). General and environmental health as the context for science education. In A. Zeyer, & R. Kyburz-Graber (Eds.), *Science | Environment | Health: Towards a renewed pedagogy for science education*. Dordrecht: Springer.

Keselman, A., Slaughter, L., Smith, C. A., Kim, H., Divita, G., Browne, A., et al. (2007). Towards consumer-friendly PHRs: patients' experience with reviewing their health records. In *Proceedings of the 2007 annual symposium of the American medical informatics association* (pp. 399–403).

Keselman, A., & Smith, C. A. (2012). A classification of errors in lay comprehension of medical documents. *Journal of Biomedical Informatics, 45*(6), 1151–1163.

Layton, D., Jenkins, E., Macgill, S., & Davey, A. (1993). *Inarticulate science? perspectives on the public understanding of science and some implications for science education*. Driffield, UK: Studies in Education.

Lemke, J. (1990). *Talking science: Language, learning, and values*. Norwood, NJ: Ablex.

Levin, D. M. (2008). *What secondary science teachers pay attention to in the classroom: Situating teaching in institutional and social systems* (unpublished doctoral dissertation). College Park: University of Maryland.

Levin, D. M. (November 2014). Promoting core practices in a middle-level mathematics and science teacher preparation program. In *Presentation at the annual conference of the association for middle level education, Nashville, TN*.

Levin, D. M., & Richards, J. (January 2010). Practices of attending to student thinking can promote collaborative conversations about science. In *Paper presented at the annual meeting of the association of science teacher education, Sacramento, CA*.

Marglin, S. A. (2008). *The dismal science: How thinking like an economist undermines community*. Cambridge, Massachusetts: Harvard University Press.

Montgomery County (Maryland) Schools. (n.d.). Draft secondary comprehensive Health education curriculum framework. Retrieved from http://www.montgomeryschoolsmd.org/uploadedFiles/curriculum/health/HealthFramework.pdf.

National Cancer Institute. (2011). *Human papillomavirus (HPV) vaccines*. Retrieved January 13, 2015, from http://www.cancer.gov/cancertopics/factsheet/prevention/HPV-vaccine.

National Research Council. (2007). *Taking science to school: Learning and teaching science in grades K-8*. Retrieved from http://www.nap.edu/catalog/11625/taking-science-to-school-learning-and-teaching-science-in-grades.

National Research Council. (2011). *A framework for K-12 science education: Practices, crosscutting concepts, and core ideas*. Retrieved from http://www.nap.edu/catalog/13165/a-framework-for-k-12-science-education-practices-crosscutting-concepts.

Neulight, A., Kafai, N., Yasmin, B. A., Kao, L., Foley, B., & Galas, C. (2007). Children's participation in a virtual epidemic in the science classroom: making connections to natural infectious diseases. *Journal of Science Education and Technology, 16*(1), 47–58.

NGSS Lead States. (2013). *Next generation science standards: For states, by states*. Retrieved from http://www.nap.edu/catalog/18290/next-generation-science-standards-for-states-by-states.

Nutbeam, D. (2000). Health literacy as a public health goal: a challenge for contemporary health education and communication strategies into the 21st century. *Health Promotion International, 15*(3), 259–267.

Osborne, J., & Dillon, J. (2008). Science Education in Europe: Critical Reflections. *A Report to the Nuffield Foundation*. London: King's College.

Paasche-Orlow, M. K., Schillinger, D., Greene, S. M., & Wagner, E. H. (2006). How health care systems can begin to address the challenge of limited literacy. *Journal of General Internal Medicine, 21*(8), 884–887.

Paasche-Orlow, M., & Wolf, M. (2007). The causal pathways linking health literacy to health outcomes. *American Journal of Health Behaviour, 31*(Suppl. 1), 19.

Paavonen, J. (2008). Baseline demographic characteristics of subjects enrolled in international quadrivalent HPV (types 6/11/16/18) vaccine clinical trials. *Current Medical Research and Opinion, 24*(6), 1623–1634.

Parker, R. M., Baker, D. W., Williams, M. V., & Nurss, J. R. (1995). The test of functional health literacy in adults: a new instrument for measuring patients' literacy skills. *Journal of General Internal Medicine, 10*(10), 537–541.

Ratcliffe, M., & Grace, M. (2003). *Science education for citizenship: Teaching socio-scientific issues*. Maidenhead: Open University Press.

Ratzan, S. C., & Parker, R. M. (2000). Introduction. [NLM Pub. No. CBM 2000-1]. In C. R. Selden, M. Zorn, S. C. Ratzan, & R. M. Parker (Eds.), *National library of medicine current bibliographies in medicine: Health literacy*. Bethesda, MD: National Institutes of Health, U.S. Department of Health and Human Services.

Roth, W. M. (2008). Bricolage, metissage, hybridity, heterogeneity, diaspora: concepts for thinking science education in the 21st century. *Cultural Studies of Science Education, 3*(4), 891–916.

Schillinger, D., Grumbach, K., Piette, J., Wang, F., Osmond, D., Daher, C., et al. (2002). Association of health literacy with diabetes outcomes. *The Journal of American Medical Association, 288*(4), 475–482.

Schulman, L. S. (1986). Those who understand: Knowledge growth in teaching. *Educational Researcher, 15*, 4–14.

Schulz, P., & Nakamoto, K. (2012). The concept of health literacy. In A. Zeyer, & R. Kyburz-Graber (Eds.), *Science | Environment | Health: Towards a renewed pedagogy for science education*. Dordrecht: Springer.

Sivaramakrishnan, M., & Patel, V. L. (1993). Role of traditional knowledge in the explanation of childhood nutritional deficiency by Indian mothers. *Journal of Nutrition Education, 25*(3), 121–129.

Smith, C., Wiser, M., Anderson, C., & Krajcik, J. (2006). Implications of research on children's learning for standards and assessment: a proposed learning progression for matter and atomic-molecular theory. *Measurement, 14*(1&2), 1–98.

Smith, S. K., Dixon, A., Trevena, L., Nutbeam, D., & McCaffery, K. J. (2009). Exploring patient involvement in healthcare decision making across different education and functional health literacy groups. *Social Science and Medicine, 69*(12), 1805–1812.

Society for Participatory Medicine. (2014). Retrieved from http://participatorymedicine.org/.

Van Driel, J. H., Verloop, N., & de Vos, W. (1998). Developing science teachers' pedagogical content knowledge. *Journal of research in Science Teaching, 35*(6), 673–695.

World Health Organization. (1998). *Health promotion glossary*. Retrieved from http://www.who.int/healthpromotion/about/HPR%20Glossary%201998.pdf?ua=1.

Zeyer, A. (2007). Naturwissenschaften unterrichten: disziplinär, interdisziplinär, integriert. [Teaching science: disciplinary, interdisciplinary, integral]. In L. Jäkel, S. Rohrmann, M. Schallies, & M. Welzel (Eds.), *Der Wert der naturwissenschaftlichen Bildung. 8. Heidelberger Dienstagsseminar* (pp. 39–53). Heidelberg: Mattes.

Zeyer, A. (2012). A win-win situation for health and science education: seeing through the lens of a new framework model of health literacy. In A. Zeyer, & R. Kyburz-Graber (Eds.), *Science | Environment | Health. Towards a renewed pedagogy for science education*. Dordrecht: Springer.

Zeyer, A., & Dillon, J. (2014). Science | Environment | Health: towards a reconceptualization of three critical and inter-linked areas of education. *International Journal of Science Education, 36*(9), 1409–1411.

Zeyer, A., & Knierim, B. (2009). Die Einstellung von zukünftigen Lehrpersonen zum Impfen vor und nach einer Informationsveranstaltung [The attitude of teacher students towards vaccination after an information event]. *Prävention und Gesundheitsförderung, 4*, 235–239. http://dx.doi.org/10.1007/s11553-009-0170-4.

Zeyer, A., & Kyburz-Graber, R. (Eds.). (2012). *Science | Environment | Health. Towards a renewed pedagogy for science education*. Dordrecht: Springer.

Zeyer, A., & Müller, T. (2007). Wie Röntgenbilder entstehen [The making of X-ray photographs]. In *EducETH - Unterrichtsmaterialien, Technik in der Allgemeinbildung* Retrieved from http://www.educ.ethz.ch/lehrpersonen/ta/roe.

Zeyer, A., & Sidler, T. (2015). Wie wirken sich Informationen zur HPV-Impfung auf den Impfentscheid junger Frauen aus? [What impact does information about the HPV vaccination have on the decision of young women to be vaccinated?]. *Prävention und Gesundheitsförderung, 2*.

Zeyer, A., & Welzel, M. (2005). Was Seife mit dem ersten Schrei des Neugeborenen zu tun hat [Soap and the first cry of a newborn baby]. *Praxis der Naturwissenschaften. Physik in der Schule, 7*(54), 40–44.

Zeyer, A., & Welzel, M. (2006). Was Viskosität und Rheuma miteinander zu tun haben [Viscosity and rheumatic disease]. Die Blutsenkung. *Praxis der Naturwissenschaft. Physik in der Schule, 7*(55), 39–44.

"You will be glad you hung onto this quit": sharing information and giving support when stopping smoking online

Marie-Thérèse Rudolf von Rohr
Department of English Linguistics, University of Basel, Switzerland; Swiss National Science Foundation Project Language and Health Online

12.1 Introduction

The Internet is a popular means to find health information in Western societies, with available sources ranging from commercial and governmental sites to medical databases (Camerini, Diviani, & Tardini, 2010; Prestin & Chou, 2014; Sillence, Briggs, Harris, & Fishwick, 2006). The Oxford English Dictionary defines information as "[k]nowledge communicated concerning some particular fact, subject,..." (see definition of "Information" OED, sense 2), but informing can occur in different shapes and forms online. For instance, Internet users may obtain information through the exchange of experiential advice in online support groups; previous research in linguistics and other fields tells us that sharing experience and lay expertise is highly valued (for a discussion centered in linguistics, see Sillence, 2010). Understanding how information is shared in online peer-to-peer formats is important in order to address concerns regarding user-generated content as well as to set up future online health interventions (Prestin & Chou, 2014). According to Harvey and Koteyko (2013), "[an] understanding of the dynamics behind such behaviors [posting in online health support groups]" is a crucial step in order to foster "successful supportive relationships online" in the future (p. 186).

Participants of the online smoking cessation forum studied here help each other when quitting smoking. They call this journey *a quit*, using the verb *quit* as a noun—*my quit*—turning the elusive activity of quitting into an object which can be owned. Participants encourage each other to keep going through their "quits" by normalizing withdrawal symptoms and giving each other advice on coping and medication. This exchange of information and advice is intricate on an interpersonal level. On the one hand, seeking advice online involves an asymmetry in power and can be threatening to the advisee's face (Locher, 2013; Sillence, 2010). On the other, giving advice involves attempting to impact someone else's decisions. Further, in the context of peer-to-peer forums, which consist only of typed (i.e., written) interaction, there are "rhetorical challenges" that online participants face to persuade others to take up their advice or to be considered help-worthy (Harvey & Koteyko, 2013, p. 165). Issues of credibility, which are negotiated through linguistic means, are at the forefront. Participants who seek help have to convince other forum members of the authenticity of their claim, while helpers need to establish their expertise to give advice or emotional support. If participants try to impact someone else's attitudes or decisions through advice or information, or if they endeavor to make contributions believable, they are essentially engaged in persuasion. This chapter focuses on the linguistic form of information/advice exchanges in an effort to understand persuasion in online interaction.

Kenneth Burke (1950) provided a useful definition of persuasion as a process. He argued that rhetoric, "the art of persuasion," should be defined as "[...] the use of words by human agents to form attitudes or to induce actions in other human agents" (Burke, 1950, p. 41). There are two points that have to be highlighted: Persuasion is an interpersonal process, which happens through discourse, understanding discourse as relating to "any instance of language-in-use" (Gee, 2011, p. 205). Persuasion can thus be studied in language structures by paying attention to relational aspects. The Aristotlean concept of persuasion involves three levels: the use of arguments (*logos*), the credibility or personal appeal of speakers (*ethos*) and, the personal involvement of the audience (*pathos*) (Cockroft & Cockroft, 2005). Credibility and personal involvement indicate why interpersonal pragmatics can contribute to the analysis of persuasion. Pragmatics "in the European tradition" deals with "language in use from a social and cultural point of view" (Locher & Graham, 2010, p. 1). Locher and Graham (2010) explain that interpersonal pragmatics analyzes the "relational aspect of interactions" (p. 2), positing that there are always informational as well as interpersonal aspects to any kind of communication. Relational work is a helpful concept to describe how interpersonal effects intersect with information. It has been defined as "the work people invest in negotiating their relationship in interaction" (Locher & Watts, 2008, p. 78). To draw the link between persuasion and relational work, we can argue that the latter facilitates linking information structures to patterns of creating personal involvement or trustworthiness.

To find out what language patterns are used when sharing information and giving support, and how these activities are linked to persuasion in this corpus of online data, this study investigates several research questions:

1. How is information shared by help-seekers and advice-givers?
2. How do help-seekers signal the authenticity of their posts?
3. How do advice-givers legitimize their contributions?

In this empirical analysis, a series of steps are undertaken to look at information exchange and support-giving with a focus on persuasion. The content structure of posts[1] is investigated by coding the discursive moves, defined as what a post contributes "to the ongoing interchange" (Miller & Gergen, 1998, p. 192). Looking at the nature, frequency, and sequencing of discursive moves makes it possible to gain insight into how participants linguistically structure information (Locher & Limberg, 2012). Participants are considered both as *initiators* (participants who have started a thread) and *respondents* (participants who react to the initial message) (Richardson, 2003). This research examines how initiators signal the authenticity of their request for help and how respondents use discursive strategies to make their contributions trustworthy. These findings are linked to linguistic patterns (i.e., recurrent language structures). For instance,

---

[1] The terms *post* and *message* are used interchangeably in this chapter.

respondents show their understanding of an initiator's problem by paralleling their experiences through comparative structures—for example, *like you*—which preface predictions of what initiators are to expect, or suggestions of what initiators should do.

This chapter is organized into several sections. First, previous linguistic work about the interpersonal aspects of advice-giving and showing support online is reviewed. Next, in the methods section, data and ethical reflections for the original study are described, after which the analytical procedure is outlined in detail. In the results section, the structure of sharing information and support is discussed by looking at entire sample messages; at the quantitative analysis of discursive moves; and at individual discursive moves (described as *assessment, advice, own experience,* and *background information*). In the conclusions, results are summarized and reflections on possible implications for health-care communication are presented.

12.2 Interpersonal aspects of advice-giving and showing support online

In the section that follows, sociolinguistic studies are introduced that have dealt with the rhetorical challenges in online health interaction, in terms of giving advice, establishing credibility or expertise, and creating authenticity. However, before treating each of these aspects of persuasion in more detail, the notion of relational work is revisited since an interpersonal pragmatic perspective is adopted.

To understand the rhetorical challenges in online health interaction, Harvey and Koteyko (2013) highlighted the utility of politeness theory when looking at online peer-to-peer communication. Relational work is an especially useful lens for the present study since it has a broader focus than early politeness theory. Relational work "refers to all aspects of the work invested by individuals in the construction, maintenance, reproduction and transformation of interpersonal relationships among those engaged in social practice" (Locher & Watts, p. 96). For instance, praising someone is a type of relational work, since one participant signals approval to another.

Participants use relational work to negotiate their "public self-image" or "face" in interaction, which is highly relevant when seeking and giving advice online. There are three different functions of relational work: it can be (1) face-enhancing, (2) face-maintaining, and (3) face-threatening.

Face-enhancing relational work—such as bonding, empathizing, or praising—furthers involvement between participants, which can be linked to *pathos* (the emotional involvement of the audience), and has been shown to be used to create support between participants (Locher, 2006).

Face-maintaining relational work tries to soften potential attacks upon one's or someone else's self-image (a process called mitigation). Examples of mitigation devices include the word *maybe* or the use of indirect structures, for example, impersonal sentences. Mitigation can help by making advice more acceptable to addressees (Ng & Bradac, 1993).

Finally, face-threatening relational work can endanger the self-image of addressees. Locher (2006) argued that "boosters," intensifiers of content, may emphasize expertise which can be face-threatening to advisees (p. 141). Locher examined advice in the online advice column "Lucy Answers," which was part of a Web site run by a large university and geared toward college students. In her analysis of Lucy's response letters, written by a professional health-care team, Locher contended that boosters, for example, *it is essential*, highlighted an asymmetry in knowledge between advisees and Lucy, the advisor.

There is no one-to-one link between linguistic form and the effect of relational work, which means that both interpretation and use of relational work depend on the norms of interaction (Page, 2012). Below, a selection of linguistic studies is discussed showing the major role played by the interpersonal, or relational, aspect when participants seek and give advice in online health support groups.

When people access online health support groups, they seem to be looking for support and advice (Harvey & Koteyko, 2013). Locher (2013) even posited that online support groups "thrive" because participants are interested in others' personal experience, which may also serve as advice (p. 343). Giving advice entails a recommendation of "a future action" (p. 343), which tends to be sequenced with assessment. Assessment can be defined as one participant addressing and evaluating another person's specific situation (Locher, 2013; Morrow, 2012). The people who are assessing need to be especially careful to take into consideration their addressee's face needs—that is, the "public self-image" the latter wants to "display and have confirmed in interaction" (Locher, 2006, p. 113)—as they are commenting on someone else's experiential ground. Hence, advice-givers tread on delicate ground, since both assessment and advice can be a tricky business on an interpersonal level. Example (1)[2] from my data set nicely shows how the advisor is careful to avoid being too assertive about how the advisee's husband is supposed to behave, as putting forward that he is not supportive enough could be seen as face-damaging.

> *(1) Is there any way your other half can agree to ensure he respects your quit while you give it a good go if poss? he prob already is, but if my other half were quitting and i still smoked i'd do my best not to put temptation in her way.*

When advice-givers assess another person's situation and give advice, as in (1), it ultimately implies that the advice-givers know what is best, potentially invading the other person's desire for self-determination. However, in (1) the advisor is careful to downtone such an impression, showing awareness of the face-threatening potential of his[3] advice.

Several studies have analyzed the impact of the communicative context on the linguistic form (i.e., the linguistic realization) of advice online, paying special attention to relational aspects. Locher (2013) investigated "Lucy Answers," an online health advice column, and found that there was a clear preference for nondirective advice-giving. Advice presented in declarative sentences, interrogative sentences (rather than imperatives), and with lexical hedges (e.g., *maybe, sort of*) underlined the "optional,"

---

[2] All examples appear verbatim, with original spelling and punctuation.
[3] The gender of participants has been deduced based on the nicknames they use. There is no guarantee that the nickname reflects real-life gender.

mitigated, character of advice (p. 346). Examples (2) and (3) illustrate advice realized as declarative (2) and interrogative (3) sentences:

> So, Lucy suggests a good cleaning at the dentist and a few extra vitamin C.
> Locher (2006, p. 64, example 5.8)

> Can you cut something out until the class is over?
> Locher (2006, p. 65, example 5.14)

Harrison and Barlow (2009) studied an online self-management program for people with arthritis to understand the effect of the interpersonal dimension of advice-giving on linguistic form. Participants had to give feedback on each other's action plans. In this context, there was a clear preference for advice in the form of declaratives recounting personal experience, as in example (4):

> I found it easier to keep up with drinking by setting out the glasses each morning on the counter. Each time I drank the required amount, I moved the glass to the dishwasher. At a glance I could see exactly how much I needed to drink.
> Harrison and Barlow (2009, p. 103, example 6)

Harrison and Barlow (2009) argued that short narratives helped advice-givers to remain unimposing, signal their experiential knowledge, and foster identification.

Morrow (2006) analyzed a discussion forum about depression and also dealt with interpersonal concerns and discourse patterns of advising. In his data, the advisory move tended to be couched within other content to mitigate possible face threats (see also Locher, 2006). In contrast to Harrison and Barlow's findings (2009), advice-givers used their own or a friend's experience to preface concrete pieces of advice. Thus, their personal experience did not contain the piece of advice itself. The interpersonal effects were similar: respondents simultaneously qualified themselves as appropriate advisors, and bonded with help-seekers. Moreover, advice-givers created a supportive environment by reassuring and sympathizing with help-seekers (Morrow, 2006). Advice could be linguistically realized in very different manners, ranging from short imperatives to mitigated, more indirect advice, including lexical hedges, such as *perhaps*, and interrogatives: "Do you know if there are any support groups for parents who's children have ADHD in your area?" (Morrow, 2006, p. 536, example from 4d). According to Morrow (2006), the supportive tone of the communicative context made short imperatives acceptable, signaling "closeness" and "intimacy" rather than being impolite (p. 544). Help-seekers did not comment on whether they followed through on advice, but they thanked others for advice given. Morrow (2006) has remarked that openly admitting to have accepted advice would leave help-seekers in a lower position.

In her study of a motherhood community blog, Kouper (2010) found that direct advice (i.e., imperatives or sentences containing *should*) was as frequently used as personal experience to give advice/recommendations. Apparently, direct advice was used in "trivial" or uncontroversial situations which were minimally face-threatening

(Kouper, 2010). Participants told their personal experience when responding to more complex problems, especially if it concerned the care of their babies. Kouper (2010) concluded that advice in the form of personal experience allowed participants to navigate between being directive and showing empathy toward their participants. This led her to hypothesize that soliciting advice online may be more about "community bonding than to obtaining information and instructions for action" (p. 16). Silence (2010) came to a similar conclusion in her study of a prostate cancer forum, in which she found that participants were interested in having their decisions confirmed and supported. Advice that did not fit their point of view was rejected or completely disregarded. Both Kouper (2010) and Silence (2010) pointed out the need for further research for how and why people engage with peers in advice-giving in online health support groups.

Forum members not only have to avoid threatening the "face" of advisees (i.e., endanger their public self-image), but they also have to come across as trustworthy and credible when giving advice. Advice-givers need to make clear why their answers should be believed, for instance, by sharing their own experience, which establishes their credibility because they have lived through and overcome a similar problem (Harvey & Koteyko, 2013). Richardson (2003) has termed strategies that aim at fostering credibility and trust "warranting strategies" (p. 172). Warranting is crucial for the uptake of a message. Silence (2010) described common warranting strategies seen in a prostate cancer forum: participants presented their key health statistics, referred to other informational sources, or told their story. Armstrong, Koteyko, and Powell (2011) observed that participants of an online diabetes community warranted their advice by displaying their knowledge of current research, by referring to their professional status or, like in Silence (2010), by referring to informational sources or sharing personal experience. As the studies cited have shown, the sharing of personal experience in the form of short narratives seems to be multifunctional. This sharing can help establish a credible identity, serve to display support, and be a mitigated (indirect) form of advice, respecting the face needs of the participant, a finding which was corroborated by Armstrong et al. (2011). Example (5) from Armstrong et al. (2011), below, shows a help-seeker explicitly interested in experiential advice, which she receives from an advice-giver who "ha[s] gone through it" and additionally possesses professional expertise (p. 11).

> [help-seeker]
> Firstly, I'm in no rush to have children (but) my ultimate worry is diabetes affecting my ability to have 'healthy' babies, ... I've just got sooo many questions and I just need anyone who has either gone through it, or knows any words of wisdom to put my little mind at rest...
> [advice-giver]
> Hi, just some input on this cos I have my own child and I work in a special care baby unit so see a lot of babies from IDDM (Insulin-Dependent Diabetes Mellitus) mothers. I found that the first 2 months of pregnancy were the worst and I couldn't believe how unstable and fragile I was, after that things just kinda fell into place and all went well until 35 wks when I delivered [child], normally...
>   Armstrong et al. (2011, p. 10–11, example Diabetes and Pregnancy)

Veen, te Molder, Gremmen, and van Woerkum's (2010) case study of a forum for celiac disease patients showed that participants also shared their personal experience to give advice and display their understanding of the help-seeker's situation. However, stories could also have a corrective or normalizing function, which contradicted, or at least relativized, the help-seeker's interpretation of her/his situation. In her analysis of a body-building forum, Page (2012) has similarly argued that while personal stories often offer ground for identification, they can also be face-threatening in other cases, especially if participants refer to their own, diverging experience that had a distancing effect.

In online peer-to-peer interaction, advice-givers show through their language use why they can legitimately give advice or why their recommendations should be adopted. In turn, novice participants have to account for the authenticity and legitimacy of their problem or help requests (Harvey & Koteyko, 2013; McKinlay & McVittie, 2011). By asking for help, participants put themselves in vulnerable positions, which may require them to use strategies to preserve face. In their review of previous studies, Harvey and Koteyko (2013) have compiled a catalog of strategies that help-seekers employ to make their claims authentic and to get support. One strategy is describing their symptoms or illness history, which allows them to identify with the online group (Harvey & Koteyko, 2013). Morrow (2006) also observed that "the description of symptoms" were widely used in problem messages in his data (Miller & Gergen, 1998, p. 539).

A second help-seeking strategy is to make direct or indirect requests for information (Harvey & Koteyko, 2013). Previous linguistic research has highlighted that such requests may require face-saving relational work, since help-seekers present themselves as needing help (Kouper, 2010; Morrow, 2006). Therefore, help-seekers may prefer to ask for descriptive experience, instead of advice as such (Veen et al., 2010). In cases where help-seekers ask for direct advice, they may opt to use "or" questions, such as in example (6) from Morrow (2006).

> *Can I really get it back to normal by forcing myself to "do" stuff, or do I have to wait till my mind heals*
>
> Morrow (2006, p. 540, example from 6a)

Such a question format minimizes the face-threatening character of seeking advice, because by using the "or" structure help-seekers present possible solutions, thus representing themselves as "a person who has some competence in dealing with the problem" (Morrow, 2006, p. 540).

A third discourse strategy used by help-seekers is the invoking of a shared background. This may be by explicitly relating themselves to what members have in common, through "statements of self-disclosure" or by telling a story (Harvey & Koteyko, 2013, p. 169). Help-seekers who share a short narrative can involve other posters and set the ground to request advice (Page, 2012).

The fourth strategy used by help-seekers to signal authenticity is the description of personal successes or by relating positive improvements (Harvey & Koteyko, 2013).

Finally, Harvey and Koteyko (2013) mention that help-seekers attempt to appeal to other participants through strategies of "being self-deprecating" and "referring to extreme behavior" (p. 169). To illustrate the latter strategy they quote an example from Eichhorn (2008): "I just swallowed many pills with beer" (p. 169).

In summary, previous linguistic research has reflected, albeit implicitly, on several aspects of persuasive mechanisms in online health peer-to-peer interaction. Within this particular context, participants seem to prefer to use advice—sometimes imparted in the form of sharing personal experience—in order to persuade others about a specific course of action, which is linked to the Aristotelian *logos*, or the arguments employed. In the study of an online smoking cessation forum described below, the focus is on how help-seekers and advice-givers linguistically structure and sequence their contributions. The aim is to find out how advice is requested and imparted in view of reinforcing the common goal of stopping smoking. Additionally, while advice-givers need to appear as credible, possessing some expertise, in order for their advice to be accepted, help-seekers have to come across as authentic in order for their requests for help to be taken up. Thus, both advice-givers and help-seekers have to deal with issues of ethos, that is, the personality of speakers (Ilie, 2006). Simultaneously, help-seekers and advice-givers try to engage each other by showing empathy, solidarity or invoking a common background (i.e., give support), which is linked to the emotional involvement of the audience (i.e., *pathos*). What kind of strategies do participants adopt in order to engage each other and deal with issues of credibility? This study builds upon the findings of previous linguistic research, but differs in its use of specific analytical lens of persuasion and interpersonal pragmatics. This analysis gives new insight into how linguistic actions, such as requesting help or giving advice, are combined with interpersonal strategies, such as showing empathy and bonding, in online interaction around health information.

12.3 Methodology

12.3.1 Data

12.3.1.1 The choice of smoking cessation

Smoking cessation is an ideal topic space to study persuasive strategies of online information sharing and support giving. First, the necessity of quitting is given, as the potential damages of smoking are well established medically. Cancer Research UK (2014) has called tobacco use "the UK's single greatest cause of preventable illness and avoidable death" ("Impacts of tobacco," Background section, para 1). Indeed, participants in a forum devoted to this topic do not have to convince each other of the dangers of smoking, but they help each other out with the challenges of stopping, trying to encourage each other when they have relapsed, and giving advice on coping with withdrawal symptoms.

Second, the UK Department of Health considers smoking cessation, helping individual smokers to stop, one of the six areas of action necessary to reduce smoking rates (Cartwright, 2008). The online support group serving as the subject of this study provides its members with a platform for self-help to stop smoking. Looking at peer-to-peer interaction online gives us crucial insight into how people make sense of quitting smoking, and how they draw on other people's experience to persevere.

12.3.1.2 The forum "SmokingisBad"

The corpus of data for this study is drawn from an online support group dedicated to smoking cessation. To safeguard the confidentiality of participants, I refer to the online support group itself using a pseudonym, "SmokingisBad."[4] "SmokingisBad" is a public forum with no commercial or governmental ties, hosted in the UK. The forum had 2379 members at the time of data collection in May 2012. While anyone can access the forum and read messages, only registered members are allowed to contribute to threads or view user profiles. The overall frame or purpose of "SmokingisBad" is to help participants through the arduous process of quitting smoking through giving advice and being supportive, as made clear on the opening Web page below the main forum title (*stop smoking help* or *quit smoking cigarettes*).

The titles of the different subforums additionally highlight the supportive character of "SmokingisBad." For example, *the Help room* is advertised as follows (spelling verbatim): "If your struggling, having a tough time, losing your quit, feel you can't carry on, need help and support right now, use this room!!"[5]

The forum is divided into four larger subsections: *Welcome*, *Your Quit Smoking Journey*, *The Help Room*, and a space for off-topic posts. The *Welcome* section is a space in which people can introduce themselves, share quotes to inspire others, and pledge not to smoke. *Your Quit Smoking Journey* is the heart of the forum, composed of multiple subforums organized into different time units, designed to chronologically accompany contributors as they quit smoking. Initially, the subforums provide a platform for each day of not smoking, culminating in a representation of the first week of quitting (*Day* 1 to *Day* 7). The rest of the first month of stopping smoking is split into larger units of weeks (*Week* 2 to *Week* 4).

Participants who have stopped smoking for a month have the opportunity of posting to different subforums every month until they reach the half-year point (subforums *Month* 2 to *Month* 6). The number of subforums thereafter decreases, with names indicating that it is an achievement to reach these stages: *The Halfway House Inn*, reached after half a year of having stopped smoking, and *The Penthouse* 1 *Year* + (for those who have been able to stop for over a year). The fact that the length of time units increases the longer the smoking cessation shows that it is mainly in the initial phase of quitting when participants need most support and the most focused opportunities to post. *The Help Room* is a place where people can request help and support in emergencies. Finally, there is a section with forums that are off-topic, in which people discuss any issue that does not pertain to smoking.

12.3.2 Sampling and deidentification

The data for this study consist of postings between March and April 2012 to specific subforums of "SmokingisBad" used by participants during the initial phase of smoking cessation. Specifically, data were collected from subforums devoted to the first three days of the "quit," *Day* 1, *Day* 2, *Day* 3, and the second week (*Week* 2), considered

---

[4] To further maintain confidentiality, the subject lines and the specific URLs of posts are not provided.
[5] Notice the use of *quit* as a noun as mentioned in the introduction.

an important early milestone. Day 4 until Day 7 show evidence of less activity, and for this reason are not included in the analysis. The rationale behind choosing these specific subforums catering to people in an early stage of quitting is that new quitters may likely seek more help, as demonstrated by the fact that *Day* 1, *Day* 2, *Day* 3, and *Week* 2 receive the most posts in the *your quit smoking journey* subsection. Thus, these four subforums provide a detailed look into the kinds of concerns participants have and share about when having just quit smoking.

Only threads that received between 10 and 20 replies have been considered for analysis, which seemed to be the typical length of threads of these subforums. A sample of 10 threads was randomly selected out of the 39 threads that fit the posting and time parameters. These 10 threads contain 138 posts overall; the number of participants posting varied. For instance, one thread had six separate participants for a total of 12 posts; on the other end of the spectrum, another thread had 13 separate participants for a total of 18 posts. Initiators of threads always posted more than once (from two to eight posts each). The length of posts varies from 2 to 359 words, with the average length being 75 words, which is a numerical indicator that participants post for different purposes. Some participants just want to congratulate someone on having decided to quit, but add their personal experience, evaluate the recent quitter's situation and give advice.

The topics cluster in three main areas:

1. Sharing experience with withdrawal symptoms/medication,
2. Announcing the start of a new attempt at quitting—"a new quit," to use the community members' term—and outlining potential pitfalls, and
3. Evaluating how the current attempt at quitting is progressing.

This study adhered to the heuristic principles outlined by the Association of Internet Researchers (Ess & the AoIR Ethics Committee, 2002; Markham, Buchanan, & the AoIR Ethics Committee, 2012). There is a personal message function for private interaction, which gives the impression that posters are aware of the public nature of their forum interaction. Additionally, my linguistic focus on form, that is, the text-based nature of my research, minimizes the possibility of linking content to a specific individual (Ess & the AoIR Ethics Committee, 2002; McKee & Porter, 2009). However, in order to minimize any potential harm to participants, screen names have been changed and any markers of location removed from the data.

12.3.3 Analytical procedure

The aim of this study was to find out how participants in an online support group for smoking cessation share information and give support when starting to quit smoking, and what practices of signaling authenticity and legitimacy (i.e., practices of constructing *ethos* and *pathos*) have been employed. Forum participants influence each other's actions and attitudes through language, which links their posting activity to persuasion (Burke, 1950). The data were analyzed using discourse analysis, a method which, in linguistics, is used to find out "the way texts are constructed" (Holmes, 2014, p. 178). Discourse analysis is, as Gee (2011) explained, "a theory of how we use language to say things, do things, and be things" (p. 3). For instance, by looking in detail at what participants *say* in their posts, it is possible to find out how they *do* (in the sense of *perform*) share information; how they *do* give support (or how they *are* supportive community members); and if there are communicative norms to these activities.

To get a sense of how each post is structured and sequenced, Miller and Gergen (1998)'s concept of discursive moves was adopted, defined by these authors as "the kind of contribution that the entry made to the ongoing interchange" (p. 192). Examining discursive moves allows description of what people *say* and actually *do* in their messages and how they exchange information/advice. This helps to identify the content structure of texts, as well as recurring patterns (Locher, 2006; Morrow; 2012; Placencia, 2012). This is especially important because even though the overall frame of most online health support groups is one of giving help and advice (Harvey & Koteyko, 2013; Locher, 2013; Morrow, 2006), participants do not restrict themselves to these activities.

Each of the 138 posts was initially read and coded using a mixture of Locher's (2006) and Morrow's (2012) catalog of discursive moves. The discursive moves employed by participants depend on the social practice under scrutiny, understanding a social practice as "a socially recognized and institutionally or culturally supported endeavor" (Gee, 2011, p. 210). Thus, the initial catalog of discursive moves was adapted after a pilot phase to do justice to this particular data set. The discursive moves are presented in Table 12.1 (see Locher, 2006; Morrow, 2012; Placencia, 2012) and further explored in the Discussion section below. These different types of discursive moves reflect what kind of language activities (i.e., discursive moves) was used when sharing information and giving advice in "SmokingisBad." They are listed in alphabetical order for ease of reference and are illustrated by examples from the data set.

While sentences were chosen as the coding unit for analysis, a discursive move could be larger than a single sentence. Each text passage received only one code. The depicted catalog of discursive moves shown in Table 12.1 was used to code each of the 138 posts systematically. A quantitative analysis was conducted, calculating the frequencies of use of discursive moves and cross-referencing them with participant roles (initiator vs respondent). Based on the frequencies of discursive moves, it became clear that initiators are likely to receive advice after sharing details about their background (but not give advice themselves) whereas respondents remained focused on initiators, giving *advice*[6] and *assessing* the latter's situation. Thus, I selected the most typical discursive moves for initiators and respondents to identify linguistic patterns. Previous research in sociolinguistics and other fields, for example, discursive psychology, has stressed the importance of sharing one's story and disclosing personal details about oneself to come across as credible. For this reason, the discursive moves *background information* and *own experience* were examined more closely, to answer how they informed practices of signaling authenticity and/or legitimacy (Armstrong et al., 2011; Cranwell & Seymour-Smith, 2012; Harvey & Koteyko, 2013; Sillence, 2010).

Three threads were double-coded by the author and another researcher to test the catalog of discursive moves. Due to the qualitative nature of this study, I opted for "subjective assessment" of intercoder agreement, Discrepancies were evaluated and resolved after discussion before the analysis was finalized[7] (Guest, MacQueen, & Namey, 2012, p. 89).

---

[6] Discursive moves are set off in italics.
[7] Percentage agreement was not included for this case study but is part of the larger PhD project (percentage agreement for the final codebook was 83%).

Table 12.1 Discursive moves employed in "SmokingisBad" in alphabetical order

Discursive move	Explanation	Example
Advice[a]	Tells someone what they should do, what is best for them	…get rid of any remaining fags you have or Nic will have you back again if you get really tempted…
Apology	Expresses sympathy for someone by saying you feel sorry	Sorry you're feeling so rubbish at the moment
Assessment	Assessment and/or evaluation of someone else; support of help-seeker	I can see you've reached your "enough is enough" point…
Background information	Initial poster provides information about themselves	Recently diagnosed with COPD and after yet another virus and chest infection and picking up a prescription for champix on Monday…
Disclaimer	Special kind of assessment, indicating insufficient knowledge	This is my first post so not sure if it's in right place
Farewell	(Optional) Closing move, saying goodbye	Love and hugs
General information	Reporting of facts and impersonal delivering of information	Smoking never did make anything better, that was the artful con of nicotine addiction…
Greeting	Greeting, (optional) first move	Hi all
Meta-comment	Text-structuring remark	Hope this makes sense lol
Official forum welcome	Welcoming of a newbie by a moderator	Welcome to the forum and well done on the decision to quit possibly one of the most important you will ever make…
Open category	A move that cannot be categorized	–
Own experience	Personal experience of advice-giver	You see I've been that occasional smoker, and I've also found myself on 40 a day due to the way I trivialized my smoking habit…
Quote	Technical feature of quoting a previous post	–
Request advice or information	Asks for guidance or information	…, are you going cold turkey?
Thank	Thanks for advice or support	Thanks for all ur lovely messages!!
Welcoming	Specifically welcomes a poster to the community or to a new subforum	Huge welcome to the forum and well done on your first day
Well-wishing	Wishes someone well	Hope you're having a good day

[a]To clearly demarcate the difference between *assessment* and *advice*, the code *advice* was only used if it there were some linguistic pointers indicating a recommendation (e.g., imperatives, interrogatives, the modal verb *should*) (Locher & Limberg, 2012).

12.4 Results and discussion

12.4.1 The structure of sharing information and giving support

The systematic coding of discursive moves in the forum threads showed that sharing information and giving support are performed through diverse discursive moves in "SmokingisBad." Recall that as discussed above, a discursive move designates "the kind of contribution that the entry made to the ongoing interchange" (Miller & Gergen, 1998, p. 192). For example, the discursive move *thank* denotes the use of a passage to express gratitude to someone. Interestingly, information exchange was by no means restricted to the discursive move *general information*, but was also performed through the discursive moves *assessment* and *advice*, confirming results from previous studies of communication in similar computer-mediated contexts (Morrow, 2012). In fact, *general information*, or the impersonal reporting of facts, was rarely found in the data. Two factors may account for this phenomenon.

First, separating *general information* from *assessment* or *advice* proved to be difficult, since administering information can be understood as advice to or as assessment of someone in the context of this forum, which explicitly aims at helping each other (Locher & Limberg, 2012). Secondly, the intimate or colloquial frame of interaction meant that participants rarely shared facts without personally addressing help-seekers (using *you*) even when talking about general stages of quitting, which fostered the interpretation of information as guidance or evaluation by help-seekers. Nevertheless, giving *advice* or the *assessment* of someone else's situation features "the communication of knowledge" (see OED definition of "information," sense 2).

In general, passing on information, in the sense of communicating knowledge of particular facts, is achieved through the combination, the composite, of several discursive moves in messages.[8]

12.4.2 Exemplar messages

The two sample messages below are both written by advice-givers, and illustrate the style of more complex posts, that is, messages containing several discursive moves. Example (7) was written in response to a direct request for advice. The help-seeker is on his sixth day of quitting and is taking medication—Champix—which helps reduce cravings and minimizes the pleasure one gets from smoking. He does not feel any different, and wonders "does this take a while with the champix."[9]

Sample message by advice-giver
<greeting>Hi [...]<greeting>

<own experience>The champix is working hun already it's made you not want so many as normal now it will start to make them taste bad at least they did for me and others I know have taken them<own experience>

---

[8] A message consists of at least one discursive move but can contain several discursive moves at once.
[9] Angle brackets feature the names of how a particular passage has been coded and appear at the beginning and the end of the passage they designate.

<own experience>They reckon anywhere between 7 and 14 days till you stop smoking with them but I have known it to take three weeks for some to stop smoking with them<own experience>

<advice>The bumph is just a guideline hun we're all different and so is the way we react to it and I suspect it also depends how many we smoked soon now you should notice they taste a bit different just go with it and don't wait for something to happen it already is just live your normal life and let the champix do it; job<advice>

<farewell>Love and Hugs
[user name]<farewell>

This advice-message contains five paragraphs, including the discursive moves *greeting* and *farewell*. In the second paragraph, the participant analyzes the help-seeker's situation, arguing that he is already experiencing the first effects of the medication. She predicts that he will soon become more aware of the influence of the medication. The advice-giver arrives at her evaluation because of her *own experience*, which gives her some expertise on the topic. In the next paragraph, the advice-giver elaborates how Champix works according to her *own experience*. The passage starts off looking like *general information*, using the personal pronoun *they*, standing for unidentified people, maybe for Champix or generally "experts," but the poster dismantles the official opinion by contrasting it with her own experiential knowledge. Finally, *advice* in the fourth paragraph acts as a summary of the previous *own experience* moves, concluding that the help-seeker should not worry about not reacting exactly as the instructions say. She changes from *we* to *you* and uses the term of endearment *hun* to create common ground and to involve the help-seeker. The core information of the advice-message does not just lie within *advice*, but is continually built up from an initial evaluation based on *own experience* to personalized *advice*, in which Champix, its effects, and the time it takes for them to become noticeable, are explained.

The psychological drawbacks of quitting are a much-discussed topic in the forum. Example (8) is about coping with feeling depressed during quitting. To understand the context of the conversation, the parts of the help-seeker's post that triggered the message by the advice-giver afterward are shown below.

sample exchange;
...
[help-seeker's message triggering advice]
<background information>Will give the ciggs away tonight but all my friends and family smoke and tbh its kinda not an issue if they are on top of the wardrobe or in the shop cause one of the huge things for me to over come is that ciggs are always going to be there...so easy to get one...<background information>

<request for advice>What do you guys do when the depression kicks in...I usually get it about 4–6 weeks in:-(<request for advice>

...

[advice-giver]
<own experience>I went through a period of some depression in my 4th month. I tried St. John's Wort (research it) which to me seemed to work, but I think back then I just gritted my teeth and figured there was many many folk in a lot worse situations than me who've been dealt much worse blows to their life...and they get through it.<own experience>

<assessment>depression isn't an easy thing...but talking about it while you are going through it helps a lot. I bet a good many folk quitting 'sigh' more than they realise too.<assessment>

<assessment>I hear what you are saying about how easy it is to get cigs, but giving them away at least means you have to go through more if temptation bites. Having to ask...or go and buy is much different to just reaching up to the top of the wardrobe.<assessment>

To answer the help-seeker's question—how to combat depression as the result of quitting smoking?—the advice-giver offers his *own experience* of how he coped with depression during his own quitting journey. The post is made up of three paragraphs. In the first, the advice-giver provides the experiential account that the initial poster wanted. He describes how herbal medicine works and how it helped him, without uttering an explicit recommendation to buy it. His comment in parenthesis, "research it," emphasizes that he does not want to impose his experience on others. In the second paragraph, the advice-giver, who has fulfilled his interactional task of recounting his *own experience*, takes a step back and makes an *assessment* of a more general nature, acknowledging that feeling depressed is difficult and is something many go through (as he himself did). In his assessment, he offers an afterthought about how talking helps, and about how widespread depression is, through which he indirectly provides a recommendation and normalizes being depressed.

In the last paragraph, the advice-giver employs *assessment* to evaluate the help-seeker's points, made in *background information*, of how easy it is to get cigarettes. The advice-giver legitimizes the help-seeker's standpoint of how owning cigarettes should not bother her because many people in her environment still smoke, but ultimately views it as too short-sighted, which can be understood as indirect advice. Even though there is neither explicit *advice* nor *general information*, this example message brings across information by moving from *own experience* to a more generalized *assessment*. Consequently, the advice-giver normalizes feeling depressed during quitting and encourages the help-seeker to stick through hard phases (on normalization through narratives, see Veen et al., 2010).

12.4.3 Quantitative analysis

After the discourse-analytical insight into how information can be wrapped up in messages, quantitative analysis offers an overview of the content/informational structure used in the selected sample. The 138 posts contained 512 discursive moves, an average of 3.71 discursive moves per post. Table 12.2, below, shows the frequency of

Table 12.2 **The frequency of discursive moves in 138 messages according to participant roles and their position in the thread in alphabetical order**

	Total	Initiator	Respondent	First move	Last move
Advice	98	2	96	21	34
Apology	4	–	4	4	–
Assessment	116	9	107	45	22
Background information	51	51	–	14	22
Disclaimer	2	2	–	2	–
Farewell	61	8	53		
General information	1	–	1	–	–
Greeting	57	7	50		
Meta-comment	12	6	6	4	1
Official forum welcome	2	–	2	–	2
Open category	9	3	6	6	1
Own experience	49	–	49	10	8
Quote	3	1	2	3	–
Request advice or information	17	9	8	4	7
Thank	13	13	–	12	–
Welcoming	12	–	12	12	–
Well-wishing	5	–	5	1	2
Total occurrences discursive moves	512	111	401	138	99

Note: *Greeting* and *farewell* are fairly fixed in their positions and have a mere interpersonal function in posts, therefore they are not considered in the analysis of sequence.
39 posts were composed of only one discursive move, which accounts for the different numbers of first and last moves. In posts featuring only one discursive move, the move is counted as occupying the initial, not the final, position (Morrow, 2012).

discursive moves in the whole corpus and their distribution according to the participant roles: initiator and respondent. Initiators are the contributors who start a thread in their search for help, while respondents react to initial posts by potentially giving advice or support. Table 12.2 indicates whether a move occurred in the first or last position of a post, in order to find out possible patterns of sequencing. Sequencing tells us about the structure of posts and whether there is a preferred order, and certain communicative norms in place, with respect to how participants contribute to threads.

Review of the distribution of discursive moves and sequences suggests the existence of communicative norms for how information is shared. The most common discursive moves, *assessment* and *advice*, are almost exclusively used by respondents who assume the role of advice-givers, tailoring their knowledge to the specific situation of initiators (see examples (7) and (8)). Conversely, when initiators start a thread, they assume the role of help-seekers, and maintain this role throughout, as they hardly ever *advise* or *assess* others. The frequency and distribution of *assessment* and *advice* shows that participants actively try to affect each other's quitting processes. They take the stated supportive purpose of the forum seriously, with the focus remaining on the initiator's problem.

The numbers of discursive moves give an impression as to how the exchange of information takes place. In their 40 posts, help-seekers often provide *background information* to receive help and only rarely *request advice/information* in their first posts. This is in accord with findings from previous studies, in which explicit requests for advice were found to be rare because of the threat it poses to a help-seeker's face (see Morrow, 2006; Page, 2012; Veen et al., 2010). Morrow (2012) convincingly argued that just the mere description of a problem in such a communicative context "could be seen as a way of soliciting advice" (p. 262). A close reading of the few instances of *request advice/information* shows that the actual number of people asking for advice is even smaller (four out of nine; the rest are bonding questions). Moreover, two of these are requests for others' experience, and one is phrased as a question using "or." Both these strategies can be considered ways to minimize risks to help-seekers' face. First, if help-seekers ask for experiential accounts, they do not explicitly state that they need advice (Veen et al., 2010). Second, if they ask for help with an "or" question, they already present two possible solutions, which shows that they are knowledgeable (Morrow, 2006). Example (9) nicely illustrates how a help-seeker works to maintain a competent identity and to establish an equal basis with other members when *requesting advice*.

(9) What do you guys do when the depression kicks in...I usually get it about 4–6 weeks in:-(

As she talks about the moment when she "usually" begins to feel depressed, she nicely positions herself as experienced in quitting, which mitigates, that is, softens, any potential advice she is going to receive. Sociolinguistic literature about advice has argued that asking for advice creates an asymmetry in knowledge between advisee and advisor (Harrison & Barlow, 2009; Locher, 2013; Locher & Limberg, 2012). Thus the initiator's pointing out of her past experience can be seen as an attempt by the initiator to show that she is knowledgeable herself.

In most cases, respondents do not immediately give *advice* or *assess* after an initiator has started a thread. First, in half of the posts they *greet* initiators (51.02% out of 98 posts), while initiators only *greet* in 17.5% out of 40 posts. If initiators are new to the community or to a particular day in their quitting journey, respondents often explicitly use the discursive move *welcoming* to admit initiators to the particular subforum right after *greeting* them. Generally, *welcoming* has a sort of ritualistic function. It often comprises either praise or an expression of affection, and is used by more experienced forum members, who have gone through the smoking cessation day in question before. Example (10) is a typical example of what *welcoming* can look like.

(10) Huge welcome to the forum and well done on your first day.

Greeting and *welcoming* have an entirely relational function, and are used to set a positive atmosphere at the beginning of a thread. If *welcoming* occurs, it is always the very first discursive move after *greeting*. If we recall the interpersonal pragmatic premise that there are always both informational and interpersonal aspects to communication (Locher & Graham, 2010), respondents move from discursive moves with a predominantly interpersonal function, to discursive moves that emphasize informational aspects, turning to *advice, assessment,* and sharing their *own experience.* Finally, respondents close their messages with a *farewell* in 54.08% of 98 posts.

In their responses, help-seekers show appreciation of advice-givers' help by *thanking* them. About a third of all posts by help-seekers (32.5%, or 13 out of 40) feature some expression of gratitude. *Thanking* is a typical follow-up move, as it never occurs in the first post of an initiator. Further, all instances of *thanking* are the first discursive move in posts. This, on the one hand, ensures a positive footing between help-seeker and advice-givers in order to ask for further advice. On the other hand, this bolsters help-seekers' authenticity, because it shows that they are genuinely interested in advice-givers' input and support. In contrast, respondents never use the discursive move *thank*, since they are the ones giving recommendations and advice.

The quantitative analysis of discursive moves allowed me to identify those that occur most frequently, as well as patterns of structuring information/content in the course of sequenced forum interaction. It revealed that there are communicative norms in place, since the content contributed varies according to a poster's role in the forum. First, it made clear that *assessment* and (often based on that) *advice* are the main discursive moves that respondents use. Further, it showed that discursive moves, particularly those with a mainly interpersonal function, tend to occur in the same sequential positions, with *welcoming* being a first move in respondents' posts and *thanking* as first move in follow-up messages by initiators. In the section below, the four most frequent discursive moves *assessment, advice, own experience,* and *background information* are discussed in detail to provide a sense of how they are composed and how they fit into the overall structure of the information exchange on the forum.

12.4.4 Discursive moves

12.4.4.1 Assessment

As we have seen in Table 12.2, *assessment* was the most frequently used discursive move. Ninety-eight posts by respondents contained 107 *assessments,* which indicates that a post could feature more than one *assessment* move. In terms of sequencing, *assessment* was the first move in 42.85% of respondents' posts (excluding *greeting*), which accounts for 42 messages out of 98. In 21 out of 98 messages, or 21.21%, it was the last move of respondents' posts (excluding *farewell*). However, *assessment* is still markedly more often used as a first than as a last discursive move. Morrow (2012) explained this sequencing with the fact that posters first needed to show their understanding of a problem before giving a piece of advice. Locher (2013) stressed that *assessments* are part of the larger speech event of advising and present a stepwise entry to *advice* (Locher, 2013; Morrow, 2012). In Morrow's study (2012), initial *assessment* overwhelmingly preceded *advice*, while he does not report whether *advice* directly followed *assessment*. In this study, respondents' messages exhibit similar tendencies in structure as in Morrow (2012) and Locher (2006, 2013), but they are not as pronounced. While *assessments* immediately presequenced *advice* in 14.28% of respondents' posts in first position, *assessments* directly presequenced *advice* in 30.84%. overall. In total, 39.79% of respondents' posts featured both *assessment* and *advice*.

In view of the personalized evaluative aspects present in many *assessments*, it has been difficult to distinguish *assessment* from *advice*. This is true not only for coding purposes, but also likely for participants of the forum themselves. Locher and Limberg (2012) have highlighted that utterances can be construed as advice even without "clear-cut linguistic pointers that mark them as such" (p. 7). Due to the supportive/help-giving frame of this forum, *assessments*, such as in example (11), can also be easily interpreted as indirect advice. In (11), the advice-giver evaluates the help-seeker's decision to immediately start quitting again after having relapsed.

(11) <assessment>I think you are putting a bit too much pressure on yourself, the goals you are setting while not unachievable are large and hard.<assessment> <advice>Maybe try to give yourself a few easy wins, get your confidence up.<advice>

Respondents also often expanded their *assessments* of help-seekers by adding informational elements. In (12), for instance, the respondent moves from a more generalized stance (including a generic *you*) to a specific evaluation of the help-seeker.

(12) The anxiety will pass, there are so many emotions wrapped up in a quit and some days you feel you are knitting fog. You will see the sky for the clouds though and you will be glad you hung onto this quit, know it doesn't seem like it at the moment though.

The respondent presents his central argument of how anxiety subsides, and how it will become easier, in a personalized form, using a rather generic *you*, addressing the help-seeker, but also any other person trying to quit. Moreover, the respondent sympathizes with the help-seeker. He writes that he understands that the help-seeker cannot believe this assertion yet, which implicitly signals the respondent's own expertise, because he is signaling that he once experienced the same stages. Locher (2006) observed in her study of the online advice column that the use of *assessment* was a way of personalizing answers. Her data revealed a clear preference for *assessments* over *general information* in the topics "emotional health" or "relationships," which Locher (2006) attributed to the fact that "giving information alone cannot solve the problem" in personal matters (p. 79). Similarly, contributors to this online group do not provide "pure" information in the sense of facts and figures, but give personalized answers which acknowledge help-seekers' individual situations, helping them to remain committed to quitting.

All instances of *assessments* were examined in order to find out how they attend to the interpersonal needs of help-seekers. Analyzing the interpersonal functions of *assessments* showed that they were most often used to praise help-seekers or to display empathy, which reassured or normalized feelings or behaviors of initiators (see Locher, 2006; Placencia, 2012). *Assessments* employed by respondents seem to be on a continuum, with normalizing *assessments* featuring a strong informational component, as they refer to the quitting process, and praise having mainly an encouraging, relational function. Pudlinski (2008) described the encouraging function of compliments in his study of calls to a peer support line. He contended that compliments, which are assessments that "focus specifically on a positive evaluation" of the participant, allowed the peer to "display affiliation" with and a "congruent understanding" of the caller (pp. 809–810).

Respondents normalized the experiences of help-seekers in *assessments* by qualifying their feelings and predicting specific future developments, behaviors described by Placencia (2012) as "deproblematizing the situation" (p. 299) (see also Locher, 2006). To normalize, respondents resorted to using the *will*-future, mostly in connection with *you*, or *it is (adjective)* constructions, as in (13) and (14) below.

(13) Sorry you still feel rough hun but all these things are normal at the start of a quit and they will pass fairly rapidly
(14) It is really common to get increased feelings of anxiety when you quit - but in the long run you will feel so much better

Respondents provide a positive outset in (13) and (14); the adjectives *normal* and *common* qualify help-seekers' emotions as nothing out of the ordinary. These qualifications occur in impersonal constructions, denoting these evaluations as perspectives that are not just personal opinions. The use of the *will*-future (instead of other modal verbs, such as *should* or *may*) frames the outcome as prediction with a great degree of certainty. Moreover, the *will*-future implies expertise in this context: Respondents know the next steps, because they have gone through the process. Some respondents explicitly mobilize their competent "quitter" identity when they boost their predictions by promising that they are true (15, 16):

(15) the first week is the pits, if you can do that you are done with the worst, promise
(16) Over time that grumpiness goes…you'll find time to smile more i promise, and soon too!!!

While normalizing *assessments* have a greater informational load than other *assessments*, because they tend to refer to the quitting process, other *assessments* are mainly used to reassure help-seekers. Normalizing and reassuring are both linked to displaying empathy and are the predominant interpersonal aspects of *assessments* seen in this data set (compare to Locher, 2006 as well as Placencia, 2012, who discussed similar aspects). When respondents reassure help-seekers that they are on the right track, they often use the present continuous of *do* with a positively connoted adverb (*great, fine, well*) as well as the modal verb *can*; see (17) and (18).

(17) You are doing great, if your head hits the pillow at night and you have not smoked a cig, then you had a successful day
(18) Well you can quit, you are well into a good quit already here and through what folk call 'hell week' already.

Reassuring *assessments* function as encouragements that reinforce the help-seeker's ability to quit, and may feature an explanation for why the respondent is positively assessing a help-seeker. In (17), the respondent's evaluation allows her to emphasize the positive aspects of the help-seeker's initial post, in which he said that he was "feeling very down [but] still not smoking." In contrast, in (18), the respondent reinforces the help-seeker's good news of having quit for a week, rather than the associated negative feeling. Finally, some *assessments* have an entirely interpersonal function, which is the case if respondents praise help-seekers. Arguably, praising strengthens the emotional link to the community, encouraging initiators to go on posting and quitting. Linguistically, these moves frequently feature the structure *well done*, as in (19).

(19) massive well done on making it through the first week

Overall, *assessments* have an overwhelmingly positive tone, through which respondents try to make their evaluations of help-seekers acceptable and keep them committed to quitting. As described by Locher (2006), *assessments* are a way of personalizing information, paying tribute to the fact that in order to continue with quitting, help-seekers need not only facts, but also to feel understood.

12.4.4.2 Advice

As mentioned in discussion of quantitative analysis above, *advice* is reserved for respondents, which implies that initiators accept their asymmetrical position of help-seekers. Since giving advice implies guidance for future actions or telling someone what is best for them, it can also function as a source of knowledge or information, especially in combination with *assessment*, as outlined above (for definitions of advice see Locher & Limberg, 2012). There are four thematic types of *advice* in the data:

1. *advice* to continue posting to the forum,
2. *advice* to persevere in quitting,
3. *advice* to take 1 day at a time, and
4. concrete pieces of *advice* (i.e., specific actions; recommendations).

The length or linguistic form of *advice* varied greatly depending on its thematic type.

Respondents advise help-seekers to frequently post to the forum and to keep the group in the loop, especially if the help-seekers are new or relatively new members. This type of *advice* is characterized by the use of the verb *post* in the imperative form (alternatively, *keep posting*), which is often combined with another imperative, the phrase *let us know*. Examples (20) and (21) are both responses to a post by a new member who has announced the start of his quit.

(20) Post often and let us know how you're doing!
(21) Post in here as much as you want/can as thats what I done and OMG it REALLY REALLY helped me sooooooooooooooooooooo much!!!!

These instances of *advice*, encouraging initiators to post often, seem to be a way of strengthening community ties. In most occurrences, respondents advise newcomers to continue posting to the forum in their identity as community members, which is indicated by the personal pronoun *us* in (20). Thus, example (21) is exceptional insofar as the respondent warrants her *advice* based on herself.

The second thematic group of *advice* has a motivational/supportive function, as respondents encourage help-seekers to continue quitting. Imperative constructions predominate in this kind of *advice*, which may be accompanied by an evaluative or well-wishing subclause. The verbs employed emphasize the nature of quitting as a process and tend to feature a time component: *stick, hang on/in there, keep, stay*. *Keep* often occurs with a verb in the gerund (*keep going, trying*) or as the phrase to *keep at it*, whereas *stay* collocates with the adjective *strong*. In example (22), *advice* simultaneously closes and is the gist of a long section that features *advice* and *assessment* on how the initiator should cope with his recent relapse.

> (22) *Keep going and I hope and wish for good fortune and better times to come your way too.*

Advice to persevere in quitting is frequently employed as a last discursive move before *farewell*, which may be so that the respondent can leave on an encouraging note. Apart from having this final bonding function, *advice* to persevere in quitting corroborates the community tenet that quitting is difficult and time-consuming. In view of the imminent danger of a relapse, it has the important function of reinforcing help-seekers' determination.

While the first two advice-types have a predominantly interpersonal function, *advice* to take 1 day at a time provides support as well as embodies a coping strategy. Respondents urge help-seekers to break their quitting journey down into manageable units, especially when they struggle with withdrawal symptoms. This type of *advice* stresses that quitting is not a linear process, but has its ups and downs. In this it resembles *advice* to persevere in quitting. Interestingly, *advice* to take 1 day at a time appears to be used more often during a later phase: when quitters are already in their second week, and not at the very start of quitting. While this could be just an idiosyncrasy of this data sample, it may indicate that respondents want to make help-seekers aware that, even though the first week is over, there are still going to be difficult moments. To "take one day at a time" seems to be the preferred coping strategy of forum members when facing withdrawal symptoms. Most of this kind of *advice* features some variation of *to take it a day at a time* (also to *take each day at a time* or *to take it day by day*) in the imperative form, in which the verb phrase occasionally is presequenced by the adverb *just*. In example (23), the respondent encourages the help-seeker, who has voiced that he is feeling low, to keep going by splitting his journey into small units.

> (23) *Just take it a day at a time and if all else fails, have a cup of tea and go to bed early*

It is notable that the first three advice groups were mostly realized with the imperative form, which gives a sense of urgency to *advice*, underscoring its supportive function. Similar to Morrow's findings (2006), imperatives seem to be signaling closeness rather than being directive.

Finally, the last group of *advice* includes more specific suggestions and recommendations for actions. While imperative sentences are also frequently used in this group, it is linguistically more varied overall. For instance, imperative clauses are more often preceded by a subclause in this than in the other three groups. The verbs *try* and *keep* come up several times as imperatives, sometimes preceded by *just*. *Try* is used with an object or another verb, which implies that respondents suggest something new for help-seekers, such as a specific coping strategy or general tips on how to visualize their quitting journey. Respondents offer ways of dealing with cravings in examples (24) and (25).

> *(24) When you get craves, try not to think of the cigarette as your friend that you miss, but rather as your enemy that wants to kill you.*

> *(25) If you find yourself facing some stress or worry, ..., just remember that things will not be helped at all by a cigarette.*

Both respondents provide coping strategies, prefacing their advice either with the temporal subclause *when* or a conditional *if*-clause in (24) and (25), which has the effect of tailoring advice to a concrete situation. Specific *advice* statements tend to display some mitigation or downtoning if they feature suggestions regarding help-seekers' daily lives. The respondent in (26) comments on a statement by a help-seeker; the help-seeker has said that she needs to learn how to live with temptation, since her husband is still a smoker.

> *(26) Is there any way your other half can agree to ensure he respects your quit while you give it a good go if poss? he prob already is, but if my other half were quitting and i still smoked i'd do my best not to put temptation in her way.*

The interrogative, further mitigated with *if poss*, suggest that the respondent seems to feel that he is on delicate territory. He adds a conditional sentence—he would support his wife in the same situation—but is careful to introduce it with a disclaimer.

Generally, forum members are more likely to hedge *advice,* or to offer more reasons to indicate the optionality of their suggestions, if their *advice* contains recommendations that affect a help-seeker's daily life and not the specific process of quitting. These findings concur with Kouper's (2010), who observed that the delicacy of the topic influenced the form of advice. In this community, experienced posters feel free to give *advice* on the quitting process as such, but are more careful when it comes to recommending actions for someone's daily life. Giving *advice* and support are strongly interrelated in this forum, since any advice or information exchange ultimately has the purpose of supporting help-seekers in their quitting journey.

12.4.4.3 Own experience and background information

The discursive moves *own experience* and *background information* are two sides of the same coin. The former refers to respondents, and the latter to initiators, each sharing their experiential world. Since initiators indicate their authenticity in *background information*, and respondents make their right to advise legitimate on the basis of their *own experience*, these two discursive moves have important communicative functions (Harvey & Koteyko, 2013; Sillence, 2010).

Several sociolinguistic studies have stressed the multifunctionality of telling one's *own experience* in online health support groups (Hamilton, 1998; Harrison & Barlow, 2009; Kouper, 2010; Morrow, 2006). There are also plenty of studies documenting this phenomenon in other fields, for example, discursive psychology (see Armstrong et al., 2011; Veen et al., 2010). This online community is no exception; *own experience* is used to signal solidarity, to warrant *advice/assessment*, and sometimes functions as indirect advice. In order to identify with help-seekers, respondents parallel their own quitting story with the one of the help-seekers, which is signaled through the use of the adverbs *too*, *also*, or structures like *exactly the same*, *like you*. The respondent in (27) reacts to a help-seeker who is frustrated by her "feeling grumpy." The respondent brings up her *own experience* to illustrate that she understands the help-seeker's position, stressing her solidarity through her repetition of *grumpy*.

> *(27) <own experience>First couple of weeks I wasn't anyone I could recognise... grumpy, irritable, short tempered and difficult [...]<own experience> <assessment>I can honestly promise you that it does get better. As each day goes by the mood swings, grumpy feelings and craves get less and less.<assessment><advice>But do be very good to yourself. Take it a day at a time and accept that some days will not feel as good as others<advice>[...]*

The respondent shares her *own experience* to illustrate that she sympathizes with the help-seeker. This also works as a warranting strategy (Richardson, 2003) that allows the respondent to show her knowledge of how quitting will continue and to give advice. On several occasions, *own experience* acquires this double function—showing solidarity and serving as a legitimation sequence, either before or after *assessment* and *advice*. Sometimes respondents warrant their *assessment* by mentioning how long ago they have stopped smoking, which serves to give their statements credibility (on sharing "key statistics" see also Sillence, 2010, p. 381). In example (28), the respondent uses herself as an example to support her previous *assessment* that "the first week is the worst part."

> *(28) i'm 10 weeks free now and am totaly over it.*

Some *own experience* moves consist of narrative sequences about how respondents resolved a similar situation as the one recounted by the help-seekers. Since respondents illustrate how they ultimately managed to cope, it can be understood as indirect advice (see example 8).

The discursive move *background information* varies greatly in its linguistic realization—although the frequent use of the verb *feel* is conspicuous—since help-seekers share details of their individual quitting journeys. Help-seekers need to be perceived as authentic for other posters to react to their initial message (Harvey & Koteyko, 2013; McKinlay & McVittie, 2011). Since all threads are taken up unchallenged by advice-givers, help-seekers are clearly successful in coming across as belonging to the group. Strategies to create authenticity largely overlap with the ones described by Harvey and Koteyko (2013). Some help-seekers initiate threads by disclosing that they have relapsed or by sharing their success in quitting as in (29).

(29) [...] I am so happy to have got through week one and COLD TURKEY!! [...]

Others evaluate steps they have undertaken to quit and how quitting has been progressing; see (30).

(30) I've finished the first week and although it wasn't very pleasant, it wasn't the week of hell that I was expecting.

Some help-seekers are self-deprecating ("I feel a total failure"), describe their withdrawal symptoms ("Always a niggle of something missing"), or discuss health problems (see example in Table 12.1).

In half of all threads in the corpus, help-seekers signal their uptake of *advice* when giving *background information* in one of their later posts. One help-seeker first *thanks* the respondent for his *advice* not to see herself as a victim of addiction and then illustrates why she considers this suggestion be useful. Alternatively, some help-seekers promise to follow *advice*, and one help-seeker describes how she adopted advice, though without mentioning the respondent. Obviously, it is impossible for a researcher to say whether the advice was indeed followed, or if the reference to administered advice is just a way of keeping conversational interaction going. Nevertheless, the fact that in half of the corpus help-seekers align themselves with advice provided by respondents seems to indicate that they appreciate guidance beyond community bonding (cf Kouper, 2010).

12.5 Conclusions

In this linguistic study, the sharing of information between initiators (help-seekers) and respondents (advice-givers) in an online smoking cessation support group was described from the perspective of persuasion. The purpose of this particular online forum is one of providing help and support to achieve the community goal of stopping smoking. Forum members actively try to have an effect on help-seekers' attitudes and actions through their posts, while facing the rhetorical challenges of being credible or authentic.

The content/informational structure analysis of discursive moves suggests that participants seem to follow a loose sequence when sharing information. Even though help-seekers rarely *request advice* explicitly, they outline their need for guidance or support by sharing their experiential world, in which they illustrate their authentic "quitter" identity. In turn, advice-givers first react by establishing a positive, interpersonal footing, involving help-seekers by *greeting* or *welcoming* them. After this, advice-givers move on to the informational core of their messages, in which they *assess* help-seekers' situation and/or give *advice*, often warranted by their *own experience*. Finally, respondents finish their messages by wishing their help-seekers *farewell*. Initiators react by *thanking* advice-givers for their input, indicating that they are authentic help-seekers and preparing the ground to continue interaction.

Thus, information is not only shared within a specific discursive move (such as *general information*) but is conveyed through the composite of a range of discursive moves. Additionally, the predominance of *advice* and *assessment* shows that information is shared in personalized form, meaning that respondents attend to the interpersonal side of interaction. This indicates that, apart from just knowledge itself, help-seekers also need motivational and supportive input to be successful quitters (see also Locher, 2006; Morrow, 2012). Since the purpose of the forum is to help participants through their quitting journey, informing is intertwined with community support.

This conclusion is further confirmed by a close reading of *assessments*. Assessments showed similar interpersonal functions to those found in previous studies (Locher, 2006; Placencia, 2012). They were used to normalize worries, to reassure, and to praise quitters in view of the common stop-smoking goal. Respondents assumed a position of expertise when they reassured and praised quitters, as well as when they commented on the typical stages of quitting, normalizing difficulties such as cravings and other withdrawal symptoms.

The analysis of the discursive move *advice* revealed that it could largely be classified into four types, two of which had a primarily relational function. *Advice* to continue posting on the forum served as means of tying newcomers to the community, whereas *advice* to persevere in quitting reminded quitters not to lose sight of the common goal. Respondents also frequently advised help-seekers to take 1 day at a time, framing quitting as a difficult process, a journey, which needed to be divided into manageable steps. While the first three types of advice work to reinforce the aspired lifestyle change and to persuade help-seekers to continue, the fourth *advice*-type contained specific guidance for individual help-seekers. Since advice-givers were careful not to impose on help-seekers, these types of *advice* were more mitigated (i.e., they were less linguistically direct, using interrogatives, lexical softeners such as *perhaps*) (Kouper, 2010). This analysis of 10 threads only offered a glimpse into discourse norms of sharing information and giving support on the "SmokingisBad" forum. However, it may serve as a basis for further linguistic research into how people help each other at the beginning of quitting smoking on a grassroots level, and how information is exchanged in online health peer-to-peer groups in general.

Harvey and Koteyko (2013) highlight how important it is to understand the "rhetorical challenges" members of online groups face, if we want to successfully use such formats for "therapeutic benefits" (p. 165). This study highlights how people try to help each other stop smoking. Some of the findings of this study could also be useful for content creation for less interactive smoking cessation Web sites, especially those that want to reinforce someone's determination to quit. First, advice to persevere in quitting could be inserted in several subsections of a Web site, alongside other reassuring moves. Second, the ups and downs of quitting could be addressed by including advice to take it 1 day at a time. Third, Web sites could feature a "when quitting" section, in which experienced quitters could talk about difficult situations of the quitting journey and how they overcame them; these could be encouraging to someone who is struggling with quitting. Finally, there could be a section of the site where successful quitters normalize difficult stages of the quitting process and share their grassroots advice. Information exchange on the *SmokingisBad* forum is tailored to individual quitters and has a reinforcing/supportive function within this online health community. These are characteristics which should inform further development of online health resources in this domain.

References

Armstrong, N., Koteyko, N., & Powell, J. (2011). "Oh dear, should I really be saying that on here?": issues of identity and authority in an online diabetes community. *Health*, 1–19. http://dx.doi.org/10.1177/1363459311425514.
Burke, K. (1950). *A rhetoric of motives*. New York: Prentice-Hall.
Camerini, L., Diviani, N., & Tardini, S. (2010). Health virtual communities: is the self lost in the net? *Social Semiotics*, 20(1), 87–102.
Cancer Research UK. (2014). *Cancer research UK Briefing: Impact of tobacco use on health inequalities*. Retrieved from http://www.cancerresearchuk.org/sites/default/files/policy_july2014_tobaccoinequalities_briefing.pdf.
Cartwright, S. (2008). *Smoking. Health and social behaviour: The effects on health of smoking and combating the Issue*. Retrieved from http://www.healthknowledge.org.uk/public-health-textbook/disease-causation-diagnostic/2e-health-social-behaviour/smoking.
Cockroft, R., & Cockroft, S. (2005). *Persuading people: An introduction to rhetoric* (2nd ed.). Houndmills: Palgrave Macmillan.
Cranwell, J., & Seymour-Smith, S. (2012). Monitoring and normalising a lack of appetite and weight loss. A discursive analysis of an online support group for bariatric surgery. *Appetite*, 58(3), 873–881.
Eichhorn, K. C. (2008). Soliciting and providing soical support over the Internet: An investigation of online eating disorder support groups. *Journal of Computer-Mediated Communication*, 14(1), 67–78.
Ess, C., & the AoIR ethics working committee (November 27, 2002). *Ethical decision-making and Internet research: Recommendations from the AOIR ethics working committee*. Retrieved from http://www.aoir.org/reports/ethics.pdf.
Gee, J. P. (2011). *An introduction to discourse analysis: Theory and method* (3rd ed.). New York: Routledge.
Guest, G., MacQueen, K. M., & Namey, E. E. (2012). *Applied thematic analysis*. Los Angeles: SAGE.
Hamilton, H. E. (1998). Reported speech and survivor identity in on-line bone marrow transplantation narratives. *Journal of Sociolinguistics*, 2, 53–67.
Harrison, S., & Barlow, J. (2009). Politeness strategies and advice-giving in an online arthritis workshop. *Journal of Politeness Research*, 5(1), 93–111.
Harvey, K., & Koteyko, N. (2013). *Exploring health communication: Language in action*. London: Routledge.
Holmes, J. (2014). Doing discourse analysis in sociolinguistics. In J. Holmes, & K. Hazen (Eds.), *Research methods in sociolinguistics: A practical guide* (pp. 177–193). Chichester: Wiley-Blackwell.
Ilie, C. (2006). Classical rhetoric. In (2nd ed.) K. Brown (Ed.), *Encyclopedia of language and linguistics* (Vol. 10) (pp. 573–579). Oxford: Elsevier.
Information, n. (2014). In *oed.com* Retrieved from http://www.oed.com/view/Entry/95568?redirectedFrom=information.

Kouper, I. (2010). In *The pragmatics of peer advice in a LiveJournal community Language @Internet*, 7 (article 1). Retrieved from http://www.languageatinternet.org/articles/2010/2464/metadata.

Locher, M. A. (2006). *Advice online: Advice-giving in an American Internet health column.* Amsterdam: John Benjamins.

Locher, M. A., & Limberg, H. (2012). Introduction to advice in discourse. In H. Limberg & M. Locher (Eds.), *Advice in discourse* (pp. 1–27). Amsterdam: John Benjamins.

Locher, M. A. (2013). Internet advice. In S. Herring, D. Stein, & T. Virtanen (Eds.), *Pragmatics of computer-mediated communication* (pp. 339–362). Berlin: Mouton de Gruyter.

Locher, M. A., & Graham, S. L. (2010). Introduction to interpersonal pragmatics. In M. A. Locher, & S. L. Graham (Eds.), *Interpersonal pragmatics* (pp. 1–13). Berlin: Mouton.

Locher, M. A., & Watts, R. J. (2008). Relational work and impoliteness: negotiating norms of linguistic behaviour. In D. Bousfield, & M. A. Locher (Eds.), *Impoliteness in language: Studies on its interplay with power in theory and practice* (pp. 77–99). Berlin: Mouton de Gruyter.

Markham, A., Buchanan, E., & the AoIR Ethics Working Committee (December 2012). *Ethical decision-making and Internet research: Recommendations from the aoir ethics working committee (Version 2.0).* Retrieved from http://aoir.org/reports/ethics2.pdf.

McKee, H. A., & Porter, J. E. (2009). *The ethics of Internet research: A rhetorical, case-based process.* New York, NY: Peter Lang.

McKinlay, A., & McVittie, C. (2011). *Identities in context: Individuals and discourse in action.* Malden, MA: Wiley-Blackwell.

Miller, J., & Gergen, K. J. (1998). Life on the line: the therapeutic potentials of computer-mediated conversation. *Journal of Marital and Family Therapy*, *24*(2), 189–202.

Morrow, P. R. (2006). Telling about problems and giving advice in an Internet discussion forum: some discourse features. *Discourse Studies*, *8*(4), 531–548.

Morrow, P. R. (2012). Online advice in Japanese: giving advice in an Internet discussion forum. In H. Limberg, & M. A. Locher (Eds.), *Advice in discourse* (pp. 255–279). Amsterdam: John Benjamins.

Ng, S. H., & Bradac, J. J. (1993). *Power in language: Verbal communication and social influence.* Newbury Park: Sage.

Page, R. E. (2012). *Stories and social media: Identities and interaction.* New York: Routledge.

Placencia, M. E. (2012). Online peer-to-peer advice in Spanish *Yahoo!Respuestas*. In H. Limberg, & M. A. Locher (Eds.), *Advice in discourse* (pp. 281–305). Amsterdam: John Benjamins.

Prestin, A., & Chou, W. Y. S. (2014). Web 2.0 and the changing health communication environment. In H. E. Hamilton, & W. Y. S. Chou (Eds.), *The Routledge handbook of language and health communication* (pp. 184–197). London: Routledge.

Pudlinski, C., (2008). Encouraging responses to good news on a peer support line. *Discourse Studies*, *10*(6), 795–812. http://dx.doi.org/10.1177/1461445608098203.

Richardson, K. P. (2003). Health risks on the internet: establishing credibility on line. *Health, Risk and Society*, *5*(2), 171–184.

Sillence, E. (2010). Seeking out very like minded others; exploring trust and advice issues in an online health support group. *International Journal of Web Based Communities*, *6*(4), 376–394.

Sillence, E., Briggs, P., Peter, H., & Fishwick, L. (2006). A framework for understanding trust factors in web-based health advice. *International Journal of Human-Computer Studies*, *64*(8), 697–713.

Veen, M., te Molder, H., Gremmen, B., & van Woerkum, C. (2010). Quitting is not an option: an analysis of online diet talk between celiac disease patients. *Health*, *14*(1), 23–40.

Health information in bits and bytes: considerations and challenges of digital health communication

Clare Tobin Lence, Korey Capozza
HealthInsight, Salt Lake City, Utah

13.1 Introduction

HealthInsight, a community-based nonprofit whose purpose is to improve the quality of health care in Utah, Nevada, and New Mexico, has expertise in communicating complex health information to consumers and patients using a variety of technology platforms, including Internet and mobile phone.

For this chapter we highlight our experience with three innovative, technology-driven efforts, all of which were targeted to consumers and patients in the nonclinical setting. We make a deliberate distinction between consumers and patients, and the difference is subtle. Here we use Gruman and colleagues' definition which defines individuals as "consumers" when they are engaged in making decisions about obtaining health care (such as choosing a health plan) and defines individuals as "patients" when they are "interacting directly with health-care providers and services about personal health concerns" (Gruman et al., 2010).

The three programs discussed in this chapter all relied upon digital technology to convey important, but sometimes complex, health information to consumers and patients. One of these uses text message technology to send health behavior messages about how to self-manage a chronic disease. The second program uses a Web site platform to provide interactive information to help consumers select among health-care providers based on objective quality data. And the third effort was a research project to investigate how patients with a difficult, poorly understood disease use social media platforms to communicate with similarly diagnosed patients. Below is a more detailed description of the programs referenced in this chapter.

13.2 The health programs

13.2.1 Care4Life

Care4Life is an interactive text messaging program for type 2 diabetes health behavior support designed to address self-management challenges outside of the clinic setting in the period between physician office visits. Our team tested Care4Life with patients identified through 18 primary care clinics in the context of a community-wide diabetes improvement project.

Care4Life is driven by a computer algorithm which sends health-related messages to patients via text (SMS) message. The program encourages patients to track their blood sugar, take their medications, and achieve exercise and weight loss goals through daily reminders and personalized messages of encouragement. The program takes advantage of the ubiquity of cell phones and the immediacy they afford, as owners tend to have them on hand at all times. Care4Life is available in both English and Spanish.

To help increase engagement with the program, the computer-generated messages come from a "virtual coach" named Paula, who addresses users by name. Patients can reply to prompts for blood sugar readings, medication-taking behavior, and progress toward weight loss and exercise goals via a structured text message reply. Three hundred patients with poorly controlled diabetes were enrolled in the program between March and December 2012, 76 of them as part of a formal research study.

The rationale for implementing this program comes from a growing body of research which has shown that health-related text messaging programs can bring about behavior change to improve short-term diabetes management and clinical outcomes (U.S. Department of Health and Human Services (HHS), 2014). Our insight into patients' interactions with the program, obtained from the data they sent into the program, yielded many useful insights which will be discussed in more detail below.

13.2.2 *UtahHealthScape*

UtahHealthScape.org was launched in July 2011 with the goal of communicating actionable information to health-care consumers in the state of Utah. Most resources currently on the site are targeted to helping consumers find and select high-quality health-care providers. The site provides a near-comprehensive directory of Utah physicians (5500) and clinics (2200). It presents performance scores on measures of health-care quality for hospitals, health plans, nursing homes, and home health agencies; these scores are determined using data collected by the Centers for Medicare and Medicaid Services (CMS) and the Utah Department of Health.

Although UtahHealthScape's quality information is similar to that found on the Medicare Compare sites (Centers for Medicare and Medicaid Services (CMS), 2014), we endeavored to create a more consumer-friendly display, achieved primarily by developing summary scores that give a high-level view of health-care provider quality, with the option for users to click through to other pages to view greater detail. For example, on the main quality page we present a summary score for hospital surgical care, but a user could click through to find the specific score for how often a given hospital administers antibiotics at the appropriate time before surgery. Other features, such as a simplified, appealing visual design, and concise explanations of common health concerns and conditions, offer additional usable information for consumers. The complete site is also available in Spanish. In addition to helping individual consumers find better care, the public, transparent nature of UtahHealthScape is expected to encourage providers to improve their care quality by creating conditions conducive to accountability.

13.2.3 Crowdsourcing social media commentary

The third project discussed in this chapter relates to aggregating patient insights from online communities to better understand coping and disease management behaviors in the real world. For this project, we looked specifically at how patients with severe atopic dermatitis (or eczema)—a recurring, chronic skin condition with a very negative impact on quality of life—used social media to get information about the disease, discuss treatments and management options, and find social support. Patient comments used for this project were anonymous, already in the public domain, and masked by the statistical method used which translated the individual inputs (or comments) into aggregated word counts and topics.

Seven Web sites were used for this project:

Inspire, www.inspire.com
Topix, www.topix.com
The Experience Project, www.experienceproject.com
Baby Center, www.babycenter.com
MedicineNet OnHealth, www.onhealth.com
MedicineNet, www.medicinenet.com
eMedicineHealth, www.emedicinehealth.com

Using natural language processing and topic modeling, a computer technique and statistical approach that allows for parsing and analyzing real-world text in an automated, reproducible way, we analyzed over 30,000 patient comments from these seven social media sites (above) and aggregated the unstructured, individually driven insights into a more powerful community-level view. The objectives of the project were to understand priorities and issues of importance to patients and caregivers and to characterize the range of treatments used by individuals with the disease. We created a software program to analyze the comments, calculate counts of prevalent words and phrases, and generate a list of dominant topics from the aggregated posts. For example, prevalent topics derived from the comments include sleep issues, psychosocial issues related to living with the disease, and concerns about steroid side effects (Capozza, 2015). This technique allowed for a revealing look into how members of a specific group of patients communicate with each other outside the clinical environment using social media to share their experience and feelings related to living with a recurring, poorly understood chronic condition.

These three programs were unique in terms of strategy employed, technology utilized, and participants targeted. Yet we found several cross-cutting themes emerge in the course of our experience implementing or observing them in the community setting. Below we discuss these themes and highlight specific examples from each program to illustrate the practical considerations and ethical questions that may be relevant to future efforts to communicate health information outside the clinical setting using digital technology.

13.3 The digital divide

I do not fear computers. I fear the lack of them.
                                    Isaac Asimov, Time Magazine (1978)

A key drawback with technology-mediated communication is the potential to exclude some consumers and patients, and possibly those most in need of assistance, due to technology access barriers. Though ownership and use of mobile phones and computers is widespread across the US population, disparities in access to devices and the Internet remain. For example, among Americans who earn less than $10,000 per year, 42% have access to broadband Internet at home; in contrast 90% of those who earn $100,000 or more do (Rainie, 2013). Age, rural residency, disability, and Spanish language preference also predict barriers to broadband Internet access (Rainie, 2013). In terms of digital health information delivery, those with a broadband connection were more likely to use the Internet for health-related information seeking and communication than those with a dial-up connection (Rains, 2008). Notably, despite a significant rural population spread across a large, mountainous geography, Utah has one of the highest rates of broadband access in the nation: only 1.8% of the total population lacks access, though not all those with access subscribe to Internet service (Federal Communications Commission, 2012, p. 77). Nonetheless, divide issues were apparent in our work on the three health information delivery tools that we describe in this chapter.

Cell phone ownership is ubiquitous at this point, with 91% of US adults reporting having a mobile phone (Duggan, 2013, p. 1). Unlike broadband access, cell phone ownership rates are high across categories of income, education, and race. However, older adults still lag behind: while 97% of 18- to 24-year olds own a cell phone, 76% of those aged 65 and over do (Rainie, 2013). This age difference was apparent in our Care4Life pilot. The program was available to adults over age 18, but those who signed up for the program were largely middle aged (average 54 years, max 78, min 25). Approximately one-third of patients invited to try the program declined participation because they did not own a cell phone. A similar proportion (one-third) indicated that they did not have access to a computer or the Internet at home. While some could have enrolled in the program using a publicly available or family member's computer, few who considered this option elected to do so, as evidenced by their enrollment status.

Among cell phone owners, access to texting technology is widespread and this functionality is highly utilized: 81% of cell phone users send and receive text messages (Duggan, 2013, p. 4). However, as with cell phone ownership in general, age differences are apparent. While 18- to 24-year olds demonstrate prolific use of this feature, older Americans are less likely to use it (see Figure 13.1) (Duggan, 2013, p. 4).

In addition to access barriers, we found that a significant number of Care4Life patients had difficulty signing up for the program due to technology proficiency barriers. Our staff members sometimes spent up to an hour on the phone helping patients navigate a basic online enrollment form that required entry of common personal information into a limited number of structured fields. We also taught several patients how

Text messaging
% of cell phone owners who send or receive text messages

All cell phone owners (n=2076)		81%
a	Men (*n=967*)	81
b	Women (*n=1109*)	81
Race/ethnicity		
a	White, Non-Hispanic (*n=1440*)	79
b	Black, Non-Hispanic (*n=238*)	85
c	Hispanic (*n=225*)	87^a
Age		
a	18–29 (*n=395*)	97^{bcd}
b	30–49 (*n=557*)	94^{cd}
c	50–64 (*n=594*)	75^d
d	65+ (*n=478*)	35
Education attainment		
a	No high school diploma (*n=144*)	71
b	High school grad (*n=565*)	77
c	Some college (*n=545*)	85^{ab}
d	College + (*n=799*)	86^{ab}
Household income		
a	Less than $30,000/yr (*n=504*)	78
b	$30,000–$49,999 (*n=345*)	80
c	$50,000–$74,999 (*n=289*)	88^{ab}
d	$75,000+ (*n=570*)	88^{ab}
Urbanity		
a	Urban (*n=711*)	82^c
b	Suburban (*n=965*)	82^c
c	Rural (*n=398*)	76

Source: Pew Research Center's Internet & American Life Project Spring Tracking Survey, April 17 – May 19, 2013. N=2076 cell phone owners. Interviews were conducted in English and Spanish and on landline and cell phones. The margin of error for results based on all cell phone owners is ± 2.4 percentage points.
Note: Percentages marked with a superscript letter (e.g., ^{a}) indicate a statistically significant difference between that row and the row designated by that superscript letter, among categories of each demographic characteristic (e.g., age).

Figure 13.1 Number of texts sent per day, by group (Duggan, 2013, p. 4).

to use the text messaging features on their phones, providing tutorials on how to use number keys to type words, and how to find the "inbox" on their screens. Notably, after patients mastered the basic steps required to send and receive text messages, they appeared to have no trouble using the program.

Using Web sites to find health information is a challenge for those without home Internet access, as it necessitates travel to a library or other source. Even for those with a home dial-up connection, Internet searching is tedious and sites like UtahHealthScape, with large databases and complex features, are slow to load, likely leading to frustration for the user. However, while the digital divide related to home Internet access is larger than that related to mobile phones, the rise of Internet-enabled mobile phones has begun to narrow that gap. Sixty percent of Americans access the Internet on their mobile phones and 34% go online primarily via mobile phone (Duggan, 2013, p. 5). Modifying UtahHealthScape for mobile view is an issue we have yet to solve, primarily due to the need to reformat the large amount of information currently provided in the Web version for a small screen, an undertaking which requires difficult decisions about which information to exclude. The ethical implications of selecting certain information for display at the expense of not showing other information is a concern we discuss at length later in this chapter.

In developing and disseminating UtahHealthScape and Care4Life, we aimed for inclusiveness in our target audiences. For our study of Care4Life, we recruited participants from primary care clinics that were known to have poorly controlled diabetes. Many of these recruits were demographically similar to populations that are more likely to lack technology access and proficiency, as they were older and in poorer health. We found that some technology-related barriers were surmountable, such as learning to use text message functionality, but others were not, such as owning a mobile phone. In promoting UtahHealthScape, we presented the tool to many groups that work with underserved populations, all of whom were excited about its potential utility. However, Internet access and knowledge of how to navigate a Web site are prerequisites to engaging with the site. Despite considerable effort spent reaching out to populations that are less likely to have technology access, we did not observe an increase in traffic as a result of these efforts, which could be related to Internet or computer access challenges faced by underserved communities.

As noted, older, low-income Americans are less likely to have access to technology as compared to their younger, higher-income counterparts; they also bear a higher burden of chronic disease, including diabetes (Centers for Disease Control and Prevention (CDC), 2013). Therefore, those most in need of support for self-managing complex medical conditions and making informed health-care decisions—the goals of Care4Life and UtahHealthScape, respectively—may be the least likely to see a benefit from technological solutions. This mismatch of need and access is an important ethical issue in the development of digital health interventions: do these tools fail to help populations with the greatest need, thereby exacerbating health disparities?

Our experience with the Care4Life program provides some hope that such tools can help ameliorate, rather than exacerbate, the digital divide. Motivated participants in the Care4Life program overcame significant initial barriers to become highly engaged users of the tool. When participants consented to join the study, but failed to fully enroll in the program by replying to a "welcome" text, we called them and, when asked, provided one-on-one coaching to help them complete this step in the enrollment process. This demonstrated that some up-front personal effort facilitated access for individuals that we might have assumed would not be capable of using Care4Life. A key challenge for future efforts will be to identify what level of assistance is required for different types of patients, in order to maximize access while conserving staff resources.

Indeed, addressing access and ability barriers should be a consideration with any health information technology intervention. For example, future cell phone-based programs should consider providing basic, low-cost phones to patients of low socioeconomic status; in the case of online resources such as UtahHealthScape, training staff in community centers and libraries to assist consumers is another opportunity, and one that we are beginning to explore.

Additionally, the digital divide continues to narrow, and over time, today's young, tech-savvy consumers and patients will be tomorrow's older users, potentially obviating a digital divide that is based on age. Differences based on demographic characteristics may persist, however. These digital divide considerations create a dilemma for program selection in terms of choosing between the latest, cutting edge technology and a platform that reaches more types of users. We confronted this issue when we were considering a mobile phone-based intervention in 2011. Although smartphones and "apps" were gaining popularity, we chose a simple, text messaging tool because it was more widely available and typically a feature included on very low-cost phones.

It will be important for developers of future tools to stay abreast of the progression of technology and how individuals with different cultural and economic backgrounds will utilize it, such as computers versus tablets versus Internet-enabled mobile phones, or whatever technology platform the future will bring.

13.4 Don't make me think

> *Real people have trouble with long division if they don't have a calculator, sometimes forget their spouse's birthday, and have a hangover on New Year's Day.*
>
> *Thaler and Sunstein (2008)*

The health behaviors we discuss in this chapter—self-managing a complex disease (Care4Life) and selecting a health-care provider based on metrics of quality (UtahHealthScape)—are not simple. Patients and consumers often struggle with these behaviors, which is why, in part, assistive technologies have emerged to address that need. However, designing and implementing an effective tool demands consideration of *why* individuals struggle with healthy behaviors.

Each of these behaviors involves decision making, whether the decision is to count carbohydrates at lunch or choose a high-quality nursing home for one's elderly parent. Making the healthy choice at a complex decision point requires (1) the ability to understand health information and (2) engagement in an effective decision-making process. The first criterion relates to "health literacy," defined as "the degree to which individuals have the capacity to obtain, process, and understand basic health information and services needed to make appropriate health decisions" (Nielsen-Bohlman, Panzer, & Kindig, 2004, p. 1). Low rates of health literacy in the US suggest that much of the population has limited ability to understand and apply information to a health decision. The other key issue, related to the second criterion, is the limited capacity of humans for rational decision making. Decision research has illuminated the fact that individuals often do not use full conscious attention to make decisions, relying instead on heuristics; although heuristics generally function as "sensible estimation procedures," they can lead individuals to make health-related choices that are not in their best interest (Gilovich & Griffin, 2002, p. 3). Effective decision making is particularly challenging for individuals when the decision involves a large number of options, which can lead to a state of cognitive overload. Many health-related decisions involve a variety of choices, making the use of heuristics even more likely—if individuals make a decision at all (Anderson, 2003, p. 158; Ariely, 2000; Olshavsky & Granbois, 1979; Peters, Dieckmann, Dixon, Hibbard, & Mertz, 2007). Both health literacy and decision-making theory considerations informed how we addressed the practical and ethical challenges of delivering health information through the Care4Life and UtahHealthScape tools.

13.4.1 Health literacy

A prominent speaker on health literacy issues illustrates the atypical use of language that is common in the health field:

> *When a clinician says, "We need to draw some blood," there likely is no need for crayons. When hearing your lab tests are "unremarkable," that is likely good. But if results come back "positive," you might have cause for concern.*
>
> Osborne (2014)

Osborne's examples demonstrate the difference between basic educational literacy and health literacy, and show how even essential health-related concepts may lead to confusion among those with low health literacy. To successfully navigate complex health decisions, patients and consumers need health literacy skills. They also need health numeracy skills. Numeracy is "the ability to comprehend, use, and attach meaning to numbers" (Levy, Ubel, Dillard, Weir, & Fagerlin, 2013, p. 107); those with low health numeracy are less able to take medications correctly (Peters et al., 2006) and to successfully manage chronic disease (Cavanaugh et al., 2008), although the exact mechanism by which this connection operates is not clear (Levy et al., 2013). Comparative research shows that individuals struggle more with health numeracy than other types of numeracy, including financial (Levy et al., 2013).

Health literacy and numeracy rates in the US are low. In 2004, the Institute of Medicine reported that 90 million Americans cannot interpret or act upon complex health information (Nielsen-Bohlman et al., 2004). A mere 12% can accomplish complicated tasks such as determining the employee share of health insurance costs from a table (National Center for Education Statistics, 2006). This issue is compounded by physicians' limited ability to assess health literacy. Evidence shows that physicians vastly overestimate the health literacy of their patients. A 2003 study found that physicians thought 89% of patients understood medication side effects after an explanation; in fact, only 57% of patients understood (Kessels, 2003).

Our awareness of low health literacy has been salient in the development and ongoing enhancement of UtahHealthScape. As we discuss further in Section 13.7, available health-care data tend to be very clinically oriented and complex even for expert audiences. Although we have received positive feedback from local community groups and endorsement from multiple federal agencies regarding the strong consumer orientation of our site design, we recognize that it poses not insignificant challenges for those with low health literacy. As such we continue to conduct focus group and usability testing to improve the site.

Our summary page for care quality ratings (see Figure 13.2) is designed to be the simplest view of the health information. The precise numeric summary score can be seen by hovering over a symbol with the mouse arrow. The summary page shows composites of multiple single measures—such as whether heart failure patients received symptom management instructions upon hospital discharge—providing an

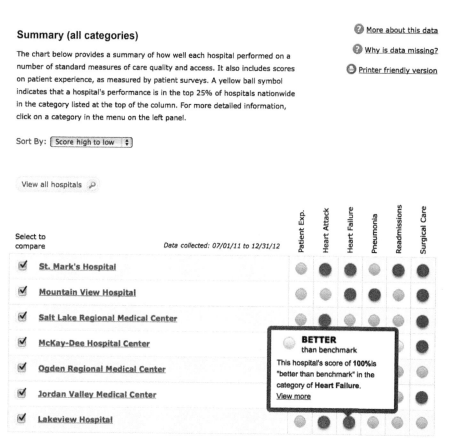

Figure 13.2 Summary page for hospital quality scores on UtahHealthScape.org, hovering to view the numeric score.

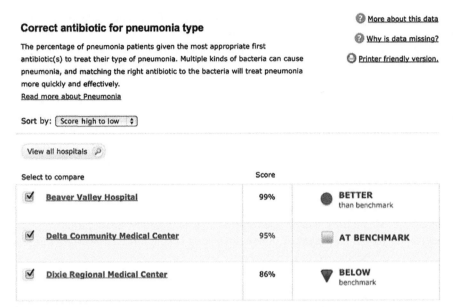

Figure 13.3 Single pneumonia measure for hospital quality scores from UtahHealthScape.org.

overall score for heart failure and other main categories like heart attack, surgical care, and patient experience (see Figure 13.3 for an example of a single measure). This summary level display strategy is supported by the quality reporting literature (Sofaer & Hibbard, 2010b; Vaiana & McGlynn, 2002). A yellow ball (shown in dark gray in Figures 13.2 and 13.3) indicates that a facility's score is in the top 25% nationwide. Thus, the yellow ball represents a high score and the absence of a yellow ball denotes a low score, effectively dichotomizing the ratings for easier comparison. All pages sort facilities from high score to low score by default, reducing the cognitive load of comparison (Sofaer & Hibbard, 2010b). Nonetheless, this display requires sophisticated chart reading skills: users must first read the explanatory paragraph to understand the meaning of the symbols, then the column labels; they must also understand how to read across and down from the hospital and measure of interest to find the correct symbol and interpret it. For many consumers, interpreting chart-based information displays is not obvious (National Center for Education Statistics, 2006).

Consumers with high health literacy who desire information beyond the summary scores can click through to other, "deeper" pages on the site. Figure 13.3 shows a page from UtahHealthScape for a single measure of pneumonia care. Along with the more detailed measure, the quality scoring is slightly more detailed as well. The yellow ball remains but at this level there are three performance categories: the top 25% (yellow ball), the middle 50% or interquartile range (gray square), and the bottom 25% (purple triangle). Although precise percentage scores are visible, the symbols serve to ease interpretation for consumers as to whether a facility's score is good, fair, or poor. The literature suggests that even consumers with high health literacy benefit from clear, simplified presentation (Hibbard, Greene, & Daniel, 2010).

For individual measures, the descriptions published by the CMS are appropriate to clinical stakeholders but not consumers (Sofaer & Hibbard, 2010a). Our team, with early input from a nurse, worked to modify the descriptions into succinct, common language while maintaining accuracy. Consistent with public reporting literature that describes the need to create context or frame a complex new idea (Hibbard & Peters, 2003), we reasoned that our interpretations of these measures should delineate why consumers should care about the measure. Our efforts are illustrated by the original and consumer descriptions below, which describe how a hospital performed on a single measure of care for pneumonia patients (see also Figure 13.3).

Original measure: Immunocompetent patients with community-acquired pneumonia who receive an initial antibiotic regimen during the first 24 h that is consistent with current guidelines.

UtahHealthScape consumer version: The percentage of pneumonia patients given the most appropriate first antibiotic(s) to treat their type of pneumonia. Multiple kinds of bacteria can cause pneumonia, and matching the right antibiotic to the bacteria will treat pneumonia more quickly and effectively.

The consumer version is clearer, but it is not simple: site users must understand percentages and the meaning of pneumonia, antibiotics, and bacteria, at a minimum. Still, many consumers have a basic understanding of these medical terms, illustrating that commonly used measures of readability—based on the text length and number of syllables in words and the number of sentences—are weak predictors of the readability of health information (Kim et al., 2007). Communicating health information clearly is best assured through consumer testing (Sofaer & Hibbard, 2010b). Although we have leveraged focus groups and user testing in the past, we have come to recognize the need for more frequent, if less formal, testing; our organization now has a standing Patient and Family Advisory Council that informs our projects, which will enable us to obtain regular feedback without the costly investment of participant recruitment.

Our efforts to develop UtahHealthScape as a usable tool even for those with low health literacy—using summary scores, dichotomized presentation, symbols to clarify the meaning of scores, default high-to-low display, and measure descriptions that are simplified and anchored in consumer concerns—provide a more understandable version of health-care data to the consumer. However, the decisions we have made to simplify and clarify the site have also obscured additional information and nuance that is present in the raw data. We have made decisions about what aspects of the data are most valuable and emphasized those aspects over others. Despite our best intentions, and a process that includes input from consumers and numerous industry stakeholders, translating health-care data for a nonexpert audience is inherently paternalistic.

The ethical concern of paternalism is also germane to the next section on decision making, where we will discuss it further.

Health literacy was a less prominent issue with the Care4Life program, which includes health content recommended by the American Diabetes Association, but written in short, simple messages at the fifth grade reading level using the Flesch–Kincaid metric (Flesch, 1948; Kincaid, Fishburne, Rogers, & Chissom, 1975). Sample messages read:

> *Take all of your medicines, even if you feel well. It can be dangerous if you skip them. Tell your doctor if you have trouble with any of your meds.*

Still, the more sophisticated features of the program, including an online personal portal, went largely unused. Numerical responses sent into the system by users were stored in this portal, where patients could view their trends over time in a graphical format, make associations between their behaviors and test results, and print their data to bring to visits with their clinical care team. Their personal blood glucose trends over time were line-graphed and classified into color-coded "high-," "ideal-," and "low-" range bands to allow for a quick, snapshot view of progress (Figure 13.4). However, only eight of the 76 users logged into the Web portal, and only two did so more than once. It is possible that this low usage was due to health literacy or numeracy obstacles, to the added time and effort involved, or to a lack of technology proficiency.

The Care4Life program also required deliberate, thoughtful translation of clinical guidelines into short reminders, feedbacks, and tips. Messages were based on the Social Cognitive Theory of behavior change (Bandura, 2001), which focuses on goal setting and self-regulation. Care4life messages were developed, reviewed, and endorsed by a team of diabetes educators and included input from physicians, including endocrinologists. Medical terminology was avoided when possible, and accompanied by an

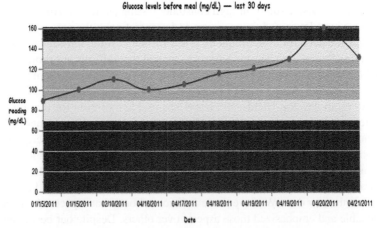

Figure 13.4 An example of blood glucose levels trended over time displayed in the Care4Life portal.

explanation when included. In most cases, each message was under 160 characters, but some more complicated ideas were extended over two, 160-character (or fewer) messages.

Crafting such messages can be similar to translating a wordy clinical guideline into a haiku poem. Distilling such information into an accurate, reader-friendly, 160-character message requires careful trade-offs between accuracy, understandability, and length, and is an exercise that requires skill and understanding of the mode of communication, the intent of clinical guidelines, and the health literacy level of users. In our experience, the Care4Life messages struck the right balance. Patients found these pithy reminders and messages informative and valuable. We administered a satisfaction survey (the Client Satisfaction Questionnaire-8) to Care4Life users at the conclusion of the study period and 94% responded that they would recommend the program to other patients with diabetes (Capozza et al., 2015; Larsen, Attkisson, Hargreaves, & Nguyen, 1979).

13.4.2 Decision making

Effective individual decision-making processes are also essential to the choice of healthy behaviors. Decision science demonstrates that individuals frequently do not employ rational decision making, even for important decisions; overcoming, and in certain cases leveraging, the ways that individuals commonly make choices was central to addressing the practical and ethical challenges we faced in developing the tools and programs discussed in this chapter.

Despite the persistence of the "rational actor" theory across disciplines and in the popular consciousness (Gilovich & Griffin, 2002, p. 1), decision science suggests that individuals generally exhibit "bounded rationality" (Simon, 1997, pp. 291–294). Although they may carefully consider certain decisions, individuals often make both incidental and complex decisions with the use of heuristics, rather than deep contemplation (Tversky & Kahneman, 1974). Decision-making heuristics are actually sophisticated processes that allow "quick and dirty" estimation (Gilovich & Griffin, 2002, p. 3); they are adaptations that allow individuals to progress through the numerous decisions of everyday life with efficiency, but they can lead to imprecise interpretations and potentially to harmful choices. One prominent theory posits that there are two tracks of thinking, one quick and automatic, the other slow and deliberate. The fast process, termed "System 1," operates continuously, but individuals can override it with the conscious reasoning of "System 2" (Stanovich & West, 2000, p. 658). System 1 relates to quick judgment through heuristics and System 2 relates to conscious, thoughtful decision making.

Individuals are more likely to engage in suboptimal decision making under certain circumstances, for example, emotional situations. Studies of brain activation patterns show that elevated activation in the amygdala, the fear center of the brain, aligns with study subjects making less reasoned choices (De Martino, Kumaran, Seymour, & Dolan, 2006). Research further shows that when presented with incomplete information in a decision context, subjects show brain activity consistent with fear and a lack of expected reward, suggesting that individuals faced with ambiguity may also be less capable of engaging in rational decision making (Hsu, Bhatt, Adolphs, Tranel, & Camerer, 2005). The health field is, unfortunately, rife with incompleteness and ambiguity: The Agency for Healthcare Research and Quality has reported that 40–80% of medical information explained to patients is forgotten immediately and half of what is retained is incorrect (Agency for Healthcare Research and Quality (AHRQ), 2011). However, too much information is similarly problematic. Cognitive overload due to an overwhelming amount of information can increase reliance on heuristics, with the ultimate effect of diminishing decision-making capacity (Ariely, 2000; Olshavsky & Granbois, 1979; Peters, et al., 2007). Too many options also increase the likelihood of complete avoidance of a decision (Anderson, 2003, p. 158). When individuals encounter situations that are emotional, ambiguous, or cognitively burdensome, they are more likely to rely on quick decision making through heuristics, or to avoid making any decision at all.

Our interventions to provide health information to patients and consumers presume a certain level of active, carefully reasoned decision making on the part of those individuals. Indeed, individuals are unlikely to find these tools without actively choosing to seek them out (crowdsourcing, UtahHealthScape) or first agreeing to participate (Care4Life). Nonetheless, the challenges to effective decision making identified by the decision literature are notable. In addition to addressing these challenges in the development of our tools, at times we also leveraged common decision-making heuristics to our advantage, with the goal of minimizing cognitive burden and facilitating effective use of the tools.

Earlier, we discussed providing context for UtahHealthScape measures in the form of an additional sentence in the description that suggests why a consumer should find the measure relevant; the example provided was "multiple kinds of bacteria can cause pneumonia, and matching the right antibiotic to the bacteria will treat pneumonia more quickly and effectively." This technique utilizes the heuristic of "anchoring," by associating something unknown with something known (Tversky & Kahneman, 1974, pp. 1128–1129). Further, the description implies that the wrong antibiotic could have negative health outcomes. Fear of potential negative health outcomes may reduce decision-making capacity, but it also encourages the consumer to consider that performance scores more seriously (Shaller, Kanouse, & Schlesinger, 2011).

Minimizing cognitive overload was crucial to UtahHealthScape's development. We recognized that too much information could overwhelm the consumer and cause her/him to abandon any efforts to glean information from the site. From our perspective as data experts, it was a challenge to avoid including confidence intervals around performance scores and a multitude of footnotes to elucidate data inconsistencies, which is a common struggle among creators of health-care quality Web sites (AHRQ, 2010). Ultimately, we balanced minimizing cognitive overload with transparency by including detailed methodology on a Web page that was separate from the quality scores but still easily accessible. The methods page also serves to enhance site credibility, even if consumers do not review the information in detail (American Association of Medical Colleges, 2014). Trust in an information source is necessary to encourage its use, and improves the likelihood of effective decision making by users by reducing the fear response that comes with uncertainty and ambiguity.

Aspects of decision theory have helped us improve the usability of UtahHealthScape, despite the inherent challenge of presenting complex information to consumers. However, our efforts to streamline the decision-making process, to guide consumers to the "best" choice of health-care provider, raise ethical concerns of paternalism. We gather input from focus group consumers and provider representatives, but ultimately we make the decisions about what information is emphasized, what is de-emphasized, and what is omitted completely. We leverage, to a small degree, consumer fears about the health-care system to encourage them to value and use the information on the site. We aim, though, for a "libertarian paternalism" that structures decisions to guide consumers toward a given choice, but does not remove any choices (Thaler & Sunstein, 2003). As an example, we present information on all health-care facilities with available data rather than presenting only the top performers.

Similarly, and as discussed above, the Care4Life program was designed to minimize cognitive strain on patients by simplifying health behavior recommendations into short, engaging educational messages, tips, reminders, and encouragement messages that could be conveyed in 160 characters or fewer. The messages follow a reliable, formulaic structure and require simple, consistent responses from the user in a numerical format. For example:

Try to exercise at least 5 days each week unless your doc has told you not to. Reply with the number of days you will exercise this week (e.g. 4).

It's time for your weekly weight check-in. Reply with your current weight in pounds (e.g. 180).

For users who desired more information, many messages included Web links and/or the option to reply to the message to get additional resources or information. For example:

No one manages diabetes perfectly. It is a learning process. Learn more about the causes & symptoms of high blood glucose at care4l.com/g02.

A before meal glucose reading of 70 or less can be dangerous. Text LOW or go to care4l.com/g03 to learn how to treat lows. Tell your Dr. about lows.

However, patients may not follow through on invitations to visit a link or contact a provider, and information that is more prominently communicated is likely to reach a wider audience. Behind the scenes, crucial decisions were made about what information to feature in core messages versus buried in a link or follow-up message, similar to the choices about what information to include and to emphasize on UtahHealthScape. A simplified communication strategy that reduces cognitive effort is essential to effectively engage consumers and patients (Hibbard & Peters, 2003), but in doing so we must be aware of and regularly reflect upon how our interpretation and actions might change the essential meaning of that which we are trying to convey. Again we note the need for "libertarian paternalism," for setting up choices in a way that guides individuals toward the healthy behavior—such as finding more information about managing blood sugar levels or communicating with their physician—but leaves them the autonomy to choose to engage in that behavior or not (Thaler & Sunstein, 2003).

13.5 Humanizing technology

I am putting myself to the fullest possible use, which is all I think that any conscious entity can ever hope to do.
Hal-the-Computer, Kubrick (1968)

Health information delivery, until recently, has been a domain governed by human relationships, especially the physician–patient relationship, and characterized by face-to-face interaction. People caring for others have always been at the heart of health care.

Only recently has the provision of health information moved beyond this model to include technology-driven options. Digital technology has enabled tools and programs that allow health consumers and patients to seek health information from various sources on their own schedule and at their personal discretion. This is a key departure from the past, when health information might have been exchanged strictly during a clinical encounter, and primarily flowed in one direction, from the physician to the patient.

Though not a replacement for human interactions, health information communication through digital media has several powerful advantages. First, it is cost effective. Even if the cost to develop a new tool is high, the cost of deploying that tool across a large population is low, especially compared with the cost of traditional disease management or education. Further, costs can be spread over large populations allowing for decreasing per person costs as wider dissemination occurs.

From the consumer and patient perspective, health information presented via technological interventions, such as information communicated by UtahHealthScape and Care4Life, and shared on social media sites, can offer a level of convenience that is not possible with traditional health-care interactions. Users of these programs can control the frequency and type of the information that they receive. They can seek the information at any time of day, in their pajamas if they wish. Instead of a 15-min in-office consultation, during which many forget the questions they wanted to ask, consumers can take their time searching, reading, and digesting information; if they forget something they learned, they can access the information again at a future time. While physician office visits must be scheduled in advance at a fixed time, tools such as UtahHealthScape, Care4Life, and social media sites can be accessed during times of need, at the patient's discretion. Given these characteristics, they are more patient-centered than face-to-face meetings with a provider, and more convenient than a trip to a physical library.

The consistency offered by technology interventions can be appealing to some consumers and patients because there is less room for bias and improvisation, which may color an interaction with a human. In general, patients value the opinion of their physician, but they may also value the objectivity provided by a standardized information source: Care4Life, UtahHealthScape, and social media sites all provide the same information to all users. Scientific evidence shows that physicians, like other humans, have biases (e.g., Fincher et al., 2004; Khan, Plummer, & Minichiello, 2008). Physicians may provide more information to patients they, consciously or unconsciously, perceive as intelligent, for example. In focus groups we conducted for UtahHealthScape, some participants expressed skepticism about the motivation behind physician referrals and a preference for more information about options when choosing a specialist.

Both UtahHealthScape and Care4Life, as technology tools, obviate the many social and emotional cues that accompany in-person health-care interactions. One would think that this aspect is a disadvantage, but this absence of typical cues can actually have a positive effect by "democratizing" relationships and reducing hierarchies that serve as barriers to communication, such as those observed in the physician–patient relationship (Kim, 2000). As Kim notes, "the relative lack of social status cues renders electronic communication more democratic, providing a voice for the voiceless, as people forget their social position, appearance, age, race and even gender. In addition, the absence of social barriers makes people express themselves more openly" (Kim, 2000, para. 4).

Indeed, digital communication may be preferred by patients because it allows for this anonymity, privacy, and discretion. For example, information shared on social media sites is surprisingly confessional and revealing. Patients may share more with each other online or with a computer program than they would with a care provider. Research on patient reported outcome measures for medication adherence, in which patients answered questions on a tablet while waiting to see their physician, suggests that self-reports on the tablet are more accurate than physician judgments of adherence, and that patients are more honest with the tablet questionnaire than with their physician (McInnes et al., 2013).

We observed this phenomenon in the Care4Life study. One patient, who had rebuffed advice from his physician and refused in-person assistance with managing his diabetes, was in contrast a very robust user of the text message program, logging blood sugar data at regular intervals. While it is hard to say why this patient responded better to the digital version of diabetes education and coaching, it seems likely that some patients prefer the neutral relationship with a digital platform to the more hierarchical, emotionally laden interaction with a human.

While the "neutrality" of digital communication can be a positive characteristic for some patients and consumers, others may find it a drawback, especially in the context of emotional or difficult situations. Patients may prefer personalized guidance, support, or empathy to the sometimes cold, impersonal interaction with a computer-driven program. For example, consumers may want the recommendation of a neighbor, communicated both verbally and nonverbally, when choosing a doctor, not just "data" such as the objective quality ratings available on UtahHealthScape (Associated Press-NORC Center for Public Affairs Research, 2014). Patients with diabetes may need to feel encouragement from a real person, who can convey caring or warmth, in order to be motivated to change health behavior. Communication theory supports this idea and posits that computer-mediated communication conveys very little "social presence"—the ability to transmit information about facial expression, direction of looking, posture, and nonverbal cues (Papacharissi & Rubin, 2000)—and thus it has a more limited impact.

However, in our experience, the advantages and disadvantages of technology-mediated versus in-person communication are much more nuanced. In some circumstances face-to-face communication is not feasible, in which case computer-driven communication can fill the need for human connection. For example, patients with rare diseases are often isolated from peers who can provide social support and guidance on coping strategies, by virtue of the uncommon incidence of their condition (Fox, 2011). In a nationally representative survey, people living with a rare disease far outpaced all other groups, including those living with chronic conditions, in accessing information from online peer networks (Fox, 2011). As one respondent pointed out: "When a disease is so rare and there are no folks in your town, and few in your state who are going through what you are going through, you need a support group that encompasses people from all over the world" (Fox, 2011, p. 3, para. 4). Interestingly, we found that the patients who engaged with the Web-based patient communities studied in the Crowdsource project expressed a strong sense of "human" connection, indicating that there is likely a distinction between technology that enables human interaction (i.e., social media and communication technology) and technology that is more unidirectional, and/or anonymous (i.e., Web information).

Social media allows such patients to connect with a large group of peers in a way that would not otherwise be possible in person. The sense of moral support and information sharing from people "who truly understand" mimics the emotional and social support that might be gleaned from an in-person interaction. For example, in our Crowdsource project which analyzes the content of patient comments in online communities associated with eczema, one individual wrote, "I'm so glad I've found this forum and can talk to people who totally understand and can relate to my problems. It makes me really upset that no one understands. I hope you guys do." Another patient comment from a different site reads, "isnt it strange how absolutely alike al our emotions and reactions to this are ... if nothing else it proves that were nothing but perfecctly normal psychologically as we all come from completely different backrounds and all react and feel the same [sic]." Another stated, "I will say that I am very moved to know that I am not alone."

Others use social media sites to get "neighborly advice" on best treatments. As one patient posted, "Are there other people out there who have experienced Elidel [a prescription treatment for eczema] over a period of time. I would be interested to read of their trials as I have just started Elidel." Another shared some advice about a particular remedy in the hope that it would help someone else: "It has really improved the skin on my arms. No more dry red patches on my skin, I haven't had this much success in 15 years of trying to treat my skin. Maybe someone can benefit from my trial and error."

And whereas computer-mediated communication may have been more impersonal in the past, that's not the case today. The explosion of mobile phone programs has created more sophisticated tools and avenues for reaching patients. As previously noted, mobile phones are personal devices, often carried by owners at all times.

Communication delivered to a cell phone is typically directed at the owner, individually, rather than a group of people more broadly, as is the case with e-mail messages which include spam, advertisements, and bulk reply messages. Further, cell phone programs (apps, text message programs) can be customized and personalized to suggest a more intimate interaction. Because of these characteristics, some users of the Care4Life program made a personal connection with "Paula" the fictional, computer-generated coach who greeted users in some of the text messages. Indeed, one user actually called our office asking to talk with Paula, indicating that she believed Paula to be real. Another user described Paula as a "friend" who "cares about me." A third user sent angry, personal text messages back to the Care4Life program when he received an error message indicating that he had entered his blood glucose reading incorrectly. It is noteworthy that these users assumed there was a human behind the program when the consent and enrollment process, written at a sixth grade reading level, clearly stated the nature of the interaction (computer-driven). It could be that mobile phones are not typically used for impersonal communication (in the way that, say, e-mail, rife with spam, might be), and therefore the idea that messages can be sent by a computer is new and unfamiliar.

This confusion between fictional and real interactions obviously poses some ethical quandaries. If patients believe they are interacting with a human, do they interpret the advice differently? Care4Life provides fairly generic coaching and tips, but the personalized messages that are specific to an individual may be misleading. However, some affinity with Paula was likely necessary in order for patients to engage with the program. Clearly, the challenge is creating communication that draws patients into an emotional connection without creating an illusion that the computer-generated persona is human.

Understanding which patients prefer technology-mediated communication and which prefer or need human-to-human communication is a topic in need of further study. Targeting human approaches only to the patients that want or need that personal attention may be a more effective strategy than an expensive, blanket approach. Borrowing from the marketing industry, this segmented, multichannel approach might make the most sense given limited resources and variability in patient needs—a topic that we will explore next.

13.6 Know your audience

No two people share exactly the same universe.
Gaelic proverb

Health information delivery tools hold promise for supporting informed health-care decision making and positive behavior change. However, organizations that develop and deploy such tools often pay little attention to individual and group differences within the audiences they seek to target. Marketers of commercial products, on the other hand, know that products appeal to a certain subset of people, to different people in different ways, and to some consumers sooner than others. In some cases, they modify the product itself for different audiences; for example, iPhones are offered at different price points depending on their available storage size. Other products are progressively marketed to a larger audience over time, as occurred with Gmail, which was marketed to early adopters first, and later to a broader consumer audience. Market segmentation allows a product to be designed more specifically to the needs of a population, increasing its value to that group, and the chances that it will be adopted. Segmentation also facilitates the concentration of promotional resources, allows development and refinement of the product with smaller scale initial production, and creates a base of satisfied customers (Moore, 2006; Rogers, 2004; Smith, 1956; Wedel & Kamakura, 2000).

The patient and consumer focus groups we have conducted over the years have yielded important insights about the perceived value—and limitations—of our interventions. Similar to the framework outlined by Longo and Woolf (2014), we found that one key way in which patients differ is in their place on the "decision journey." For example, some consumers and patients may be trying to understand the basic terms and context of a condition (diabetes) or medical decision (assessing providers on quality) and may not be ready for a more active role. Some may ultimately decide to defer to their physician rather than engage in a complex contemplation of trade-offs.

Some of these distinctions became apparent in the Care4Life program. The intervention was targeted to a specific population: adult patients of primary care clinics with poorly controlled diabetes. Clinic staff recruited this group using data from their electronic health records, and therefore, the intervention could be precisely targeted to users' needs. Yet we still observed widespread variation in interest and ability to use Care4Life. Several patients declined to participate because they felt their condition was already well managed, even though their lab results had indicated otherwise. Several others simply stated that they were not interested in focusing on their diabetes at the moment and/or able to change their behavior due to competing priorities, including other health issues. From a market segmentation perspective, it makes sense to target the intervention and its promotion to individuals who are ready to change their behavior, but by excluding those who are not ready to change we may be overlooking groups that would most benefit from help. This is the danger in pursuing the receptive and "ready" market segment—resources and effort may be diverted from understanding the needs and motivating factors associated with the groups that are less willing, or able, to change behavior.

In the case of social media sites, audience segmentation is less of a concern, because users effectively "self-segment." Individuals with a specific diagnosis or health concern can find a community of peers that share very particular questions and insights of interest to a narrow group. For this reason, they are sometimes referred to as "empathic communities," because of the social cohesion and affinity demonstrated by the group (Preece, 1999). The nature of information sharing on these sites is highly specific. For example, one diabetes-related blog included a very detailed thread about how to hide an insulin pump while wearing a wedding dress. In our Crowdsource study of social media sites related to eczema, there are threads related to product reviews, dating issues, makeup tips, and coping with stress. Given this self-segmentation, online

Select to compare		Quality Award	Accepts Medicaid	Accepts VA Benefits	Pediatric Services	Certified Wound Specialist
☑	**Access Home Care** 74 West 100 North Logan, UT 84321 (435) 787-4990		M	VA	☻	✎
☑	**Alpha Home Health Care Llc** 250 West Center Street Orem, UT 84057 (801) 225-1080	▧	M			✎
☑	**Applegate Homecare** 1492 E Ridgeline Drive Ogden, UT 84405 (801) 393-2760	▧	M	VA		

Figure 13.5 Home health care "badges" featured on UtahHealthScape.org.

patient communities are viewed by product marketers as ready-made targets—some sites allow commercial advertising, while others do not.

Depending on the intervention, audience segmentation can be more or less realistic. For example, building multiple versions of a complicated Web site like UtahHealthScape for multiple audiences is not practical. Although we aim to make the site accessible to as many consumers as possible, we have recognized that the complex nature of health-care quality data will likely put the site out of reach of less health-literate populations, despite our best efforts at a clear display strategy. Nonetheless, one way that we attempt to reach consumers with different abilities and interests is to include "badges" in the directory listing (see Figure 13.5). Badges are snapshot indicators that convey information such as which providers speak other languages, which clinics have weekend office hours, which health plans offer a 24-hour nurse hotline, and which nursing homes have a dedicated dementia care unit. UtahHealthScape also has a section that provides basic information about common health conditions and concerns, written in plain language with the oversight of a nurse. As we continue to develop and improve the site, we aim to provide more educational materials that are appropriate to consumers with different interests, needs, and capabilities.

13.7 Data dilemmas

Torture the data, and it will confess to anything.
Ronald Coase, Nobel Prize Laureate, Warren Nutter Lecture, 1982.

Coase (1988)

The primary function of UtahHealthScape is to provide information on health-care quality to consumers, with the aim of informing their choice of health-care provider. The process of obtaining, re-forming, and displaying this information creates numerous practical and ethical challenges related to making health-care data actionable for consumers. In this section, we highlight several of these key challenges and describe how we sought to address them (see Figures 13.2, 13.3, and 13.5 for reference to UtahHealthScape's visual displays).

The data we use for UtahHealthScape are obtained primarily from the Centers for Medicare and Medicaid Services (CMS), which collects data on care quality and patient experience for hospitals, nursing homes, home health agencies, and a few large physician groups as of 2014 (CMS, n.d.b). Although there are other sources of data on health-care quality, the CMS data have several advantages: the information can be downloaded directly from their site; it is free; the history of reporting provides a level of consistency over time; providers are required to submit data if they receive Medicare reimbursements; and providers are accustomed to and relatively accepting of the results (Hatry, 2010, p. 243–261).

However, data from CMS have several weaknesses. Primarily, the data are oriented to clinical practice and not consumer utility. They are also targeted to the Medicare population, and focus on some aspects of care which are less relevant metrics for younger populations—for example, conditions like heart attack, heart failure, and pneumonia (HHS, 2011). Also, a heart attack is an emergency situation, thus the consumer utility of measures related to that condition—such as whether patients receive a statin prescription upon discharge from the hospital—is questionable considering that no one would search for information on a site like UtahHealthScape during a heart attack. Nonetheless, we report heart attack data as a proxy for the quality of hospital care more generally, information which consumers would likely find useful prior to seeking health-care services. The time delay of CMS data release, usually about three-quarters of a year plus the time to process it for display on UtahHealthScape, further diminishes the relevance for consumers.

Different types of measures within the CMS data, and other data sources, have different levels of utility for consumers. Key types are care quality and patient experience measures. Care quality encompasses both process and outcomes measures. Process measures assess whether accepted standards are followed in the provision of care. They are the easiest measure type for data gathering, but tend to be meaningless to the consumer, ranging from the obscure (percutaneous coronary intervention with 90 min for heart attack patients) to the bizarre (body actively warmed during surgery) to the counterintuitive (shorter length of hospital stay). Outcomes measures, like rate of mortality or readmission to the hospital within 30 days after discharge, are results-oriented; they are more intuitive to the consumer but can be seen as more controversial among providers, since they depend largely on the underlying health of the population served by a facility (CMS, n.d.a; Rubin, Pronovost, & Diette, 2001). Currently, we provide both process and outcomes measures on UtahHealthScape when available but expect to have a greater focus on outcomes in the future.

In our work on UtahHealthScape and in the literature, we have found that consumers are most attuned to patient experience measures (Hibbard & Jewett, 1996). These measures include, for example, whether other consumers would recommend a home health agency to their friends or family and how clean they considered their hospital room. The literature is mixed as to whether performance on patient experience measures aligns with performance on measures of care quality, which assess actual processes and outcomes that support patient safety (Timian, Rupcic, Kachnowski, & Luisi, 2013). Consumers may be more apt to utilize information about patient experience than care quality, but while this information may drive them to a more satisfying experience, it may not improve their chances of a safe health-care encounter nor encourage health-care facilities to improve their care quality. As a compromise, we provide both care quality and patient experience measures on UtahHealthScape.

In addition to the practical challenges presented by the characteristics of the data available to us, we needed to address how to best display the information. The key elements of our particular display strategy were identified in the section of this chapter on health literacy: the use of summary scores, dichotomized symbols on the summary page indicating high versus average or low performance, symbols to show the meaning of percentage scores at the individual measure level, and default presentation ordering by performance score. These elements were informed by our own experience and the body of research related to optimal display of performance data for consumers (HHS, 2011; Hibbard & Peters, 2003; Peters et al., 2007).

Beyond practical challenges, the ethical challenges of delivering health-care data to consumers are significant. As discussed in Sections 13.4.1 and 13.4.2, provision of this information raises important concerns of paternalism. On one hand, we are providing information to consumers that would be difficult for them to obtain elsewhere, but on the other, we are controlling that flow of information. We have selected which data source to use, which data to present and which to withhold, how to calculate summary scores, what criteria designate a provider as high or low performing, and we have made myriad other decisions that impact the ultimate results. To ensure as much accuracy and objectivity as possible, we ground our work in best practices from the literature, extensive input from subject matter experts, and feedback from provider and consumer representatives.

To use the information on UtahHealthScape to make a decision, consumers must trust in our methods and execution. Even then, our values may diverge from that of the consumer, leaving her/him with recommendations for choosing a provider who is not aligned with her/his values. For example, focus group participants have described to us the importance of location over some aspects of quality. We allow users of UtahHealthScape to filter providers by location, but we are unable to facilitate a precise decision-making calculus tailored to every individual user's values.

In considering how best to provide health-care quality information to consumers, we must also consider how to ensure fairness to health-care providers, which adds a layer of complexity to decisions about data and display. Misrepresenting providers' true performance quality has the potential to damage their reputation and livelihood unjustly. We attempt to address this concern by providing a private portal where providers can review their data prior to publication, if those data are not yet publicly available on the Medicare Compare sites (CMS, 2014). Although we have robust provider involvement in various small advisory committees we convene, few providers have availed themselves of this review opportunity through the portal. Nonetheless, we have created the mechanisms through which providers can provide input on and improve the accuracy of the information on UtahHealthScape, and we continue to pursue a strategy of transparency and partnership with the health-care community.

In keeping with the "libertarian paternalism" concept discussed earlier (Thaler & Sunstein, 2003), we regard our role as stewards of consumer decision making with awareness and care, with an eye toward fairness to providers as well. We work continually to evaluate how we are guiding consumers toward our best interpretation of the information available, while still allowing them the freedom to make their own choice.

13.8 Conclusions

This is an exciting time for consumer and patient engagement. Changes in how the government, private insurance companies, and individuals pay for health care are creating incentives and resources that will help patients take a more active role in decision making—from choice of treatment, to provider selection, to how best to manage a chronic condition. Increasingly, entities and individuals that pay for health care are asking for more accountability from health-care providers as well, and are tying payment to improvements in health outcomes. At the same time, providers and health insurance plans are requiring more personal responsibility from individual patients and consumers. For example, enrollment in high-deductible health plans, or insurance coverage that requires greater out-of-pocket spending than traditional insurance plans, is on the rise. In 2013, 38% of workers with employer-sponsored coverage had a deductible of at least $1000, compared to 10% of workers in 2006 (Kaiser Family Foundation, 2013, p. 110). With this added incentive to use health-care services judiciously, consumers and patients will likely be more attuned to both provider choice decisions and how they manage their own health outside of the clinical setting.

This new era has also ushered in a period of innovation in technology-based health information tools that can facilitate the expanding role of patients in decisions about their own health and health care. As we highlighted in this chapter, amid the promise of these tools are important practical and ethical challenges that should be considered when communicating health information to patients and consumers via digital health platforms outside of the clinical environment.

First, large percentages of the public have limited ability to understand and interpret health-related information. Further, consumers' and patients' interest in and ability to engage in complex decision making are equally varied. While these realities present important challenges, in some cases they also present opportunities for innovations that will reach broader populations and communicate health information in a more accessible way.

Second, as we attempt to make health information easier to consume, we are manipulating the nature of that information. In the case of both text messaging and quality data displays, we need to balance the need for simplification to drive use and understanding, with the ethical imperatives of accuracy and fairness. In our manipulation of health-care quality information, are we removing the consumer's autonomy to make decisions based on her/his own personal values? These questions are particularly important given that some health information, such as that found on UtahHealthScape, is difficult for patients and consumers to gather independently.

Third, digital communication innovations have allowed for more humanlike interactions, and the possibility of engaging consumers and patients in ways that mimic person-to-person connections. This is an exciting advancement that is, in the case of online patient communities, creating new avenues for finding information about specific health conditions and connecting with a supportive group of peers. The "humanizing" of computers can improve consumers' and patients' affinity for and engagement with health information delivered via digital platforms. But the notion that a computer can behave like a human is likely new to many patients—especially those less familiar with technology—and the confusion that may arise as a result should be continually examined and addressed as newer technology continues to blur the line between "human" and "nonhuman" communication.

Finally, consumers and patients have differing levels of access to and proficiency with technology. There is a risk that sophisticated tools will disproportionately aid the well educated and those already better served by the health-care system. Access to technology, in particular, raises ethical questions: Do the type of programs discussed in this chapter exacerbate the existing "digital divide," by providing more information and support to individuals who already have more resources available to them? Not all people will be able to engage in health care in this way, and those occupying lower socioeconomic positions in our society already suffer disproportionately from limited access to health-care services and poor health outcomes (CDC, 2013).

Consumer- and patient-oriented tools have an important role to play in improving health and health care, but as we have demonstrated in this chapter, successful adoption and utilization requires careful consideration of the audience and its capacity to use them.

References

Agency for Healthcare Research and Quality. (2010). *Model public report elements: A sampler.* Rockville, MD: AHRQ.
Agency for Healthcare Research and Quality. (2011). *Training to advance physicians' communication skills.* Rockville, MD: AHRQ.
American Association of Medical Colleges. (2014). *Guiding principles for public reporting of provider performance.* Retrieved from https://www.aamc.org/download/370236/data/guidingprinciplesforpublicreporting.pdf.
Anderson, C. J. (2003). The psychology of doing nothing: forms of decision avoidance result from reason and emotion. *Psychological Bulletin, 129*(1), 139–167.
Ariely, D. (2000). Consumers' decision making and preferences. *Journal of Consumer Research, 27*(2), 233–248.
Associated Press-NORC Center for Public Affairs Research. (2014). *Finding quality doctors: How Americans evaluate provider quality in the United States.* Retrieved from http://www.apnorc.org/projects/Pages/finding-quality-doctors-how-americans-evaluate-provider-quality-in-the-united-states.aspx.
Bandura, A. (2001). Social cognitive theory: an agentic perspective. *Annual Review of Psychology, 52*(1), 1–26.

Capozza, K. L. (2015). Clinical Research Abstracts (Living with atopic dermatitis: Patient priorities and needs expressed in social media fora). *Journal of Investigative Dermatology, 135*, S28–S48. http://dx.doi.org/10.1038/jid.2015.69.

Capozza, K., Woolsey, S., Georgsson, M., Bello, N., Lence, C. T., Oostema, S., et al. (2015). Going mobile with diabetes support: a randomized study of a text message-based personalized behavioral intervention for Type 2 diabetes self-care. *Diabetes Spectrum, 28*, 83–89.

Cavanaugh, K., Huizinga, M. M., Wallston, K. A., Gebretsadik, T., Shintani, A., Davis, D., et al. (2008). Association of numeracy and diabetes control. *Annals of Internal Medicine, 148*, 737–746.

Centers for Disease Control and Prevention. (November 22, 2013). CDC health disparities and inequalities report. *Morbidity and Mortality Weekly Report, 62*(Suppl. 3). Retrieved from http://www.cdc.gov/mmwr/pdf/other/su6203.pdf.

Centers for Medicare and Medicaid Services. (n.d.a). 30-Day death and readmission measures data. Retrieved from http://www.medicare.gov/hospitalcompare/data/30-day-measures.html.

Centers for Medicare and Medicaid Services. (n.d.b). Hospital compare: Data sources. Retrieved from http://www.medicare.gov/hospitalcompare/Data/Data-Sources.html.

Centers for Medicare and Medicaid Services. (2014). *Medicare hospital compare*. Retrieved from http://medicare.gov/hospitalcompare/search.html.

Coase, R. H. (1988). *How should economists choose. Ideas, their origins and their consequences: Lectures to commemorate the life and work of G. Warren nutter*. Washington, DC: American Enterprise Institute for Public Policy Research.

De Martino, B., Kumaran, D., Seymour, B., & Dolan, R. J. (2006). Frames, biases, and rational decision-making in the human brain. *Science, 313*(5787), 684–687.

Duggan, M. (2013). *Cell phone activities 2013*. Washington, DC: Pew Research Internet Project, Retrieved from http://www.pewinternet.org/fact-sheets/mobile-technology-fact-sheet/.

Federal Communications Commission. (2012). *Eighth broadband progress report* (pp. 1–181). Washington, DC.

Fincher, C., Williams, J. E., MacLean, V., Allison, J. J., Kiefe, C. I., & Canto, J. (2004). Racial disparities in coronary heart disease: a sociological view of the medical literature on physician bias. *Ethnicity & Disease, 14*(3), 360–371.

Flesch, R. (1948). A new readability yardstick. *Journal of Applied Psychology, 32*, 221–233.

Fox, S. (2011). *Peer-to-peer health care*. Pew Research Center's Internet and American Life Project, Retrieved from www.pewinternet.org/Reports/2011/P2PHealthcare.aspx.

Gilovich, T., & Griffin, D. (2002). Heuristics and biases: then and now. In T. Gilovich, D. Griffin, & D. Kahneman (Eds.), *Heuristics and biases: The psychology of intuitive judgment* (pp. 1–18). Cambridge: Cambridge University Press.

Gruman, J., Rovner, M. H., French, M. E., Jeffress, D., Sofaer, S., Shaller, S., et al. (2010). From patient education to patient engagement. *Patient Education and Counseling, 78*, 350–356.

Hatry, H. P. (2010). Using agency records. In *Handbook of practical program evaluation* (3rd ed.) (pp. 243–261). San Francisco: Jossey-Bass.

Hibbard, J. H., Greene, J., & Daniel, D. (2010). What is quality anyway? Performance reports that clearly communicate to consumers the meaning of quality of care. *Medical Care Research and Review, 67*(3), 275–293.

Hibbard, J. H., & Jewett, J. (1996). What type of quality information do consumers want in a health care report card? *Medical Care Research and Review, 53*, 28–47.

Hibbard, J. H., & Peters, E. (2003). Supporting informed consumer health care decisions: data presentation approaches that facilitate the use of information in choice. *Annual Review of Public Health, 24*, 413–433.

Hsu, M., Bhatt, M., Adolphs, R., Tranel, D., & Camerer, C. F. (2005). Neural systems responding to degrees of uncertainty in human decision-making. *Science, 310*(5754), 1680–1683.
Kaiser Family Foundation. (2013). *Employer health benefits 2013 annual survey: Employee cost sharing* (pp. 98–132). Retrieved from http://kff.org/private-insurance/report/2013-employer-health-benefits/.
Kessels, R. P. (2003). Patients' memory for medical information. *Journal of Social Medicine, 96*(5), 219–222.
Khan, A., Plummer, D., & Minichiello, V. (2008). Does physician bias affect the quality of care they deliver? Evidence in the care of sexually transmitted infections. *Sexually Transmitted Infections, 84*, 150–151.
Kim, J. Y. (2000). Social interaction in computer-mediated communication. *Bulletin of the American Society for Information Science, 26*(3). Retrieved from http://www.asis.org/Bulletin/Mar-00/kim.html.
Kim, H., Goryachev, S., Rosemblat, G., Browne, A., Keselman, A., & Zeng-Treitler, Q. (2007). Beyond surface characteristics: a new health text-specific readability measurement. *AMIA Symposium Proceedings, 2007*, 418–422.
Kincaid, J., Fishburne, R., Rogers, R., & Chissom, B. (1975). *Derivation of new readability formulas (automated readability index, fog count and flesch reading ease formula) for navy enlisted personnel.* [Research Branch Report 8–75]. Springfield, VA: National Technical Information Service.
Kubrick, S. (1968). *2001: Space odyssey [motion picture].* United States: Metro Goldwyn Meyer.
Larsen, D., Attkisson, C., Hargreaves, W., & Nguyen, T. (1979). Assessment of client/patient satisfaction: development of a general scale. *Evaluation and Program Planning, 2*, 197–207.
Levy, H., Ubel, P. A., Dillard, A. J., Weir, D. R., & Fagerlin, A. (2013). Health numeracy: the importance of domain in assessing numeracy. *Medical Decision Making: An International Journal of the Society for Medical Decision Making, 34*(1), 107–115.
Longo, D. R., & Woolf, S. H. (2014). Rethinking the information priorities of patients. *JAMA, 311*(18), 1857–1858.
March, J. G. (1994). *Primer on decision making: How decisions happen.* New York: Simon and Schuster.
McInnes, D. K., Hardy, H., Goetz, M. B., Skolnik, P. R., Brewster, A. L., Hofmann, R. H., et al. (2013). Development and field testing of an HIV medication touch screen computer patient adherence tool with telephone-based, targeted adherence counseling. *Journal of the International Association of Providers of AIDS Care, 12*(6), 397–406.
Moore, G. (2006). *Crossing the chasm.* New York: HarperBusiness.
National Center for Education Statistics. (2006). *The health literacy of America's adults: Results from the 2003 national assessment.* Washington, DC.
Nielsen-Bohlman, L., Panzer, A. M., & Kindig, D. A. (Eds.). (2004). *Health literacy: A prescription to end confusion.* Washington, DC: Institute of Medicine.
Olshavsky, R. W., & Granbois, D. H. (1979). Consumer decision making: fact or fiction? *Journal of Consumer Research, 6*(2), 93–100.
Osborne, H. (May 2014). *Health care homonyms, jargon, and other confusing words.* Retrieved from http://healthliteracy.com/tips.asp?PageID=11878.
Papacharissi, Z., Rubin, A. M. (2000). Predictors of internet use. *Journal of Broadcasting & Electronic Media, Vol. 44*, No. 2, Spring.
Peters, E., Dieckmann, N., Dixon, A., Hibbard, J. H., & Mertz, C. K. (2007). Less is more in presenting quality information to consumers. *Medical Care Research and Review, 64*(2), 169–190. Retrieved from http://dx.doi.org/10.1177/10775587070640020301.

Peters, E., Västfjäll, D., Slovic, P., Mertz, C. K., Mazzocco, K., & Dickert, S. (2006). Numeracy and decision making. *Psychological Science, 17*(5), 407–413.

Preece, J. (1999). Empathic communities: balancing emotional and factual communication. *Interacting with Computers, 12*(1), 63–77.

Rainie, L. (2013). *The state of digital divides. [Slide presentation]*. Pew Research Internet Project, Retrieved from http://www.pewinternet.org/2013/11/05/the-state-of-digital-divides-video-slides/.

Rains, S. A. (2008). Health at high speed: broadband internet access, health communication, and the digital divide. *Communication Research, 35*, 283–297.

Rogers, E. M. (2004). A prospective and retrospective look at the diffusion model. *Journal of Health Communication, 9*(1), 13–19.

Rubin, H. R., Pronovost, P., & Diette, G. B. (2001). The advantages and disadvantages of process-based measures of health care quality. *International Journal for Quality in Health Care: Journal of the International Society for Quality in Health Care/ISQua, 13*(6), 469–474.

Shaller, D., Kanouse, D., & Schlesinger, M. (2011). *Meeting consumers halfway: Context-driven strategies for engaging consumers to use public reports on health care providers. [Slide presentation.]*. Retrieved from http://www.ahrq.gov/news/events/conference/2011/shaller2/index.html.

Simon, H. A. (1997). Bounded rationality. In *Models of bounded rationality* (Vol. 3) (pp. 291–294). Cambridge, MA: MIT Press.

Smith, W. (1956). Product differentiation and market segmentation as alternative marketing strategies. *Journal of Marketing, 21*(1), 3–8.

Sofaer, S., & Hibbard, J. (2010a). *Best practices in public reporting no. 1: How to effectively present health care performance data to consumers*. Retrieved from http://archive.ahrq.gov/professionals/quality-patient-safety/quality-resources/tools/pubrptguide1/pubrptguide1.html.

Sofaer, S., & Hibbard, J. (2010b). *Best practices in public reporting no. 2: Maximizing consumer understanding of public comparative quality reports: Effective use of explanatory information*. Retrieved from http://archive.ahrq.gov/professionals/quality-patient-safety/quality-resources/tools/pubrptguide2/pubrptguide2.html.

Stanovich, K. E., & West, R. F. (2000). Individual differences in reasoning: Implications for the rationality debate? *The Behavioral and Brain Sciences, 23*(5), 645–726 .

Thaler, R., & Sunstein, C. (2003). Libertarian paternalism. *American Economic Review, 93*(2), 175–179.

Thaler, R., & Sunstein, C. (2008). *Nudge*. New York: Penguin Books.

Time Magazine. (February 1978). *The miracle of chips*. 44–45.

Timian, A., Rupcic, S., Kachnowski, S., & Luisi, P. (2013). Do patients "like" good care?: Measuring hospital quality via Facebook. *American Journal of Medical Quality: The Official Journal of the American College of Medical Quality, 28*(5), 374–382.

Tversky, A., & Kahneman, D. (1974). Judgment under uncertainty: heuristics and biases. *Science, 185*(4157), 1124–1131.

U.S. Department of Health and Human Services. (2011). *Public reporting as a quality improvement strategy: A systematic review of the multiple pathways public reporting may influence quality of health care*. Retrieved from http://effectivehealthcare.ahrq.gov/ehc/products/343/763/CQG-Public-Reporting_Protocol_20110817.pdf.

U.S. Department of Health and Human Services, Health Resources and Services Administration. (2014). *Using health text messages to improve consumer health knowledge, behaviors, and outcomes: An environmental scan*. Rockville, MD: U.S. Department of Health and Human Services.

Vaiana, M., & McGlynn, E. (2002). What cognitive science tells us about the design of reports for consumers. *Medical Care Research and Review*, 59(1), 3–35.
Wedel, M., & Kamakura, W. (2000). *Market segmentation: Conceptual and methodological foundation.* New York: Springer.

Does specialization matter? How journalistic expertise explains differences in health-care coverage

14

Michael W. Wagner
School of Journalism and Mass Communication and Department of Political Science, University of Wisconsin-Madison, USA

14.1 Introduction

Americans have regarded health care as one of the most important problems facing the nation for several years (Brown, 2013). Outside of their own experiences with the health-care system, most people come to understand what health care is like, what the latest cutting-edge research is, and how to interpret the increasingly complex and interconnected webs of research, new treatments, insurance rules, and the like through what they learn from the mass media. While "interest in health news is as high as it's ever been," the rapidly changing media environment is placing new and challenging pressures on newsrooms to do more with less and on journalists to present their work across multiple platforms (Schwitzer, 2009, p. 16).

In other words, more people are seeking news about health care—an issue they find more vital than ever—at the same time it is becoming much more difficult for journalists to provide quality health-care coverage. Since the quality of health-care coverage can greatly affect what people know about health-related issues and the kinds of behavior people will engage in to address their own health needs, it is crucial to understand how journalists approach covering health-care issues (Tian & Robinson, 2008; Viswanath et al., 2006).

Unfortunately, the picture of health-care coverage in the United States that is painted by the journalists who know the most about the issue is not a pretty one. Indeed, a 2008 survey revealed that 65% of members of the Association of Health Care Journalists (AHCJ), the premier journalistic organization on the globe dealing with health-care issues, believed health-care coverage in the US was either fair or poor (Schwitzer, 2009). Only 1% believed that coverage was excellent. Schwitzer's (2009) comprehensive review of health-care journalism revealed that AHCJ members lamented about the rise in "quick hit" stories (i.e., stories that are reported quickly, without much depth, and not followed up in future coverage). Quick hits provide little time for in-depth investigation, the interviewing of multiple sources, and engagement with the evidence on which the most important elements of the stories rested.

The continuing diminishment of the amount of space (in print) and time (on radio and television) that health-care journalism is afforded in news coverage, coupled with the fact that shrinking newsrooms make it more difficult for journalists to develop complex, time-consuming stories—the very stories that can provide crucial information to citizens about the state of health care—raise potentially frightening questions about the quality of health-care reporting that can be produced in such an environment. This is likely to be especially true for the legions of reporters who have little to no training in health-care journalism. After all, they are the ones doing most of the health-care reporting in the US.

As one veteran reporter described what was happening at his newspaper to health journalism scholar Gary Schwitzer (2009), "A reporter with no specialized knowledge in health/science news was named as health/science editor. He thought we could simply use AP wires on research stories" (p. 4). Wire service news agencies like the Associated Press (AP) are organizations of journalists who sell their stories to subscribing newsrooms. They often provide quick, reliable coverage of breaking news and daily events, but they do not do much in the way of investigative work. Colloquially, journalists view newsrooms that rely on wire services to fill their pages as weaker than newsrooms that generally cover the news on their own.

Recent evidence from Australia reveals that "specialists" in health-care journalism score higher on a range of measures taken from their reporting—including covering the novelty and availability of new health interventions, the evidence supporting the intervention, the benefits, harms, and costs of the intervention and independent sources' views of the research—than reporters without any particular affinity for or training in health coverage (Wilson, Robertson, McElduff, Jones, & Henry, 2010).

This study compares journalists who specialize in health-care coverage with those who do not explain the willingness of journalists to cover advances in health-care delivery in the United States; the style of coverage journalists favor; and the process of reporting journalists follow when covering these advances. The specialists are journalists who are members of a professional organization dedicated to improving coverage of health care-related news. This organization puts on conferences, workshops, web-based seminars, and maintains a listserv where reporters can ask questions and share tips about covering health care. These specialists have been exposed to in-depth information about a variety of health-care issues—including cutting-edge research about the effects of complicated regulations, new research methods, peer-learning opportunities, and short courses adding to their reportorial toolkit in ways that nonspecialists (general assignment reporters) have not.

The research questions explored in this chapter are (1) Does journalists' level of specialization affect whether they view a health-care topic as newsworthy? (2) Does specialization affect the kind of story (hard news vs feature) journalists believe a health-care topic merits? Hard news is generally defined as timely stories about serious topics while feature stories generally refer to human-interest stories that are not always tied to a timely event. (3) Does specialization affect the process by which journalists would seek to write a story about health-care advances?

To answer these questions, a detailed survey of journalists was conducted, some of whom are health-care specialists and some of whom who are not. The survey asks the journalists to decide how they would respond if they learned of the development of a new mobile phone web application ("app") that would connect cancer patients to their doctors and to other cancer patients. The survey also asks open-ended questions that allow for the reporters to explain their choices in more detail. The analysis reveals major, crucial differences in how specialist and generalist journalists approach health-care coverage in the United States.

14.2 Why specialization should matter

Despite lofty expectations that reporters act as independent, fair, accurate, open, investigative, truth-telling, multiple perspective-giving, alarm-sounding watchdogs who defend the public interest, journalists often end up reporting what is said by official sources in power with far greater frequency than almost anything else (Bennett, 1990; Cook, 2005; Hayes & Guardino, 2010; Janowitz, 1975). This behavior is rooted in the training journalists receive in journalism schools and in newsrooms to produce balanced, "objective" stories containing conflict between authoritative sources about novel topics (Groeling & Baum, 2009; Schudson, 2001; Shoemaker & Reese, 1995; Sigal, 1973).

The most prominent explanation of how media coverage operates in the United States is W. Lance Bennett's indexing hypothesis (1990). Indexing argues that news coverage of political issues tends to chronicle debates—and a limited range of views—between official sources. For example, rather than covering how health-care policy decisions affect individuals' health, finances, and families, news reporting is more likely to cover what public officials like lawmakers and bureaucrats say about health-care policy (Bennett, 1990, 1996, 2012). While there is impressive evidence supporting the indexing hypothesis, there is a considerable amount of evidence that reporters can act as more than indexing stenographers, more closely fulfilling the democratic requirement that a free press provides a variety of critical perspectives on issues (Althaus, 2003; Hayes & Guardino, 2013; Sartori, 1987). However, journalists often fall short of this standard (Baker, 2007; Dahlgren, 2009; Patterson, 1994). Indexing studies demonstrate that Republican and Democratic elected officials are regularly used as sources in political news coverage because their general style of debate provides the opportunity for reporters to tell "both sides of the story" (Cook, 2006; Tuchman, 1972; Wagner & Gruszczynski, in press).

Indeed, Bennett argues that a "small set of rules account for a large share of political content in the news" (1996, p. 374). The rules include creating stories based on official or authoritative views and indexing those views with respect to the magnitude and content of conflict. Indexing is almost always considered in the context of political news coverage. Indeed, its development was conceived as a theoretical addition to the understanding of press–state relations. However, the chronicling of official positions on medical breakthroughs, health-care provision strategies, and the like are also quite likely to be indexed (see Schwitzer, 2009).

Advances in health-care research are not routinely accompanied by opposition voices in the way that Democratic proposals are regularly answered with Republican counterproposals and vice versa. The powerful voices indexed in health coverage are likely to be those of doctors, bureaucrats, and scholars. Since journalists are not generally experts in health care, research, or statistics, they are likely to defer to their sources, failing to ask critical questions about the generalizability of research findings. This often results in sensational, overblown stories promising readers and viewers the moon—frustrating the very sources of the stories in the process (Shuchman, 2002; Voss, 2002). The doctors, nurses, and administrators responsible for delivering health care to the citizenry, and the scientists who play the major role in medical advances, regularly complain about the coverage they receive in the news media. They claim the coverage is simplistic—oversimplifying research, making claims that go beyond what the research evidence suggests, and sensationalizing stories about patient care.

What is more, several organizations such as political interest groups, medical industry representatives, and the like seek to take advantage of journalists' lack of health expertise and the increasing pressures of producing multiple stories on tighter and tighter deadlines in a 24-hour news world. Companies like Ivanhoe Broadcast News sell newsrooms canned (preproduced) stories about health care promoting particular health-care "breakthroughs," treatments, and the like so that media outlets will print or air the stories in their entirety. The stories are presented in a way that makes it look like they are locally produced when they are not. These kinds of stories are desirable from an economic perspective both because health news "sells," and because the stories are already completed, and thus will not take the time of the already limited (and shrinking) newsroom staff (Hamilton, 2005).

In addition to these pressures, journalists themselves have not expressed confidence in their ability to report health news—largely because of a lack of understanding of how to interpret statistics. When Midwestern journalists were asked about four skills that are required for quality health reporting—understanding key health issues, putting news about health care into the proper context, writing balanced stories while still meeting a deadline, and interpreting health statistics—Voss (2002) found that between 66% and 85% of journalists said these skills were difficult or to master (Voss, 2002). Thus, they are not likely to want to cover health-care stories in much depth, nor are they likely to want to engage with the research that produced the advance that is newsworthy in the first place.

14.3 Methodology

14.3.1 Hypotheses

This study seeks to test three hypotheses about how specialization affects how journalists judge what they should cover when it comes to the topic of health care.

1. Journalists who specialize in health-care coverage will be more likely than other journalists to find advances in health-care provision to be newsworthy. Those who do not have any special training in health-care journalism should be less likely to see advances as newsworthy because they are not as likely to recognize them as advances and because they are less aware of potential sources for the story.
2. Journalists who specialize in health-care coverage will be more likely than other journalists to believe that an advance in health-care provision merits a feature-length treatment. Even if nonspecialty reporters believe the topic merits coverage, they should be more likely to see it as a "one-day story"—something requiring minimal effort and not returned to later

for updates. Thus, they should be less likely to do the extra work involved with writing a lengthier "feature" story that typically contains highly personal stories from sources who have been or are likely to be affected by the topic of the story.
3. Journalists who specialize in health-care coverage will be more likely than other journalists to believe it is important to read the research that produced the advance about which they are reporting. If my first two hypotheses are correct, nonspecialty journalists should be less likely to do the work with respect to reading the research that preceded the health-care innovation they are going to cover. They ought to be more likely to trust the information on the press release or ask a few simple follow ups to a source rather than read the research for themselves and follow up on areas they personally felt needed clarification, amplification, or outright challenging.

14.3.2 Participant recruitment

In order to test these hypotheses, I conducted an original web-based survey, programmed in the web-based survey platform Qualtrics. The participants in the survey came from two separate groups. One came from a sample of working journalists in the largest 150 media markets in the United States. Nielsen Media Research identifies media markets as locations that receive the same (or nearly the same) television and radio stations. They are generally named for the largest city/cities in the market. New York is the number one market, boasting over 7.4 million television households (over 6% of the US population). In the middle is the Omaha, NE market. It has over 420,000 television households and makes up 0.36% of the country. The 150th market is the Odessa/Midland, TX market. It has over 155,000 television households and comprises 0.135% of the United States. Over 90% of the American people reside in these 150 markets, though there are a total of 210 markets in the country.

The sample was created by creating a database of every reporter who covered any kind of political or public affairs news for a major newspaper or the local NBC, CBS, ABC, and FOX television affiliates in each market in the fall of 2013. Decisions about which employees of a newspaper or television station might cover politics were made by visiting the Web site for each news outlet. For those outlets that had staff biographies, a team of research assistants read them and created a list of reporters who covered politics. These were mostly local television stations. For newspapers and other outlets that did not include staff biographies, research assistants combed the Web site for any information that revealed where the employees worked. Most newspaper Web sites had information about the "desks" (topic areas such as city politics, sports, crime, etc.) to which reporters were assigned. For outlets that provided no help with respect to the work done by their reporters, research assistants clicked on the tabs related to political news and acquired the e-mail addresses of the bylines of reporters who had produced a story within the previous 7 days.

The generalist, or nonspecialist, group of reporters received an e-mail in the spring of 2014 at her or his station/newspaper e-mail address that invited them to participate in a study of how news reporters approach their jobs. The full database consisted of 6733 reporters. I e-mailed a random sample of 2000, introducing them to the project by explaining that the survey was seeking to understand how reporters approach their jobs and by inviting them to participate. Ninety-four e-mails were bounced back to me immediately because the e-mail address was incorrectly inputted or because the reporter had left the station or newspaper. Over the course of 4 weeks, links to the survey were sent to the journalists who had agreed to participate. In addition, three follow-up e-mails were sent to those who did not respond to the first invitation e-mail. In total, 631 completed the study for a response rate of 31%. Each participant who completed the survey received a $5 gift card to Starbucks as a small token of appreciation.

The second group was comprised of members of a professional organization of journalists that was dedicated to news coverage of health-care issues. The organization boasts more than 1000 members. I e-mailed an invitation to participate to a random sample of 300 members, 102 of whom elected to participate for a response rate of 34%. Of the 631 journalists in the first sample, seven answered the survey questions asking if they were members of a professional organization related to health care in the affirmative. Thus, the final sample of specialty journalists grew from 102 to 109 and the final sample of nonspecialist journalists dropped to 624.

14.3.3 Characteristics of the sample

Before the survey began, the participants answered a battery of questions relating to demographics, political attitudes, and their judgments of the American political system. Males made up 59% of the sample while females comprised 41%. Ninety percent of the sample was white; 35% were Democrats while 22% were Republicans. In terms of ideology, 40% of journalists were identified as liberal (including those who "leaned to the left") while 27% identified as conservative (or leaning to the right); see Table 14.1. These numbers are highly comparable to the last two major national surveys (Weaver, Beam, Brownlee, Voakes, & Wilhoit, 2006; Wilnat & Weaver, 2014) of journalists in the past decade. The median (average) media market size a respondent worked in was 71.

There was one statistically discernible difference in the demographics of the two samples. As shown in Table 14.2, the health-care specialists were far more likely to be women (61% female in the specialists sample as compared to 41% female in the generalists sample). Otherwise, Table 14.2 shows the two groups to be remarkably similar. The two major differences between my samples of journalists and Wilnat and Weaver's (2014) sample, the most recent national sample of journalists, was that my sample had slightly higher percentages of Republicans and women. It is important to know the political background of the reporters because their personal beliefs might influence whether they cover stories, especially those that promote beliefs consistent with the reporters' own political beliefs.

Participants were asked several questions about their attitudes regarding which political party was best able to handle a variety of health care-related issues (including medical device taxes, legal reform, prescription drug prices, medical research, and health insurance coverage). Participants were also asked to estimate which political party "conventional wisdom in America" would say is better at handling each of the issues.

As shown in Table 14.2, 61% of the sample believed Democrats were best able to handle health insurance coverage; 69% of the sample also believed that conventional

Table 14.1 **Characteristics of the two samples of journalists**

Journalist type	Female	Liberal	Moderate	Conservative	Democrat	Independent	Republican
Specialist	61%	40.3%	29.3%	30.2%	36.7%	40.3%	22.9%
Generalist	41%	39.4%	34.1%	26.4%	34.6%	43.5%	21.8%

Note: Percentages of ideological and partisan groups may not add up to 100% due to rounding.

Table 14.2 **Journalists' beliefs about which party would better handle health-care issues**

	Party handling			
	Health insurance best		Medical devices best	
	Democrats	Republicans	Democrats	Republicans
Journalists' own attitudes	61%	39%	32%	68%
Journalists beliefs about "conventional wisdom" in the US	69%	31%	30%	70%

Note: Participants were forced to choose between each party and were not allowed to say "both" or "neither."

wisdom put the Democrats on top (see also Petrocik, 1996; Petrocik, Benoit, & Hansen, 2003). Sixty-eight percent of the sample believed that Republicans were best able to handle medical device taxes while 70% of the sample believed that conventional wisdom gave the advantage on that issue to the GOP.

The most important part of the survey for our purposes was the introduction of a (fictitious) web application designed for mobile phones called Connect2FightCancer. New communication technologies have been shown in scholarly research to play important roles in treatment for several diseases (Han et al., 2010; Johnson, Isham, Shah, & Gustafson, 2011; Noh et al., 2010). Respondents were told that the Connect-2FightCancer app was not real, but that it was similar to new developments in both academic research and the health-care field more generally. The app was described as one that was developed after years of research from scholars at multiple universities.

The reporters were asked to imagine that scholars had learned that some patients were less likely than others to follow their doctor's recommendations to treat and manage their cancer. Research suggested that a quick link to doctors and to other cancer patients was a promising way to encourage patients to follow doctor's orders. The reporters were told that the app had performed well in pilot studies and was about to be released for general use by cancer patients nationwide. Then, they were asked a series of questions about their interest in the story and how they would choose to cover it if assigned to write a story about it. These questions are described in Section 14.3.4.

The Connect2FightCancer app was designed for a variety of reasons. First, it was important that generalist journalists would plausibly find the app's release to be newsworthy. As such, the app met most of the conventional elements of newsworthiness: it was timely, proximal (happening in the journalists' community), and had the potential to impact the reporters' audience. Second, the scenario could not be so appealing that there would not be any variance in the specialist journalists' attitudes about whether the app's release was newsworthy. Thus, the story did not include another conventional element of newsworthiness: conflict. Moreover, the app's release did not promise revolutionary advances in cancer care. Thus, the specialists might have been more likely to see the app's release as similar to previous cancer care advances and judge the app's release as unoriginal and not worth covering. Finally, the scenario is a realistic example of the press releases newsrooms receive to announce the implementation of a variety of health-care advances. As a consequence, any results stemming from the analyses would be more likely to generalize to other aspects of health-care coverage.

14.3.4 Questions, coding, and analysis

The major dependent variables I used in the analysis were *Newsworthiness*, *Feature Story*, and *Read Related Research*. *Newsworthiness* was measured on a 7-point scale. The question respondents answered was:

> On a scale ranging from 1 to 7, where 1 represents "not at all newsworthy" and 7 represents "extremely newsworthy," how newsworthy is the national launch of the Connect2FightCancer app?

Here, the "national launch" referred to the story noting that the app had successfully advanced through a test market-testing phase and was being made available nationwide.

The second major dependent variable, *Feature Story*, was dichotomous. It was measured by asking:

> If you were assigned to cover this story, do you think it would be best presented as a lengthy feature or a short, hard news story?

Responses were coded as 1 if the respondent preferred a feature story strategy and a 0 if the journalist chose a hard news story.

The third major dependent variable was *Read Related Research*. It was measured along a 7-point scale. The question respondents answered was:

> On a scale ranging from 1 to 7, where 1 represents "not at all important"
> and 7 represents "extremely important" if you were to cover the launch of the Connect2FightCancer app, how important would it be to read the research studies that were used to develop the information you have learned about the app?

In order to systematically test how specialist and generalist journalists approached the coverage of the Connect2FightCancer app, we included several independent variables in our statistical models. The major concept we are testing in this study is whether the way journalists approach health-care coverage differs based on whether the reporter is a health-care specialist or a generalist. Thus, *Specialist* is coded as 1 if the reporter was a member of a professional organization of health-care journalists and 0 otherwise. *Years Experience* is a categorical measure of how many years the respondent has been a journalist (0–5 years, 6–10 years, 11–15 years, more than 15 years) as reporters with more experience may approach stories differently than "cub reporters" just getting their start.

Television is coded as 1 if the reporter is primarily a television reporter and 0 if the reporter is primarily a newspaper or magazine reporter. Television reporters have far less time to tell their stories; they also have the added requirement that their coverage play well on television. Indeed, all of the words said on a typical 30-min newscast would fit on the front page of the *New York Times* with plenty of room to spare. As such, television reporters may be less likely to want to cover an issue without an obvious video component, and they may prefer a quick, "hard news" story to a lengthy feature that takes up valuable airtime.

Journalists increasingly wear multiple hats across formats when doing their work. To arrive at the dichotomous measure used in the *Television* variable for the generalist sample, I coded 1 for television reporters if they worked at a television station and 0 if they worked at a newspaper. For the specialty sample, I first asked a question about what kind of newsroom the reporter worked in (television, newspaper, magazine, web-only, or other). If the reporter chose "Web-only" or "other" I asked if their reporting was primarily presented as a print-style story or a broadcast-style story. Six percent of the sample fell into the web-only or other categories and the results of the analysis are robust when these journalists are included and when they are not. Recall that both samples are merged together for the analysis reported below.

In Schwitzer's (2009) survey of members of the AHCJ, "having the time to do research" was one of the top two elements specialists said was necessary for quality reporting. Thus, *Time* was measured by asking respondents to agree or disagree with the following question statement on a 7-point scale where 1 represented strongly disagree and 7 represented strongly agree: "In general, I have enough time to do the research that I feel is necessary to tell a compelling, accurate story."

Advertising pressure was a variable intended to capture the influence that advertisers sometimes have on the news decisions made in the newsrooms where the individual journalists worked. It was measured on a 4-point scale (1 = never, 2 = rarely, 3 = sometimes, 4 = frequently) that was used when answering the question, "how often, if at all, does your news organization allow sponsors or advertisers to influence the content of your health care coverage?" (adapted from Schwitzer, 2009). *Ideology* was measured along a 7-point scale from very liberal to very conservative. *Female* was coded 1 if the respondent self-identified as a female and 0 if the respondent self-identified as a male. *White* was coded 1 if the respondent self-identified as white and 0 otherwise.

For the *Newsworthiness* and *Read Related Research* models, ordered probit regression analysis was the appropriate statistical strategy. For the *Feature Story* model, logistic regression analysis was the best strategy.

In addition to the survey questions, the respondents were asked open-ended questions. These were invitations for them to explain their responses on the numbered scales. Respondents were permitted up to 200 words per response. By explaining their views about the app and their professional reactions to it, participants can provide more depth and context to their responses and bring up issues that I did not think to ask about in the survey (Hibbing & Theiss-Morse, 2002). Moreover, allowing journalists to describe their attitudes in their own words casts a revealing light on the process of how they form opinions and the way they characterize their own professional identity (Walsh, 2012).

14.4 Results

14.4.1 Descriptive analysis

Figures 14.1 and 14.2 reveal the average responses that health-care journalists and generalist journalists gave to the questions used as the three dependent variables described above. Recall that the values in two of the three variables are arrayed along a 7-point scale and one question was a dichotomous "yes/no" question (Figure 14.1). In terms of *Newsworthiness*, health-care journalists saw the potential story about the health-care communication app as highly newsworthy ($M = 6.1$, $SD = 1.1$), whereas generalists found the same topic about as newsworthy as not ($M = 4.4$, $SD = 1.5$).

In other words, health-care journalists were certain that they saw a story; generalists barely saw one, if at all. Second, when it came to the amount of work that reporters felt would be necessary to write an accurate story, health-care specialists believed that they needed to read the original research on which the delivery of the health-care communication app was based ($M = 5.8$, $SD = 1.1$), whereas generalists were less likely to believe that judging the scholarship for themselves was necessary ($M = 3.9$, $SD = 1.4$).

Figure 14.2 reports the results for how specialist and generalist journalists would approach a story about the app's launch if they were assigned to cover it. The results are nearly the inverse of each other. Specialists overwhelmingly preferred to cover the story as a feature (68%) as compared to 32% who thought a shorter, "quick hitting" hard news story was best. On the other hand, only 37% of generalists believed a lengthy feature would be the way to go with a story about the Connect2FightCancer app. Sixty-three percent favored a traditional hard news story.

Across a range of measures, then, journalists who were part of a professional organization that provides training and professionalization opportunities related to the conduct of health-care reporting appear to have behaved in a fundamentally different way than did nonspecialist reporters. The raw data show that specialists saw the development of a new health-care delivery strategy, the Connect2FightCancer app, as newsworthy, worthy of feature-length coverage, and a topic on which it was important to read the studies that lead to the development of the app and the expectation that it would affect health outcomes for cancer patients.

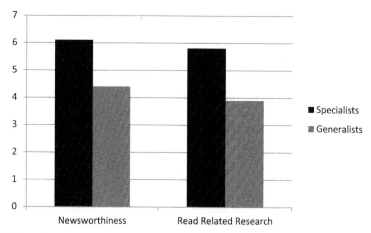

Figure 14.1 Differences in how specialist and generalist journalists approach health-care coverage.

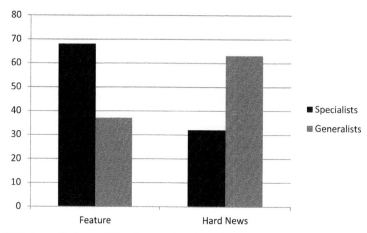

Figure 14.2 Journalistic specialization and decisions about story type.

14.4.2 Quantitative results

It is important to establish that the populations of specialist and generalist journalists reported in Figures 14.1 and 14.2 are statistically different from one another. Thus, Table 14.3 reveals the results to two-ordered probit regressions (for which *Newsworthiness* and *Read Related Research* are the dependent variables) and one logistic regression model (for which *Feature Story* is the dependent variable).

The pattern revealed in Table 14.3 is striking. Across all issues, journalists who were health-care specialists were more likely to find a story newsworthy, believe the Connect2FightCancer app was worth a feature story, and agree with the idea that it was important to read the research that led to the development of the app.

Table 14.3 **Journalistic specialization and reportorial decisions making**

	Newsworthiness	Feature story	Read related research
Specialist	0.491** (0.181)	0.813** (0.101)	0.544** (0.201)
Years of experience	0.036 (0.120)	0.040 (0.097)	0.032 (0.120)
Ideology	0.001 (0.0003)	−0.001 (0.009)	0.001 (0.0003)
Television	−0.240** (0.008)	−0.127 (0.088)	−0.261** (0.02)
Time	0.002 (0.01)	0.004** (0.001)	**0.003 (0.001)
Advertising pressure	0.177 (0.121)	0.218** (0.020)	0.122 (0.046)
Female	0.125 (0.11)	0.012 (0.057)	0.213 (0.11)
White	0.022 (0.026)	0.141 (0.194)	0.024 (0.031)
Constant		−1.533** (0.271)	
Cut 1	−1.900		−1.881
Cut 2	−0.611		−0.432
χ^2	73.54**	80.34**	71.32**
Log-likelihood	−1433.12	−1143.22	−1695.87
Observations	733	733	733

Standard errors in parentheses. **$p < 0.01$, *$p < 0.05$.

The left-hand column of Table 14.3 reports the results of an ordered probit model in which *Newsworthiness* is the dependent variable. The results demonstrate that being a specialty journalist had a positive, statistically significant ($p < 0.01$) effect on believing that the Connect2FightCancer app was newsworthy. Additionally, television reporters were less likely than newspaper reporters to find the app to be newsworthy ($p < 0.01$). The years of experience the reporter had as well as the time she or he had to commit to doing quality work were not related to judgments related to the news value of the app's launch. Moreover, the effects of the demographic controls were also not different than zero.

In the *Feature Story* model, reported in the middle column of Table 14.3, specialty journalists had a positive, significant relationship with the dependent variable ($p < 0.01$). This time, being a television reporter was not meaningfully associated with the desire to cover the app as a feature story. As the journalists in our sample reported that their newsrooms were increasingly open to the influence of *Advertising pressure*, they were more likely to favor covering the story as a feature ($p < 0.01$). Additionally, the desire to cover the app as a feature was positively and significantly associated with believing that one had the time to do quality reporting ($p < 0.01$). Once again, the length of time one had been a journalist was not statistically related to wanting to cover the app story as a feature or hard news story. The demographic controls were not significant predictors of the dependent variable either.

The right-hand column of Table 14.2 reports the results of an ordered probit regression examining the factors associated with the belief that it was important to *Read Related Research* about the app before printing or airing a story about it. Once again, being a specialty journalist had a positive, significant ($p < 0.01$) effect on the dependent variable. Television journalists were less likely to indicate that it was important to read the research about the app ($p < 0.01$) while those who said that they had time to do good work believed it was important to read the scholarly evidence. Advertising pressure and years on the job had no statistical relationship with the belief that reading the research was important.

As was the case in the first two models, *Ideology* and *White* had no effect on the dependent variable. However, *Female* was marginally significant in the model ($p < 0.10$). This finding should be interpreted with caution as it does fall outside of conventional levels of statistical significance. It does suggest, however, that future research might examine differences between male and female journalists with respect to the importance of different kinds of evidence (e.g., scholarly) that is important to understand during the reporting process.

Taken together, the data revealed in Figures 14.1 and 14.2 and the statistical analyses reported in Table 14.3 demonstrate with impressive clarity that specialty journalists approach their work in a fundamentally different way than do their nonspecialist counterparts. Indeed, even when controlling for structural matters such as whether one is a television or print reporter, the power of advertisers, the time one has to produce one's work, and the years on the job, as well as demographic factors like political ideology, gender, and ethnic background, the importance of specialization shines through.

14.4.3 Qualitative results

Of course, survey responses arrayed across 7-point scales and the like have their limits with respect to explaining individual behavior. Most notably, they often fall short in helping to elucidate the process people go through when making decisions. Additionally, respondents, of course, cannot comment on issues that were not asked about by the survey researcher. As noted in the discussion of research design above, in order to further clarify the differences in how specialty and generalist journalists approach health-care coverage, respondents were given an opportunity to explain their views—in their own words—about whether and how they would go about covering the Connect2FightCancer app and whether they would cover it at all.

After respondents answered the *Newsworthiness* question, they were asked:

> *In the space below, please share why you selected the option you did when you answered the previous question about the newsworthiness of the Connect2FightCancer app.*

Regardless of whether the respondents were specialty journalists or generalist journalists, respondents who answered on the positive side of the midpoint discussed that the app was a "good story." However, there were differences in how each type of reporter explained what it was that made the app a good story.

One generalist described how the app fit his narrative about the increasing ubiquity of mobile devices in modern life:

> Generalist: *This sounds like a story, a good one in fact. People are using mobile devices for everything now, so why not to battle cancer? I think our audience would be interested in something like this.*

Another generalist highlighted the novelty of the app:

> Generalist: *It's new and it's relevant. That's why I'd want to cover it.*

A female generalist with more than 10 years of experience covering the news took the value of the topic as a given, spending her time describing how the story would not take much effort to put together.

> *It should be pretty easy to find people to interview. The company who made the app will want to promote it, which will help find users to talk to. At a minimum, a hospital's PIO (public information office) will talk though I'd rather have a doc who uses the thing too.*

In contrast, specialists—health-care journalists—who thought the app was newsworthy focused more on the science behind the app's development and the directions that the story might go. A health-care journalist who worked for a newspaper wrote:

> Specialist: *This has the potential to be a great story. The fact there is a university research team behind the app's development makes me less worried about conflicts of interest and shilling for the app. They (the researchers) should be able to talk benefits and harms.*

Another experienced health-care specialist reporter showed evidence that she had considered the scientific relevance of the app and potential consequences—both positive and negative—of its use, noting that:

> Specialist: *This isn't a phase one medical study that people rush to cover as the latest and greatest when it is really just making sure that a real study is safe to conduct. The app addresses what should theoretically facilitate cancer patients to be more likely to follow doctor's orders. I wonder how people will feel if they hear that they're screwing up from other cancer patients. One thing I've learned from psychology professors is that people don't like it when they're told they're wrong.*

These quotations illustrate that, in judging the app's potential newsworthiness, generalist journalists thought about the potential story's impact on the audience and how difficult the story would be to cover. Health-care specialists, on the other hand, were more interested in what benefits and harms might come from the app's use, considered the quality of the scientific research ahead of their newsworthiness judgment, and asked whether they would be promoting a product instead of covering the news.

Similar patterns of difference appeared in the comments made by those who did *not* find the app to be newsworthy; the comments below were made by those who chose 3 or less when answering the 7-point scale newsworthiness question.

One television-oriented generalist journalist thought that the story would not be a good fit for her medium.

> Generalist: *It's a cool idea, but I don't know that people sitting around fiddling with an app plays well on TV.*

Another generalist worried about the impact the story would have on readers, musing that:

> Generalist: *This doesn't seem like it gets us any closer to a cancer cure or anything like that. I'm all for people listening to their doctor, but I don't see where there is a story here.*

Health-care specialists, on the other hand, were less concerned about the medium they worked in or the audience impact when explaining why they scored the app as being less than newsworthy. "This may well be a big deal," one specialist newspaper reporter wrote:

> Specialist: *but I can't judge it without reading a study or two about why it has to be this app. I mean, why can't doctors just call patients or have them go to support groups or do other things that we know works? Do we really know this is better? If so, I'd be interested in telling that story but as it stands, it's thin.*

The health-care journalists who thought the app was a newsworthy development described how the app's development appeared to have gone through a lengthy research-oriented development process. In contrast, skeptics indicated a desire to see how the app was an improvement over other areas of research with which the journalist was familiar. The specialists' skepticism was more closely hemmed to audience expectations rather than scientific concerns.

With respect to whether the story should be a short, hard news story or a lengthy feature, additional differences between specialists and generalists revealed themselves. Turning first to those who thought the app merited a feature story, both generalists and specialists thought that examples of how the app were used would be key components of the feature. One generalist wrote:

> Generalist: *I could see a nut graf [the paragraph that refers to the major point of the story] or a pull quote [a quote from the story that is also printed in larger type and set apart, from a design perspective] that was actually the texting back and forth between a patient and her doctor or two cancer patients chatting about whether to listen to their doctor's advice. You need room (a long feature) to have the time to put something like that in there.*

Similarly, a health-care specialist wrote:

> Specialist: *It would be really great to find a way to show how this worked in the story itself. If people who used the app would let us put a screen shot of how they*

talked with their doctor or something. You know, so long as the patient and doctor sign off on revealing that type of relationship… but you could really show how it works and pull out examples from a real person to add color to the theoretical stuff that researchers used to develop the app in the first place. It'd be especially great if people used the app in ways the researchers didn't imagine. Are they sharing recipes? Advice about what to do with their teenagers? You know, that kind of stuff. I'd want to go back and ask the study team if they foresaw other uses of the app or if they think it would be better if the users focused on all cancer treatment all the time.

While the health-care specialist's response was similar to the generalist's response from a "structure of the story perspective," once again, the additional context of the response reveals a key difference in the thinking of specialists and generalists concerning their approach to health-care coverage. The specialists consistently wrote about putting the quotes they might gather from their reporting in context with the research–research they had already indicated was important to read. On the other hand, generalists knew instinctively how giving an example of how the app works would improve the story, but they did not speculate about how the colorful examples of app use might reflect, confirm, or contradict the research that led to the app's development.

Finally, respondents were asked:

Is there anything else you would like to comment on regarding whether and how you would cover a story about the Connect2FightCancer app?

Most respondents (70%) did not answer this question. Far more specialists (60%) answered the question than generalists (24%), but keep in mind that generalists were a much larger proportion of the sample overall (85% of the total respondents). Generalists who answered this question focused on questions they had to which an affirmative answer would make them more interested in covering the story. These questions largely related to the availability of interviews and the likelihood the app would be widely used (e.g., "Do we know people are really going to use this? How?").

Health-care specialists wanted to know more about potential benefits and harms of using the app, what methods were used in the research that led to concluding the app would be of value, how available the app would be to the average cancer patient (for what kinds of cancer would patients have to use the app), and would other researchers be familiar enough with the evidence to offer independent comment upon it. Indeed, the health-care specialist comments reflect the AHCJ Statement of Principles about health-care coverage, including the sections of "Professionalism, Content, and Accuracy," "Independence," and "Responsibility" (Schwitzer, 2004).

14.5 Discussion

Scientists and doctors have long been skeptical of journalists' ability to understand statistics and the scientific method well enough to cover adequately health-care advances, questions facing health-care provision, and political battles over health care (Shuchman, 2002).

The analysis reported here suggests that there is a subset of journalists who do understand these matters quite well. These specialists are members of a professional organization dedicated to improving health-care coverage and thus receive information on listservs™, attend conferences, receive training, and the like to help them tell the most accurate, interesting stories possible about health care in the United States. Specialists who join organizations dedicated to improving health-care coverage gain access to a yearly national conference, regular regional meetings on a variety of topics ranging from the technical—statistics, "big data," and arcane new health-care policy rules from a national or state government—to web-based tutorials about specific concepts—covering hospitals—to workshops and seminars where reporters can engage in hands-on learning with experts and other reporters interested in health care. These specialists approach the decision to cover an advance in health care, what kind of story to write, and what legwork needs to be done when reporting the story in fundamentally different ways than nonspecialists.

On the one hand, this is good news. Specialists take accuracy, integrity, and the scientific method seriously in their work. They understand what kinds of questions to ask scholars, doctors, and patients; they are not prone to the most sensational framing of the story that is possible. The evidence presented here reveals that journalists who are health-care specialists can provide the kind of quality reporting needed to facilitate the creation and provision of important health information to local and national audiences.

On the other hand, the analyses revealed here show that a much larger group of journalists are not as prepared to cover health news. In fact, they are less inclined to think that developments in health-care delivery (in the context of the study conducted here) are even worth reporting in the first place. If they had to cover a health-care development, they generally prefer a quick, one-day story that is easy enough to tell so that they can meet their deadline. They do not want to spend the time on understanding the science that was used to develop the health innovation, and are thus more likely to report major innovations using the same tone as they do for minor ones and worse, for *potential* ones—the studies that have not yet passed a level of empirical scientific scrutiny matching the euphoria that is often present in the coverage.

The good news is that journalists can be trained to provide excellent health coverage. First, journalists have to participate in the training and seek to join associations that provide it. Second, newsrooms have to be willing to spend scarce resources on funding their reporters so that they can join relevant associations, travel to conferences and workshops, and spend the time learning from the web-based tutorials that teach new skills, share helpful techniques, and connect those interested in health-care journalism with fellow travelers. Newspapers could take advantage of the space and technological innovations on the web to produce "special series" coverage of major health-care issues. Going beyond the traditional print, these special series could include videos and audios that tell the stories of doctors, researchers, nurses, administrators, patients, families, and even politicians to cover an issue from all angles.

Journalism programs at universities have a role to play as well. "J-schools" can invite experts in health-care coverage to guest lecture and meet privately with students, as well as develop Masters-level curriculum for specialty reporting on topics like health-care news reporting.

The bad news is that newsrooms are facing tighter budgets and more structural constraints than they have in generations. Thus, it will be difficult to provide generalists covering the health beat with the training they need to do excellent work. This places the onus on the scholars developing new health innovations and the doctors, nurses, and allied health professionals administering them to patients to be able to explain advancements in clear, accurate terms: highlighting the importance of the development while still emphasizing the limits, time horizon, and potential harms that it carries.

14.6 Conclusions

The increasingly polarized political environment (Carmines, Ensley, & Wagner, 2012; McCarty, Poole, & Rosenthal, 2006) has tended to boil down the vast, complex issue of health care to whether one supports or opposes the Affordable Care Act. Comprehensive health-care coverage is more important now than ever. Journalists who specialize in health care have the training, ambition, and ability to tell quality stories to a hungry public. Whether there are enough of these reporters to meaningfully affect public opinion and individual behavior—and whether news organizations can be convinced that it is worth their while to produce more of these kinds of journalists—are crucial questions that we will continue to face in the next decade.

References

Althaus, S. L. (2003). *Collective preferences in democratic politics: Opinion surveys and the will of the people.* Cambridge: Cambridge University Press.
Baker, C. E. (2007). *Media concentration and democracy: Why ownership matters.* Cambridge: Cambridge University Press.
Bennett, W. L. (1990). Toward a theory of press-state relations in the United States. *Journal of Communication, 40*(2), 103–127.
Bennett, W. L. (1996). An introduction to journalism norms and representation of politics. *Political Communication, 13*, 373–384.
Bennett, W. L. (2012). *News: The politics of illusion* (9th ed.). New York: Longman.
Brown, A. (2013). Americans mention healthcare as top problem in U.S. Retrieved from http://www.gallup.com/poll/165848/americans-mention-healthcare-top-problem.aspx.
Carmines, E. G., Ensley, M. J., & Wagner, M. W. (2012). Who fits the left-right divide? Partisan polarization in the American electorate. *American Behavioral Scientist, 56*, 1631–1653.
Cook, T. E. (2005). *Governing with the news, second edition: The news media as a political institution.* Chicago: University of Chicago Press.
Cook, T. E. (2006). The news media as a political institution: Looking backward and looking forward. *Political Communication, 23*(2), 159–171.
Dahlgren, P. (2009). *Media and political engagement.* Cambridge: Cambridge University Press.
Groeling, T., & Baum, M. A. (2009). Journalists' incentives and media coverage of elite foreign policy evaluations. *Conflict Management and Peace Science, 26*(5), 437–470.
Hamilton, J. T. (2005). The market and the media. In G. Overholser, & K. H. Jamieson (Eds.), *Institutions of American democracy: The press* (pp. 351–371). Oxford: Oxford University Press.

Han, J. Y., Shah, D. V., Kim, E., Namkoong, K., Lee, S. Y., Moon, T. J., et al. (2010). Empathic exchanges in online cancer support groups: distinguishing message expression and reception effects. *Health Communication, 26*, 185–197.
Hayes, D., & Guardino, M. (2010). Whose views made the news? Media coverage and the march to war in Iraq. *Political Communication, 27*, 59–87.
Hayes, D., & Guardino, M. (2013). *Influence from abroad: Foreign voices, the media, and U.S. public pinion*. New York: Cambridge University Press.
Hibbing, J. R., & Theiss-Morse, E. (2002). *Stealth democracy: Americans' beliefs about how government should work*. New York: Cambridge University Press.
Janowitz, M. (1975). Professional models in journalism: the gatekeeper and the advocate. *Journalism Quarterly, 52*(4), 618–626.
Johnson, K., Isham, A., Shah, D. V., & Gustafson, D. (2011). Potential roles for new communication technologies in treatment of addiction. *Current Psychology Reports, 13*(5), 390–397.
McCarty, N., Poole, K. T., & Rosenthal, H. (2006). *Polarized America: The dance of ideology and unequal riches*. Cambridge, MA: MIT Press.
Noh, J. H., Cho, Y. J., Nam, H. W., Kim, J. H., Kim, D. J., Yoo, H. S., et al. (2010). Web-based comprehensive information system for self-management of diabetes mellitus. *Diabetes Technology & Therapeutics, 12*(5), 333–337.
Patterson, T. E. (1994). *Out of order*. New York: Vintage.
Petrocik, J. R. (1996). Issue ownership in presidential elections, with a 1980 case study. *American Journal of Political Science, 40*(3), 825–850.
Petrocik, J. R., Benoit, W. L., & Hansen, G. H. (2003). Issue ownership and presidential campaigning, 1952–2000. *Political Science Quarterly, 118*(4), 599–626.
Sartori, G. (1987). *The theory of democracy revisited*. Chatham, NJ: Chatham House Publishers.
Schudson, M. (2001). The objectivity norm in American journalism. *Journalism, 2*, 149–170.
Schwitzer, G. (2004). A statement of principles for health care journalists. *The American Journal of Bioethics, 4*(4), W9–W13.
Schwitzer, G. (2009). *The state of health journalism in the U.S.* Report to the Kaiser Family Foundation. Retrieved from https://kaiserfamilyfoundation.files.wordpress.com/2013/01/7858.pdf.
Shoemaker, P. J., & Reese, S. D. (1995). *Mediating the message: Theories of influence on mass media content*. New York: Longman Trade/Caroline House.
Shuchman, M. (2002). Journalists as change agents in medicine and health care. *Journal of the American Medical Association, 287*(6), 776.
Sigal, L. V. (1973). *Reporters and officials*. Lexington, MA: D.C. Heath.
Tian, Y., & Robinson, J. D. (2008). Incidental health information use and media complementarity: a comparison of senior and non-senior cancer patients. *Patient Education and Counseling, 71*, 340–344.
Tuchman, G. (1972). Objectivity as strategic ritual: an examination of newsmen's notions of objectivity. *American Journal of Sociology, 77*(4), 660–679.
Viswanath, K., Breen, N., Meissner, H., Moser, R. P., Hesse, B., Steele, W. R., et al. (2006). Cancer knowledge and disparities in the information age. *Journal of Health Communication, 11*(Suppl. 1), 1–17.
Voss, M. (2002). Checking the pulse: midwestern reporters' opinions on their ability to report health care news. *American Journal of Public Health, 92*(7), 1158–1160.
Wagner, M. W., & Gruszczynski, M. When framing matters: how partisan and journalistic frames affect public opinion and party identification. *Journalism & Communication Monographs, 18*(1), in press.

Walsh, K. C. (2012). Putting inequality in its place: rural consciousness and the power of perspective. *American Political Science Review, 106*(3), 517–532.

Weaver, D. H., Beam, R. A., Brownleee, B. J., Voakes, P. S., & Wilhoit, G. C. (2006). *The American Journalist in the 21st Century*. New York: Lawrence Erlbaum Associates, Inc.

Wilnat, L., & Weaver, D. H. (2014). *The American Journalist in the Digital Age*. Key Findings. Accessed at https://larswillnat.files.wordpress.com/2014/05/2013-american-journalist-key-findings.pdf.

Wilson, A., Robertson, J., McElduff, P., Jones, A., & Henry, D. (2010). Does it matter who writes medical news stories? *PLoS Medicine, 7*(9), 1–5.

Afterword

One surprising and interesting finding from our work on this book occurred very early in the editing process—so early, in fact, that the editors had nothing yet to edit. Our Call for Chapters was distributed via numerous outlets—professional listservs™ frequented by practitioners, educators and researchers in medical informatics, consumer health informatics, medical librarianship, and consumer health librarianship; and Web-only venues where similar people in the fields of journalism and linguistics congregate. We also used our personal connections in these and related fields, such as health communications, based on our years of experience in our own fields of practice and research. So the community we sought to reach was by very definition a large and diverse one, united by one common attribute: health information communication, provision, or both, by humans to other humans who were not health-care professionals.

In that Call for Chapters announcement, we said: "The book will address the challenges and ethical dilemmas concerning the delivery of health information to the general public in a wide variety of non-clinical settings. Instead of patient education or patient communication in hospitals and clinics, our interest is the challenges and successes of presenting health information outside of patient care by non-clinicians." To our surprise, we received a great deal of interest from people with interesting, worthwhile research to contribute, who had somehow managed to miss the "nonclinical" piece of the instructions. We received many queries about work with patients in hospital settings, for example. The strong identification of "health information" with physicians, nurses, patients, and the clinical setting was apparently *so* strong that it was, for some people, actually blinding.

We wonder if the widespread amalgamation of "consumer" and "patient" as synonymous terms is also part of the problem. As author Nancy Seeger expresses it in her chapter on the Consumer Health Library, "When patients become library patrons, [health information concerns] do not automatically disappear." If every member of the general public is seen as a patient, it is quite easy to assume the boundaries of the health-care model—in which the only provider of information is a physician, nurse, or allied health professional. It becomes difficult for practitioners and researchers alike to understand the special challenges of information provision when the provider does not herself/himself have training as a health-care professional. Recognition of these challenges was our principal motivation for writing and editing this book.

What do our assembled chapters tell us about these challenges, and the degree to which they are shared across health information provision communities?

We began the volume with two overview chapters to provide a context for understanding health information provision. Kreps and Neuhauser focused on health communication and population-specific issues; conversations around health information are impacted by cultural distances between the consumer and the information provider. "Those most in need of health communication may receive the least benefit from it," they write, because traditional health communication fails to be aligned with the needs and life situations of specific audiences. When there is no shared frame of reference, the information communicated is "foreign" to the audience, as well as overly technical. The distance *between* the experience of the information provider and of the consumer audience turns out to be critically important. Devoting resources to understanding the audience is, potentially, an effective way for health information providers to understand the *gaps* between themselves and that audience.

The second overview chapter, by Logan, examines health literacy—as a concept, and as the subject of research, including its evolution from a "narrow discipline" to a topic attracting widespread and diverse attention from health-care professionals, organizations, public health, insurance and pharmaceutical companies, health foundations, and the government. It is noteworthy but hardly surprising that every single chapter in this book raises health literacy as an issue. The challenge that Logan describes, however, arises from this rapid acceleration of interest in different domains of inquiry. He quotes one researcher who warned that "the field's credibility might be compromised by its current diversity and disagreements about underlying definitions, conceptual models, instruments, and methods." The positive side of the problem is that there are many interesting research questions posed by a call for "conceptual transparency" and the need for basic research and validation of the numerous instruments used in health literacy work. For the editors of this volume, an important contribution of Logan's work is its candid and revealing description of health literacy research—warts and all. Nonclinicians engaged in health information provision in *any* setting see the effects of health literacy issues on their client population. Understanding that health literacy has not only nuances, but its own controversies and dilemmas as a field, is extremely important to understanding how health literacy works in the world.

The five chapters in the *Libraries* section are grouped together because of their overarching common characteristic. The ethical dilemmas outlined by McKnight are real-life occurrences in any library setting. Medical terminology is everywhere: it appears in Catherine Arnott Smith's chapter as one of the oldest barriers to effective health information provision ever recorded; McKnight also raises the problem in the context of information access, Flaherty as a facet of preprofessional education for public librarianship, and Seeger and Chernaik in the context of services to patrons in consumer health and community college libraries, respectively. Community college and public libraries are described by Chernaik as "closely aligned"; this author points out that each of these library types is defined by open access to all, and each has a close tie to its community and thus must understand and meet the information needs of its community. As Flaherty and Chernaik make clear in their respective chapters, shared challenges involving health information include staff expectations, the dilemma of specialist information needs addressed by generalist training, and the presence of popular and authoritative health information in the same collection. However, community college libraries, unlike public libraries, have a user base that explicitly includes academics—students, staff, and faculty—and must adhere to standards set by educational accreditors. Does this make their task more difficult, or less? We suggest that there is a case to be made for both.

Seeger's portrait of a consumer health library—serving the needs of the general public but embedded inside a hospital setting—makes it clear that the public's expectations are critical components of all health information processes. The public likes symptom checkers, for example. These computerized decision support tools are widely available online. They make self-diagnosis very easy, which is not the same thing as "correct"; this is a problem for librarians because, as Catherine Arnott Smith's historical review makes clear, the librarian's fear of enabling dangerous self-diagnosis is even older than her fear of dangerous medical terminology. In a library setting this is difficult because the librarian's information may be taken as clinically credible, "even if the patron is warned of the symptom checker's limitations through the librarian's own written and verbal disclaimers." Seeger acknowledges her patrons' expectations and advises readers to meet their patrons where librarians are most comfortable—rather than "outright prohibition" of symptom-checking, they can "use their skills" to help patrons do better searching for symptom information, and direct them to quality resources instead of poor ones.

The seven chapters in the *Contexts* section are not library-specific, but even here, themes found in libraries recur. Medical terminology, for example, is referenced by Lence and Capozza as a design consideration for Care4Life, health behavior support via interactive text messaging—medical terminology was "avoided when possible." Health information as a sensitive topic is mentioned by several authors. Sayed and Weber describe the prevailing cultural norm in the Arabian Gulf Region, where one does not talk about bad news. The result is that a serious prognosis provokes a "wide range of conflicting emotions" which stigmatize the topic of cancer and lower public understanding of the disease—thus presenting a minefield for health education and health information provision. Linguist Von Rohr's findings are a little different. She examined peer-to-peer support in an online bulletin board for people quitting smoking, and describes the importance of "public self-image" or "face" in these digital interactions, as well as participants' awareness that they must appear "trustworthy" and "credible" during the advice-giving process. It is fascinating that in Von Rohr's analysis, community members' recommendations for coping with "daily life" were considered more sensitive topics, requiring such strategies as linguistic "hedging," than were specific suggestions about quitting smoking. So while "the delicacy of the topic influenced the form of advice," context appears to be everything, and health information was less sensitive than other kinds.

Zeyer, Levin, and Keselman's context is the science classroom as a venue for improving critical and culturally appropriate health literacy. These authors, no less than a number of the authors in the *Libraries* section, stress the importance of scientific understanding on the part of the professionals doing the information provision—which these authors see as a "collaborative endeavor" between science education researchers, teachers, administrators, school librarians, and resource developers in science education.

Two authors in this section focused on special populations within the larger circle of the general public. Kay Hogan Smith examines information provision considered specifically in the context of the older adult which means challenges posed by sensory and cognitive declines as well as information access. Older adults' nonadoption of information technology means that the information provider faces additional barriers with this age group. The interaction of sensory deficits and technology deficits means that the information provider must pay increased attention to meaningful, usable *formats* for information delivery. Dalrymple and Zach report on information seeking in an urban ethnic minority population; in contrast to other research, these authors did not select their audience based on specific diagnoses, but on their residence in a public housing neighborhood typifying health disparities. The work of these authors makes explicit how important it is to understand information needs that drive information seeking among the public. Information needs expressed by disadvantaged populations may not translate well to the traditional information-seeking model of research. This has implications not only for the conducting of research itself—who is recruited and from whom can we generalize??—but also for our understanding of information needs and provision of resources to meet those needs—in any medium and any professional context.

Lence and Capozza's scope of interest is digital health, and they cite the benefits of exchanging health information through digital media—it is both cost-effective and convenient, and the convenience itself makes technology the more person-centered of the information delivery options, since the person is in control of both how and when they access digital information. Lence and Capozza call our attention to ethical problems of delivering health-care data this way: the inherent paternalism of information control; the inability of any "calculus" to meet the individual values of individual people; the overarching context of health literacy in which information providers and consumers engage with each other.

Michael Wagner's examination of journalist expertise reminds the editors of the challenges discussed in the *Libraries* section of our book. The public and community college library settings place heavy demands on paraprofessionals and professionals who have generalist training—in information resources about everything from auto repair to chemotherapy—but have information clients with very specialized needs. The journalists surveyed by Wagner were significantly more likely to find a hypothetical health-care story "newsworthy," promotable as a feature, and worthy of research time if the journalists were health-care specialists than if they were generalists. Training, Wagner concludes, is the answer as well as the "good news."

Taken together, the 14 chapters collected in this book illustrate the multiplicity of situations involving health information provisions in nonclinical contexts, as well as the diversity of information providers' professional backgrounds, thus underscoring the importance and the challenging nature of such information provision. We hope that this volume speaks to researchers who are interested in setting agendas for health information provision and education research in their fields. We also hope that it provides specific, practical tips to professionals, from librarians struggling with the dilemmas inherent in their training, setting, and client base, to journalists and public health professionals looking for most effective ways to present information, to science teachers wishing to make their lessons useful in real-life situations, to eHealth applications developers, and ultimately, to everyone else involved in the challenging and worthy task of supporting lay individuals seeking health information.

Catherine Arnott Smith
Alla Keselman

Index

Note: Page numbers followed by "f", "t" and "b" indicates figures, tables and boxes, respectively.

A

AACC. *See* American Association of Community Colleges (AACC)
Academy of Health Information Professionals (AHIP), 80
Access, 55–58, 151–153, 160
Accessibility, 82
Accreditation Commission for Education in Nursing (ACEN), 154
ACEN. *See* Accreditation Commission for Education in Nursing (ACEN)
Acquisitions and collection development, 154–155
Adult literacy, 20–21
 national assessment, 21–23
Advanced library users, 152
Advice, 283–285
Advice-giving, interpersonal aspects of, 265–270
Age-related hearing loss. *See* Presbycusis
Agency for Healthcare Research and Quality (AHRQ), 134t
Agency for Healthcare Research and Quality, 303–304
AHCJ. *See* Association of Health Care Journalists (AHCJ)
AHIP. *See* Academy of Health Information Professionals (AHIP)
AHIP Trade Association. *See* America's Health Insurance Plans Trade Association (AHIP Trade Association)
AHRQ. *See* Agency for Healthcare Research and Quality (AHRQ)
ALA. *See* American Library Association (ALA)
Algorithmic knowledge, 246
AMA. *See* American Medical Association (AMA)
America's Health Insurance Plans Trade Association (AHIP Trade Association), 134t
American Association of Community Colleges (AACC), 144–145
American Library Association (ALA), 78
American Medical Association (AMA), 134t
American Social Hygiene Association, 46
Analytical procedure, 272–273
Anecdotal reports, 174
Anxiety, 120
AP. *See* Associated Press (AP)
Approximations of practice, 254
Arabian Gulf health information literacy, 172, 184–185
Aspects of quality collections, 69–70
Assessment, 280–283
Associated Press (AP), 322
Association of Health Care Journalists (AHCJ), 321
Audience analysis, 8–9
Audiences, 309–311
Audiovisual formats, 195
Authority, 81

B

"Bad" books, 70–71
Bay Area, 63
Biomedical knowledge in information processing, 242–244
Bounded rationality, 303
"Build and evaluate loops" approach, 9–10, 12–13

C

"Calgary Charter", 26, 28
CAM. *See* Complementary and alternative medicine (CAM)
Cancer information delivery outside clinical setting, 183–186
 role of medical libraries, 185–186
Cancer Strategy, 171
Cancer survivors, 182–183

CAPHIS. *See* Consumer and Patient Health Information Section (CAPHIS)
Care4Life, 291–292, 302f
Caregiver, 222
CBC. *See* Cultural border crossing (CBC)
CDC. *See* Centers for Disease Control and Prevention (CDC)
Centers for Disease Control and Prevention (CDC), 134t, 199, 201
Centers for Medicare and Medicaid Services (CMS), 292
Chemotherapy, 180–181
CHI. *See* Consumer health information (CHI)
CHIN. *See* Community Health Information Network (CHIN)
CHIRR. *See* Consumer Health Informatics Research Resource (CHIRR)
City colleges. *See* Community college(s)
CJCLS. *See* Community and Junior College Libraries Section (CJCLS)
Classic typologies, 221–222
"Clear communication" design criteria, 6
Clinical programmatic faculty, 154
Clinical trials, 136–137
"Closed to Public", 41–44
CMS. *See* Centers for Medicare and Medicaid Services (CMS)
Cognitive disorders, 201–202
Cognitive issues, 201
  communication tips, 202
  environment tips, 202
  formatting tips, 202–203
Collaboration, 102
Collection development, 119, 144, 154–155, 161–162
Collection librarians, 155
Communication
  channels, 12
  tips
    cognitive issues, 202
    hearing problems, 200
    vision problems, 198
Community and Junior College Libraries Section (CJCLS), 149, 163
Community college(s), 143, 143b, 153
  challenge for, 151
  future, 157–163
    access, 160
    collections, 161–162

    electronic resources, 161
    health literacy instruction, 162–163
    materials, 160
    services, 161
    staff training, 159–160
  health information needs, 147–149
  librarians, 144, 153
  libraries, 144, 146–147, 152
  setting, 144–147
  health literacy, 156–157
  students, 146
Community Health Information Network (CHIN), 61
Complementary and alternative medicine (CAM), 137–139
Complementary message strategies, 11
Computer-mediated communication, 307–309
Confidentiality, responsibility for, 83–85
Confusion. *See* Insecurity
Connect2FightCancer app, 327–328
Consumer and patient engagement, 306
Consumer and Patient Health Information Section (CAPHIS), 155, 80
Consumer health, 149
Consumer Health Informatics Research Resource (CHIRR), 22–23, 27, 29
Consumer health information (CHI), 44–45, 118, 136, 191, 209. *See also* Ethical health information; Medical information for consumer
  American Social Hygiene Association, 46
  comprehension among older adults, 197
  cognitive issues, 201–203
  hearing problems, 199–201
  vision problems, 197–199
  format considerations, 194
    audiovisual formats, 195
    demonstration models and materials, 196
    games, 196–197
    in-person presentations, 195–196
    online formats, 196
    print format, 194
    puzzles, 196–197
    visual materials, 195
  format summary, 197
  future, 157–163
  health sciences libraries users, 48

issues in health information provision, 149
 accessing, 151–153
 acquisitions and collection
  development, 154–155
 electronic resources, 153
 library services, 154
 organization of materials, 156
 staffing, 149–151
medical information users, 48–49
medical libraries in action, 46
medical questions, 50–51
medical subjects, 49–50
medical terms and topics, 51
needs at community college, 147
 consumer health, 149
 medical academic, 148
 nonmedical academic, 148–149
provision, 99–100
quality, 135–136
resources quality, 105–106
seeking, 210–211
 findings from survey on internet access, 219
 focus groups on, 220–223
 "on-ramp" to health information, 218
 patterns and themes, 223–225
 survey, 218–219
services, 161
sources in GCC, 173–175
Surgeon General's Library, 47
technology, 32
tools, 314
World Wide Web, 51
Consumer health librarians, 119, 122, 129, 133–134, 136
 experimental treatments, 136–139
 health literacy, 133–136, 134t
 integrative medicine, 136–139
 reference transactions, 119–126
 self-diagnosing, 126–132
 symptom-checkers, 126–132
Consumer health libraries, 51, 123, 133–134. *See also* Public libraries
Consumer-directed health care, 117–118
Core practices, 254
County colleges. *See* Community college(s)
Credibility, 12
Critical literacy, 237
Cross-cultural understanding, 252

Crowdsourcing social media commentary, 293
Cultural border crossing (CBC), 250–251
Cultural clash, 252
Cupping (bloodletting). *See Hijama*
Currency, 82

D

Data
 dilemmas, 311–314
 display issues, 314
 inconsistencies, 304–305
DDC. *See* Dewey Decimal Classification (DDC)
Decision making, 303–306
Deidentification, 271–272
DeLib. *See* Distributed *e*Library (DeLib)
Delirium, 202
Demonstration models and materials, 196
Depression, delirium, and dementia (Three D's), 192
Descriptive analysis, 330
Design sciences, 12–13
Dewey Decimal Classification (DDC), 50–51
Diabetes mellitus, 30
Dictionary for Library and Information Science, 121
Digital divide, 294–297
Digital technology, 291, 293, 306
Discomfort. *See* Anxiety
Discourse analysis, 272–273
Discursive moves, 273, 274t, 278t
 advice, 283–285
 assessment, 280–283
 own experience and background information, 285–287
Diseases in Qatar, 172
Distributed *e*Library (DeLib), 172, 178, 185
Doctor of philosophy (PhD), 81
Doctor–patient relationship, 117, 120, 133–136

E

"eHealth", 216
Electronic resources, 153, 161
Embarrassment, 121
Embedded consumer health librarians in Delaware, 110–112

"Embedded" librarians, 150–151
Empathic communities, 310–311
Empirical vacuum, 242
Environment tips
　cognitive issues, 202
　hearing problems, 200
　vision problems, 198–199
Equitable access
　advocacy for information access, 88
　barriers to information access, 86–87
　fair and, 85
　intellectual property rights and access to information, 87
Ethical health information. *See also* Consumer health information (CHI)
　conflicting values, dilemmas, and tough decisions, 90–91
　fair and equitable access, 85–87
　information ethics, 77
　information *vs.* giving advice
　　disclaimers, 89
　　don't play doctor, 89
　　patient education and information prescriptions, 90
　learning, 91–92
　responsibility for confidentiality, 83–85
　responsibility for information service, 78b
　　answer, 80–81
　　evaluation, 81–83
　　question, 78–79
　　right to privacy, 83–85
Evaluating Health Information, 136
Exemplar messages, 275–277
Experiential knowledge, 246
Experimental treatments, 136–139

F

Face-enhancing relational work, 265
Face-maintaining relational work, 265
Face-threatening relational work, 265–266
"Face" in interaction. *See* Public self-image
Fear of unknown, 121
Feedback mechanisms, 13
Fibroendoscopy, 250
Fiduciary
　ethical model, 120
　relationship, 120
Final Exit, 91

Focus groups
　findings from, 221–223
　on health information seeking, 220–221
Formal liaison role, 149–150
Format summary, 197
Formatting tips
　cognitive issues, 202–203
　hearing problems, 201
　vision problems, 199
Framework model, 246, 246f
FTE students. *See* Full-time equivalent students (FTE students)
Full-time equivalent students (FTE students), 150
Functional literacy, 236–237

G

Games, 196–197
GCC. *See* Gulf Cooperation Council (GCC)
General public, 151
Generalist librarians, 151
Group Health Research Institute (GHRI), 134t
Gulf Cooperation Council (GCC), 170
　CHI sources in, 173–175

H

Hakims, 172, 174
Hamad General Hospital, 172
Hamad Medical Corporation (HMC), 169
Harmful books, 71–72
Hayat Cancer Support group, 180
Health, 268
　consumers, 129–130, 133–134
　disparities, 214–215
　health-care coverage, 321
　health-related information behavior, 214–217
　programs
　　Care4Life, 291–292
　　crowdsourcing social media commentary, 293
　　UtahHealthScape, 292–293
Health communication, 3, 19, 21
　barriers, 4–7
　evaluation, 12–13
　improving for vulnerable populations, 7–8
　practice implications, 13–14

strategic communication, strategies to developing, 8–12
traditional, 4
Health education, 240. *See also* Science education
  collaborative nature, 256–257
  in sciences outside biology, 249–250
"Health Happens in Libraries", 102
Health information. *See* Consumer health information (CHI)
Health literacy, 5, 133–136, 134t, 297–303
  in community college setting
    librarians, 157
    patrons, 156–157
  deconstruction, 236–237
  framework model, 246f
  HPVs, 237–239
  instruction, 162–163
  principles, 6
  science classroom, 239–248
  single pneumonia measure for hospital quality scores, 300f
  summary page for hospital quality scores, 299f
Health Literacy Innovations (HLI), 134t
Health literacy research, 19
  current needs and frontiers, 30–33
  definition and conceptual underpinnings, 26–29
  milestones in, 20
    and adult literacy, 20–21
    adult literacy national assessment, 21–23
    interventions, 23–24
    outcomes, 24–26
  range and vitality, 29–30
Health-Line, 56
HealthInsight, 291
"Healthy living" activities, 218
Healthy People 2010, 236–237
Hearing problems, 199
  communication tips, 200
  environment tips, 200
  formatting tips, 201
Heart diseases, 29
Help-seeking strategy, 269
Hijama, 172
HLI. *See* Health Literacy Innovations (HLI)
HMC. *See* Hamad Medical Corporation (HMC)

Home health-care badges, 311, 311f
HPVs. *See* Human papillomaviruses (HPVs)
Human papillomaviruses (HPVs), 237–239
Humanizing technology, 306–309

I

IBSE. *See* Inquiry-based science education (IBSE)
ICT. *See* Information and communications technology (ICT)
Imams, 174
IMLS. *See* U.S. Institute for Museum and Library Services (IMLS)
In-person presentations, 195–196
Indexing hypothesis, 323
Individual "deficit" model of health literacy, 27
Information, 263–264, 267–269
  behaviors, 210–217
    health-related information behavior, 214–217
    information needs, 211
    information scientists study, 214
    information seeking, 211–213
    information use, 213–214
  providing information *vs.* giving advice
    using disclaimers, 89
    personal advice, 89
    prescribed patient education and information, 90
Information access, advocacy for, 88
Information and communications technology (ICT), 171
Information grounds, 213–214
Information needs, 119, 121–122, 126, 129–130, 133, 211, 224
Information seeking, 211–213
Inquiry-based science education (IBSE), 250–251
Insecurity, 120–121
Institute of Medicine (IOM), 27
Institutional Review Board (IRB), 220
Instrument proliferation, 28
Integrative medicine, 136–139
Intellectual access to collection, 162
Intellectual property rights, 87
Intended audience, 82–83
Interactive literacy, 237

I

Internet, 135, 263
  self-diagnosis, 129
Interprofessional Education model, 170
IOM. *See* Institute of Medicine (IOM)
IRB. *See* Institutional Review Board (IRB)

J

Joint Commission International (JCI), 172
Joint Review Committee on Education in Radiologic Technology (JRCERT), 154
Journalism, 323
  programs at universities, 337
Junior colleges. *See* Community college(s)

K

Khaleeji (Gulf) Arabs, 171
Knowledge
  biomedical knowledge in information processing, 242–244
  decision-making influences, 244–248
  development, 255
  economy, 170–171
  scientific knowledge impact on daily life and health, 241–242

L

"Lay public", 54–55, 62
Leading discussion, 254
Learning, 91–92
LEP. *See* Limited English proficiency (LEP)
Libertarian paternalism, 305–306, 314
LibGuides, 159, 185
Librarians, 78–81, 157. *See also* Consumer health librarians; Public libraries
  at community colleges, 144
  interpersonal challenges, 58–61
  roles, 62–63
Libraries, 52, 159. *See also* Consumer health libraries; Public libraries
  access to collections, 55–58
  services, 154
  staff perception, 101–102
  terminals, 153
  types, 52–55
Library and information science (LIS), 210–211
Limited English proficiency (LEP), 6

Limited health literacy, 121
Linguistics, 263, 272–273
LIS. *See* Library and information science (LIS)
Lucy Answers, 265–267

M

Macular degeneration, 197–198
Manageability, 247
Master's degree in library and information science (MLIS), 78
Material collections, 147, 160
Meaningfulness, 247
Medical academic, 148
Medical imaging program, 154
Medical information for consumer. *See also* Consumer health information (CHI)
  "closed to public", 41–44
  content, 67
    appropriate for audience, 69
    aspects of quality collections, 69–70
    "bad" books, 70–71
    harmful books, 71–72
    terminology, 67–68
  librarian
    interpersonal challenges, 58–61
    roles, 62–63
  libraries, 52
    access to collections, 55–58
    types, 52–55
    patron, 63–67
Medical libraries, 43–44, 46, 55–56, 65
Medical libraries in Qatar, 169, 178
  role of, 185–186
Medical Library Association (MLA), 41, 80, 109, 159–160
Medical schools in Arabian Gulf, 169
Medical Subject Headings (MeSH), 29–30
Medical terminology, 67–68
"Medical terms and topics", 51
MENA region. *See* Middle East and North Africa region (MENA region)
MEPS. *See* US Medical Expenditure Panel Survey (MEPS)
MeSH. *See* Medical Subject Headings (MeSH)
mHealth. *See* mobile health (mHealth)
Middle East and North Africa region (MENA region), 169

Index 353

MLA. *See* Medical Library Association (MLA)
MLIS. *See* Master's degree in library and information science (MLIS)
mobile health (mHealth), 195, 216
Muftis, 174
Myers-Briggs Type Indicator, 221–222

N
NAAL. *See* US National Assessment of Adult Literacy (NAAL)
National Cancer Institute (NCI), 134t
National Cancer Strategy, 171, 180, 185
National Center for Biotechnology Information (NCBI), 159
National Center for Cancer Care and Research (NCCCR), 172
National Center for Complementary and Alternative Medicine (NCCAM), 137–138
National Institutes of Health (NIH), 159
National launch, 328
National Network of Libraries of Medicine (NN/LM), 24, 55, 80, 108, 159–160
NCBI. *See* National Center for Biotechnology Information (NCBI)
NCCAM. *See* National Center for Complementary and Alternative Medicine (NCCAM)
NCCCR. *See* National Center for Cancer Care and Research (NCCCR)
NCI. *See* National Cancer Institute (NCI)
Neurocognitive disorders, 201–202
Newest Vital Sign (NVS), 22–23
NGOs. *See* Nongovernmental organizations (NGOs)
NIH. *See* National Institutes of Health (NIH)
NLM. *See* U.S. National Library of Medicine (NLM)
NN/LM. *See* National Network of Libraries of Medicine (NN/LM)
Nonauthoritative health information, 106–107
Nonfiction health information, 106–107
Nongovernmental organizations (NGOs), 170
Nonmedical academic, 148–149
Nursing programs, 154
NVS. *See* Newest Vital Sign (NVS)

O
Obligations, 120, 122
Older adults, 191. *See also* Health information
  graying of America, 192
  settings, 193–194
"On-ramp" to health information, 218
Online
  formats, 196
  health support groups, 266, 273
  showing support, 265–270
Ooredoo, 180–181
Organization of materials, 156
Organization of Petroleum Exporting Countries, 170
Own experience, 285–287

P
Parental nutrition, 157
Participant recruitment, 325–326
Participatory
  care, 235–236
  design, 12–13
  medicine, 235
Partners in National Cancer Strategy, 185
Paternalism, 301–302
Patient education, 44
Patient Protection and Affordable Care Act, 102
Patient readiness, 120
Patient-centered care, 117–118
Patron(s), 63, 156–157
  challenges, 64
  competition for time, 65
  expectations, 65–66
  self-diagnosis, 66–67
Pedagogical content knowledge (PCK), 255
Persuasion, 264
PhD. *See* Doctor of philosophy (PhD)
Plain language
  design criteria, 6
  tools, 134t
Practice-based teacher education, 253–255
Presbycusis, 199
Print format, 194
Privacy, 83–84
  policies, 84
Privileges of reference, 97–98

Public libraries, 43–44, 52–54, 58–59, 97, 143. *See also* Consumer health libraries
　challenges, 100
　differences in training opportunities and staff motivation, 108–109
　educational requirements for staff, 104–105
　financial and community support, 103
　health information resources quality, 105–106
　library staff perception, 101–102
　nonauthoritative health information, 106–107
　nonfiction health information, 106–107
　public and medical library practitioner communities, 109–110
　public perceptions, 102–103
　skill sets of public library staff, 107–108
　embedded consumer health librarians in Delaware, 110–112
　and health information provision, 99–100
　privileges of reference, 97–98
　staff skill sets, 107–108
　in US, 98–99
Public perceptions, 102–103
Public reporting, 301
Public self-image, 265
Public understanding of science, 244
PubMed MEDLINE, 172
Puzzles, 196–197

Q

Qatar, 170–172
　demography, 171
　health care, 172
　　barriers to, 175–176
　Qatar Vision 2030, 170–171
　truth-telling, 176
Qatar Cancer Society (QCS), 169, 177
　cancer information delivery outside clinical setting, 183–186
　role of medical libraries, 185–186
　CHI sources in GCC, 173–175
　dual language pamphlet on colorectal cancer, 179f

　financial support and fund-raising, 183
　methods, 172
　outreach to community, 180–181
　　targeting high-risk groups, 181
　　visits to local schools, 181
　patient and family support, 181–182
　promotion and prevention information, 177–180
　questionnaire, 187
　services, 177–180
　support groups, 182
　utilizing cancer survivors for support services, 182–183
Qatar Foundation institutions, 185
Qatar National Cancer Research Strategy, 171
Qatar National Cancer Strategy, 183
Qatar National Library, 186
Qatar National Research Fund, 170–171
Qatar Science and Technology Park (QSTP), 170–171
Qatari nationals, 171
QCS. *See* Qatar Cancer Society (QCS)
QSTP. *See* Qatar Science and Technology Park (QSTP)
Quality assurance, 69–70
Quantitative analysis, 273, 277–280
"Quick and dirty" estimation, 303
Quranic medicine, 172

R

Rapid Estimate of Adult Literacy in Medicine (REALM), 22–23
"Rational actor" theory, 303
REALM. *See* Rapid Estimate of Adult Literacy in Medicine (REALM)
Reference, 151, 160
　interviews, 78–79, 121–122, 159
　materials, 153
　services, 50–51, 57, 100–101, 154
　transactions, 119–126
Reference and User Services Association within the American Library Association (RUSA), 119, 121
Relational aspect of interactions, 264
Relational work, 265
Reliability, 81–82
Reporting, 322–323, 329, 333
"Representations of practice", 254

Resource centers. *See* Consumer health libraries
Responsibility for information service, 78b
   answer, 80–81
   evaluation, 81–83
   question, 78–79
RUSA. *See* Reference and User Services Association within the American Library Association (RUSA)

S

Sampling, 271–272
SCH. *See* Supreme Council of Health (SCH)
Science classroom, 239–248
   making room for health, 248–250
      health education in sciences outside biology, 249–250
   "scientific practices" and "practice-based" teacher education, 253–255
   structuring student opportunities, 250–253
Science education, 251, 239, 245
Scientific practices teacher education, 253–255
Self-censorship, 66
Self-diagnosing, 126–132
Self-diagnosis, 66–67, 129, 133
Sense-making, 211–212
Sensible estimation procedures, 297–298
"Serious" video games, 196–197
SES. *See* Socioeconomic status (SES)
Shame. *See* Embarrassment
Shared health care decision-making, 119
Sharing information, structure of, 275
Sheikh (feminine *Sheikha*), 174
Signal authenticity, 269
Smoking cessation, 270
"SmokingisBad" forum, 271
Social determinants, 19–20
Social media, 291, 293, 308
Social presence, 307–308
Socialized and preventive medicine, 49–50
Socioeconomic status (SES), 6–7, 216
Sore throat, 131
Specialization, 323–324
   methodology
      characteristics of sample, 326–328, 327t
      hypotheses, 324–325

      participant recruitment, 325–326
      questions, coding, and analysis, 328–330
   results
      descriptive analysis, 330
      journalistic specialization and reportorial decisions making, 332t
      qualitative results, 333–336
      quantitative results, 331–333
Staff training, 159–160
Staffing, 149–151
State of Qatar, 170–172
Strategic health communication, 7–8
   development strategies, 8
      audience analysis, 8–9
      communication channels, 12
      complementary message strategies, 11
      credibility, 12
      criteria for strategic message design, 9
      tailored communication, 11
      user-centered design, 9–10
Strategic message design, criteria for, 9
Sufferer, 222–223
Sunni Muslims, 171
Support, 264–265
Supreme Council of Health (SCH), 171, 180
Surfer, 221–223
Survey respondents, 155–156
Symptom-checkers, 126–132
Symptom-checking tools, 127t–128t

T

Tailored communication, 11
Talking science, 251
Technical colleges. *See* Community college(s)
Test of Functional Health Literacy in Adults (TOFHLA), 22–23
Themes, 223–225
Time constraints, 120
TOFHLA. *See* Test of Functional Health Literacy in Adults (TOFHLA)
Traditional Healers in Riyadh, 174
Training, 101, 104–105
   differences in training opportunities and staff motivation, 108–109
   staff training, 159–160
Truth-telling, 176
Typology of users, 221–222

U

U.S. Institute for Museum and Library Services (IMLS), 99–100
U.S. National Library of Medicine (NLM), 80, 106, 159
UAE. *See* United Arab Emirates (UAE)
Underserved population, 216
Understandability, 247
United Arab Emirates (UAE), 170
"Universal precautions" approach, 198
US Institute of Medicine, 5
US Medical Expenditure Panel Survey (MEPS), 192
US National Assessment of Adult Literacy (NAAL), 20
Usability testing, 10
User-centered design, 9–10
UtahHealthScape, 292–293, 312
  consumer version, 301b

V

Virtual services, 150–151
Virtue ethics, 77

Vision problems, 197–198
  communication tips, 198
  environment tips, 198–199
  formatting tips, 199
Visual materials, 195
Vulnerable populations, improving health communication for, 7–8

W

Warranting strategies, 268
WCMC-Q. *See* Weill Cornell Medical College in Qatar (WCMC-Q)
Webinars mechanism, 108
Weill Cornell Medical College in Qatar (WCMC-Q), 178, 185
Weill Medical College (WMC), 170–171
White-coat silence, 235
WHO. *See* World Health Organization (WHO)
WMC. *See* Weill Medical College (WMC)
World Health Organization (WHO), 5, 27, 170, 237
World Wide Web, 51

Printed and bound by CPI Group (UK) Ltd, Croydon, CR0 4YY
08/06/2025
01896872-0004